Characterization & Control of Interfaces for High Quality Advanced Materials

T0328412

Journal of the American Ceramic Society

www.ceramicjournal.org

With the highest impact factor of any ceramics-specific journal, the *Journal of the American Ceramic Society* is the world's leading source of published research in ceramics and related materials sciences.

Contents include ceramic processing science; electric and dielectic properties; mechanical, thermal and chemical properties; microstructure and phase equilibria; and much more.

Journal of the American Ceramic Society is abstracted/indexed in Chemical Abstracts, Ceramic Abstracts, Cambridge Scientific, ISI's Web of Science, Science Citation Index, Chemistry Citation Index, Materials Science Citation Index, Reaction Citation Index, Current Contents/ Physical, Chemical and Earth Sciences, Current Contents/Engineering, Computing and Technology, plus more.

View abstracts of all content from 1997 through the current issue at no charge at www.ceramicjournal.org. Subscribers receive full-text access to online content.

Published monthly in print and online. Annual subscription runs from January through December. ISSN 0002-7820

International Journal of Applied Ceramic Technology

www.ceramics.org/act

Launched in January 2004, *International Journal of Applied Ceramic Technology* is a must read for engineers, scientists,and companies using or exploring the use of engineered ceramics in product and commercial applications.

Led by an editorial board of experts from industry, government and universities, *International Journal of Applied Ceramic Technology* is a peer-reviewed publication that provides the latest information on fuel cells, nanotechnology, ceramic armor, thermal and environmental barrier coatings, functional materials, ceramic matrix composites, biomaterials, and other cutting-edge topics.

Go to www.ceramics.org/act to see the current issue's table of contents listing state-of-the-art coverage of important topics by internationally recognized leaders.

Published quarterly. Annual subscription runs from January through December. ISSN 1546-542X

American Ceramic Society Bulletin

www.ceramicbulletin.org

The *American Ceramic Society Bulletin*, is a must-read publication devoted to current and emerging developments in materials, manufacturing processes, instrumentation, equipment, and systems impacting the global ceramics and glass industries.

The *Bulletin* is written primarily for key specifiers of products and services: researchers, engineers, other technical personnel and corporate managers involved in the research, development and manufacture of ceramic and glass products. Membership in The American Ceramic Society includes a subscription to the *Bulletin*, including online access.

Published monthly in print and online, the December issue includes the annual *ceramicSOURCE* company directory and buyer's guide. ISSN 0002-7812

Ceramic Engineering and Science Proceedings (CESP)

www.ceramics.org/cesp

Practical and effective solutions for manufacturing and processing issues are offered by industry experts. CESP includes five issues per year: Glass Problems, Whitewares & Materials, Advanced Ceramics and Composites, Porcelain Enamel. Annual subscription runs from January to December. ISSN 0196-6219

ACerS-NIST Phase Equilibria Diagrams CD-ROM Database Version 3.0

www.ceramics.org/phasecd

The ACerS-NIST Phase Equilibria Diagrams CD-ROM Database Version 3.0 contains more than 19,000 diagrams previously published in 20 phase volumes produced as part of the ACerS-NIST Phase Equilibria Diagrams Program: Volumes I through XIII; Annuals 91, 92 and 93; High Tc Superconductors I & II; Zirconium & Zirconia Systems; and Electronic Ceramics I. The CD-ROM includes full commentaries and interactive capabilities.

Ceramic Transactions
Volume 146

Characterization & Control of Interfaces for High Quality Advanced Materials

Proceedings of the International Conference on the Characterization and Control of Interfaces for High Quality Advanced Materials (ICCCI 2003), Kurashiki, Japan, 2003

Edited by

Kevin Ewsuk
Sandia National Laboratories

Kiyoshi Nogi
Osaka University

Markus Reiterer
Sandia National Laboratories

Antoni Tomsia
Lawrence Berkeley Laboratory

S. Jill Glass
Sandia National Laboratories

Rolf Waesche
Federal Institute for Materials Research and Testing (BAM)

Keizo Uematsu
Nagaoka University of Technology

Makio Naito
Osaka University

Published by
The American Ceramic Society
735 Ceramic Place
Westerville, Ohio 43081
www.ceramics.org

Proceedings from the International Conference on the Characterization and Control of Interfaces for High Quality Advanced Materials (ICCCI 2003) in Kurashiki, Japan, 2003

For information on ordering titles published by The American Ceramic Society, or to request a publications catalog, please call 614-794-5890, or visit our website at www.ceramics.org.

4 3 2 1-07 06 05 04

ISSN 1042-1122

ISBN 1-57498-170-6

Contents

Nanoparticle Design and Suspension Control

High Temperature Interfaces

Particulate Materials

Novel Processing

Microstructure

Hot Gas Cleaning Technology

Interface Control

Nanotechnology

Preface

Interfaces are important in many different areas of materials science and technology, and to a broad spectrum of industries. The International Conference on the Characterization and Control of Interfaces for High Quality Advanced Materials (ICCCI 2003) was organized and held in Kurashiki Japan in September 2003 to provide a multidisciplinary forum for international scientists and engineers to discuss interface science and technology. The conference addressed the influence of interface structure and composition on joining and materials properties, on controlling interfaces in materials synthesis and processing, and on interface characterization. Over 100 scientists and engineers from 18 different countries attended ICCCI 2003. The Proceedings of ICCCI 2003 features 58 peer-reviewed papers on interface science and technology that provide a unique and state-of-the art perspective on interface characterization and control. The articles address interface control, high temperature interfaces, nanoparticle design, nanotechnology, suspension control, novel processing, particulate materials, microstructure, and hot gas cleaning technology. This unique volume will serve as a valuable reference for scientists and engineers interested in interfaces, particulate materials, and nanotechnology.

Kevin Ewsuk
Kiyoshi Nogi
Markus Reiterer
Antoni Tomsia
S. Jill Glass
Rolf Waesche
Keizo Uematsu
Makio Naito

Nanoparticle Design and Suspension Control

SUPERCRITICAL HYDROTHERMAL SYNTHESIS OF NANOPARTICLES

Tadafumi Adschiri, Seiichi Takami, Mitsuo Umetsu, Satoshi Ohara,
and Takao Tsukada
Institute of Multidisciplinary Research for Advanced Materials
Tohoku University, 2-1-1 Katahira, Sendai 980-8577, Japan

ABSTRACT

Some important aspects of supercritical hydrothermal synthesis of nanoparticles are (i) nanoparticle formation, (ii) single crystal formation, (iii) the ability to control particle morphology to some extent with pressure and temperature, and (iv) the ability to create a homogeneous reducing or oxidizing atmosphere by introducing gases or additional components (O_2, H_2). In this study, another important aspect of supercritical hydrothermal synthesis was discovered; that is, in-situ surface modification of nanoparticles with alcohols, aldehydes, or carboxylic acids during crystallization. This paper also describes the simulation of the supercritical hydrothermal synthesis process, based on fluid dynamics, kinetics, solubility, nucleation, particle growth, and particle coagulation. Simulation results are compared with experimental results to elucidate the effect of mixing on the nucleation and growth of particles. Based on the simulation results, the mixing method and solution flow rate have been optimized to continuously produce ZnO whiskers with high crystallinity and good opto-electronic properties.

INTRODUCTION

We are developing a continuous process of hydrothermal crystallization at supercritical conditions.[1-11] In the proposed method, a metal salt in aqueous solution is mixed with high temperature water to rapidly increase the temperature of the metal salt solution, as shown in Figure 1. This minimizes reactions and crystallization during the heat-up period. A homogeneous phase is formed using this gas-supercritical water system, and in most cases, the particles produced are single crystals. O_2 or H_2 gas is introduced into the system to produce an oxidizing or reducing atmosphere as

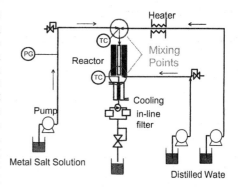

Figure 1. Experimental Apparatus used to make nanoparticles.

required for synthesis.   Using this method, we have succeeded in developing a continuous process to rapidly produce nanocrystals.

Nanoparticle formation, and changes in particle morphology are influenced by the solubility of the metal oxide, as well as the kinetics of the hydrothermal synthesis.   Both vary significantly around the critical point due to changes in the properties of water.[3,4]   In supercritical hydrothermal synthesis, the reaction rate increases by about two orders of magnitude above the critical point, while the solubility of a metal oxide in supercritical water significantly decreases.   This results in an extremely high supersaturation at the point of mixing, which gives rise to a high nucleation rate, and thus nanoparticle formation.   The variation in particle morphology around the critical point also is attributed to changes in the properties of water.[5,6]

Supercritical hydrothermal synthesis can be used to produce a variety of powders for various applications, including magnetic material $(BaO_6Fe_2O_3)$[7], phosphors $(Tb:YAG)$[8], metallic Ni nanoparticles, and Li ion battery materials $(LiCoO_2, LiMn_2O_4)$[9,10,11], as shown in Table I.

A critical issue for nanoparticle technology is particle handling.   In hydrothermal synthesis, the recovery of the nanoparticles from the water is essential.   Additionally, for spraying or coating applications, a stable dispersion of nanoparticles in an organic solvent or resin is required.

Surface modification is an effective way to control the dispersion of nanoparticles.   The conventional means to control surface characteristics is with the use of surfactants; however surfactants may not always be effective with particles that are already agglomerated.   Another method of controlling particle interactions is via surface chemical reactions.   If nanoparticles can be recovered in an organic solvent, silane coupling provides one way to introduce various

Table I.   Supercritical Hydrothermal Synthesis of Nanoparticles.

Metal Salt Aqueous Solution	Product	Particle Size [nm]	Reducing /Oxidizing	Ref.
$Al(NO_3)_3$	$AlO(OH)$	80-1000		7, 11
$Ce(NO_3)_3$	CeO_2	$\sim$100	Ox (NO_3)	3
$Co(NO_3)_3$	Co_3O_4	$\sim$50	-	7
$Fe(NO_3)_3$	αFe_2O_3	$\sim$50	-	7
$FeCl_3$	αFe_2O_3	$\sim$50	-	7
$Fe(SO_4)_3$	αFe_2O_3	$\sim$50	-	7
$Fe(NH_4)_2H(C_6H_5O_7)_2$	Fe_3O_4	$\sim$50	Red (CO)	7
$Fe(NO_3)_3 + Ba(OH)_2$	$BaO \cdot 6Fe_2O_3$	50$\sim$1000	-	8
$LiOH + Co(NO_3)_3$	$LiCoO_2$	40$\sim$200	Ox (O_2)	10
$LiOH + Mn(NO_3)_2$	$LiMn_2O_4$	10-20	Ox (O_2)	10
$Ni(CH_3COO)_2$	Ni	100	Red (H_2)	-
$Ni(NO_3)_2$	NiO	$\sim$200	-	7
$Ti(SO_4)_2$	TiO_2	$\sim$20	-	7
$TiCl_4$	TiO_2	$\sim$20	-	7
$Y(NO_3)_3 + Al(NO_3)_3 + TbCl_3$	$Al_5(Y+Tb)_3O_{12}$	20$\sim$600	-	9
$ZrOCl_2$	ZrO_2	$\sim$20	-	7

functional groups on the particle surface. An in-situ method to modify nanoparticle surfaces has been reported by Ziegler, Johnston, and Kogel.[12] They introduced surface modifiers during the chemical fluid deposition of nanoparticles, which reacted during the thermal decomposition of the metal salts in supercritical toluene to modify the particle surface.

In this study, we attempted to use an in-situ surface modification process during supercritical hydrothermal synthesis. At low temperatures, phase separation takes place for water-organic systems, such that effective surface modification cannot occur. However, a homogeneous phase can be expected at higher temperatures.

In the process shown in Figure 1, the mixing region inevitably affects the particle size, size distribution, and crystal structure of the product. Computer simulations of the supercritical hydrothermal synthesis process can be used as a tool to understand the mixing effect on crystallization.

The first objective of this study was to demonstrate in-situ surface modification during supercritical hydrothermal synthesis. The second objective of this study was to develop computer simulation technology for the supercritical hydrothermal synthesis process. ZnO whisker formation will be demonstrated on the basis of the simulated results.

EXPERIMENTAL

In this study, we performed flow experiments using the apparatus shown in Figure 1. Details of the experimental method are shown elsewhere.[1] In this study, we used two types of mixing junctions (shown in Figure 2). The reactant used was a 0.02 M concentration aqueous $Zn(NO_3)_3$ solution. Using a feed rate of 0.02 m/s at room temperature, this solution was mixed with supercritical water at 733 K fed at 0.5 m/s. After mixing, the solution temperature was 673 K. The pressure was controlled at 30 MPa using a back-pressure regulator.

For the in-situ surface treatment experiments, several organic reagents including hexanol, hexanoic acid, hexanal, and octanethiol were used to modify the surface of metal-oxide nanopaticles under the supercritical conditions. Experiments were performed in a pressure vessel with an inner volume of 5 cm³. A mixture of water, reagents, and oxide particles were added to the vessel and heated up to 400°C. The vessel was rinsed with water to collect the oxide particles. The particles obtained were dispersed in water or chloroform to determine their affinity to either water or chloroform.

SIMULATION

To simulate the process shown in Figure 1, a thermo-fluid model was needed to link with an estimation of the metal oxide solubility, the kinetics of the chemical reaction, and nucleation and crystal growth over a wide range of temperatures around the critical point, where thermophysical properties change significantly.

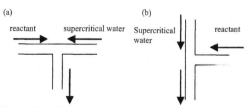

Figure 2. The two types of mixing junctions used.

Thermo-fluid Dynamics Model

In the present work, a computational fluid dynamics program for supercritical fluids in the Research and Development for Applying Advanced Computational Science and Technology (ACT-JST) Software Library[13] was employed to calculate the flow and thermal fields in the reactor. In this code, the continuity equation, momentum equation, equation of state, and energy equation are solved by the CIP (Cubic-Interpolated Propagation) method.[14]

Nucleation and Crystal Growth Model

To estimate the size distribution of the particles produced during supercritical hydrothermal synthesis, conventional nucleation theory and a particle growth model were used. The local monomer concentration was calculated from kinetics and transport equations. The degree of supersaturation was calculated from the solubility of the monomer, and the local monomer concentration. The nucleation rate and the nuclei radius were calculated using conventional nucleation theory. Crystal growth occurs by consuming monomers. The size distribution of the particles precipitated and transported by the flow was determined from the moment equations of the population balance using the finite difference numerical method, based on the control volume method. [15-18]

Metal Oxide Solubility and Kinetics

The solubility of the metal oxide in high temperature water was estimated using a simplified HKF model.[19-21] The parameters required for the estimation were obtained from the database for the HKF model.[19] The concentration of the chemical species in sub- and supercritical water was calculated by solving a set of nonlinear equations (a chemical equilibrium and a charge balance equation) using an iterative method. The calculated solubility was determined from the total concentration of the dissolved metal species. The estimated solubility of the metal oxides under acidic conditions decreases with increasing temperature up to the critical point, and then decreases significantly.

The reaction rate used in the simulation was determined experimentally using the apparatus shown in Figure 1. The results are shown in the literature.[1-3] The first order rate constant was determined from an Arrhenius plot. The rate constants fall on a straight line in the subcritical region, but the reaction rate deviates from a straight line to higher values above the critical temperature.

RESULTS AND DISCUSSION
In-situ Surface Modification

In most cases, hydroxyl or ether groups terminate the surfaces of the oxide particles formed in hydrothermal conditions. Therefore, the oxide particles are hydrophilic, and disperse well in water and other solvents whose dielectric constant is high. One possible way to modify the surface of the oxide particles is to use

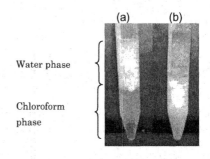

Water phase

Chloroform phase

Figure 3.   Dispersive properties of: a) as prepared; and b) surface modified oxide particles in a water/chloroform binary system.

reagents that react with the hydroxyl group. We examined various kinds of reagents including hexanol, hexanoic acid, hexanal, hexylamine and octanethiol, and we succeeded in modifying the surfaces of the oxide particles. Figure 3 shows a typical result of surface modification. On the left side, where the surface was not modified, the particles are dispersed only in the water phase. In contrast, the surface modified particles on the right side are dispersed only in the chloroform. We modified TiO_2 particles with hexanal to put alkyl groups on the particle surface to make the particles hydrophobic. This result indicates that in-situ surface modification is a convenient and rational method to control the properties of particles.

Simulation and Reactor Design
     We performed experiments using the two types of junctions shown in Figure 2. Using side injection, the mean size of the particles formed was smaller, and the particle size distribution was narrower than with the T junction (Figure 4).
     Figures 5 and 6 show the computer simulation flow pattern (a), and the distribution of total number concentration of particles (b) in the reactor. The reactor width was of 2.0×10^{-3} m, where the inlet velocity of reactant at 293 K was 0.1m/s, and of the 700K hot water was 0.4 m/s (nominally the same conditions as the experiment of Figure 3, although a 2 dimensional simulation was performed). In Figure 5a, there appears to be a stagnant zone at the mixing point, and a back-mixing region after the joint. This clearly influences particle growth and the particle size distribution, because the particles in these regions grow more than those in the main flow stream. On the other hand, for the side injection reactor, there is a relatively uniform flow pattern, and uniform particle growth occurs (Figure 6). This result agrees well with the experimental results shown in Figure 4. This suggests that simulations may enable us to rationally design reactors for hydrothermal synthesis under supercritical conditions.
     Based on our understanding of the hydrothermal reaction, experiments were performed to produce ZnO whiskers. Nanosize ZnO of high crystallinity has a wide band-gap with ultraviolet lasing action that is suitable for blue opto-electronic applications.[22, 23]

To produce high crystallinity whiskers, higher water density is desirable, since anisotropic crystal growth is expected at a higher dielectric constant. The temperature and pressure were set at 673 K and 30 MPa, respectively. A computer simulation was employed to optimize the flow rate to eliminate back-mixing in the reactor.

(a)   (b)
T   junction   Side injection

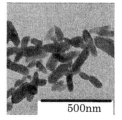

500nm   500nm

Figure 4.   Particles produced with the different mixing junctions shown in Figure 2.

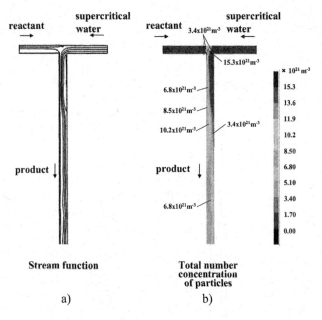

Figure 5.   Simulation results for the Figure 2a mixing configuration:
a) flow pattern; and b) particle concentration.

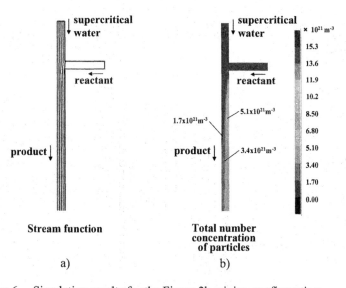

Figure 6.   Simulation results for the Figure 2b mixing configuration:
a) flow pattern; and b) particle concentration.

Figure 7 shows a SEM photograph of ZnO whiskers produced by supercritical water synthesis at 673 K and 30 MPa. As shown in this figure, well-developed crystalline whiskers with hexagonal end faces were produced.

Figure 7. SEM photo of ZnO whiskers.

CONCLUSIONS

In-situ surface modification of particle surfaces using an organic modifier was found to be possible during supercritical hydrothermal synthesis of nanoparticles. Effective organic surface modification can be attributed to the homogeneous reaction atmosphere for the organic modifiers, and to the high temperature water.

Computer simulation technology was developed and applied to supercritical hydrothermal synthesis. It is based on a thermo-fluid model, nucleation theory, kinetics, and a model of solubility estimation. The simulation results qualitatively explain the experimental results of fluid dynamics and particle formation in the reactor. Based on simulation results, the reactor conditions were refined to produce smaller nanoparticles with narrower particle size, and to produce ZnO single crystal nanowhiskers.

ACKNOWLEDGMENTS

The authors wish to acknowledge the Ministry of Education, Science, Sports and Culture, Japan, Genesis Research Institute, and New Energy, Industrial Technology Organization (NEDO) for financial support of this research. This research was also partially supported by the Ministry of Education, Culture, Sports, Science and Technology, and a Grant-in-Aid for the COE (center of excellence) project, Giant Molecules and Complex Systems, 2003.

REFERENCES

[1]T. Adschiri, M. Kanazawa and K. Arai, "Rapid and Continuous Hydrothermal Crystallization of Metal Oxide Particles in Supercritical Water," *J. Am. Ceram. Soc.*, **75** [4] 1019-1022 (1992).

[2]T. Adschiri and K. Arai, "Hydrothermal Synthesis of Metal Oxide Nanoparticles Under Supercritical Conditions," pp. 311-325 in *Supercritical Fluid Technology in Materials, Science and Engineering: Synthesis, Properties, and Applications*, Edited by Ya-Ping Sun, Marcel Dekker, Inc., 2003.

[3]T. Adschiri, Y. Hakuta, K. Sue and K. Arai, "Hydrothermal Synthesis of Metal Oxide Nanoparticles at Supercritical Conditions," *J. Nanoparticle Research*, **3** [2-3] 227-235 (2001).

[4]Y. Hakuta, S. Onai, H. Terayama, T. Adschiri, and K. Arai, "Production of Ultra-fine Ceria Particles by Hydrothermal Synthesis Under Supercritical Conditions," *J. Mat. Sci. Letters*, **17** 1211-1213 (1998).

[5]T. Adschiri, M. Kanazawa and K. Arai, "Rapid and Continuous Hydrothermal Synthesis of Boehmite Particles in Sub Critical and Supercritical Water," *J. Am. Ceram. Soc.*, **75** [9] 2615-2618 (1992).

[6]Y. Hakuta, T. Adschiri, H. Hirakoso, K. Arai, "Chemical Equilibria and

Particle Morphology of Boehmite (AlOOH) in Sub and Supercritical Water," *Fluid Phase Equilibiria*, **158-160** 733-742 (1999).

[7]Y. Hakuta, T. Adschiri, T. Suzuki, T. Chida, K. Seino, and K. Arai, "Flow Method for Rapidly Producing Barium Hexaferrite Particles in Supercritical Water," *J. Am. Ceram. Soc.*, **81** [9] 2461-2464 (1998).

[8]Y. Hakuta, K. Seino, H. Ura, T. Adschiri, H. Takizawa and K. Arai, "Production of Phosphor (YAG:Tb) Fine Particles by Hydrothermal Synthesis in Supercritical Water," *J. Mat. Chem.*, **9** 2671-2674 (1999).

[9]K. Kanamura, A. Goto, R.Y. Ho, T. Umegaki, K. Toyoshima, K. Okada, Y. Hakuta, T. Adschiri, and K. Arai, "Preparation and Electrochemical Characterization of LiCoO$_2$ Particles Prepared by Supercritical Water Synthesis," *Electrochemical and Solid-State Letters*, **3** [6] 256-258 (2000).

[10]T. Adschiri, Y. Hakuta, K. Kanamura, and K. Arai, "Continuous Production of LiCoO$_2$ Fine Crystals for Lithium Batteries by Hydrothermal Synthesis Under Supercritical Condition," *High Pressure Research*, **20** [1-6] 373-384 (2001).

[11]T. Adschiri, Y. Hakuta, and K. Arai, "Hydrothermal Synthesis of Metal Oxide Fine Particles at Supercritical Conditions," *Industrial & Engineering Chemistry Research*, **39** [12] 4901-4907 (2000).

[12]K.J. Ziegler, R.C. Doty, K.P. Johnston, and B.A. Korgel, "Synthesis of Organic Monolayer-Stabilized Copper Nanocrystals in Supercritical Water," *J. Am. Chem. Soc.*, **123** 7797-7803 (2001).

[13]http://act.jst.go.jp/index_e.html

[14]K. Oka, K. Amano and I. Enbutsu, "Development of a Method of Analysis for Flow, Heat and Chemical Reactions in Supercritical Fluid," *Trans. JSME B*, **66** 2823-2830 (2000).

[15]M.L.J. van Leeuwen, O.S.L. Bruinsma and G.M. van Rosmalen, "Influence of Mixing on the Product Quality in Precipitation," *Chem. Engineering Sci.*, **51** 2595-2600 (1996).

[16]H. Wei and J. Garside, "Application of CFD Modelling to Precipitation Systems," *Trans. IChemE.*, **75** 219-227 (1997).

[17]L. Falk and E. Schaer, "A PDF Modeling of Precipitation Reactors", *Chem. Engineering Sci.*, **56** 2445-2457 (2001).

[18]D.L. Marchisio, R.O. Fox, A.A. Barresi, M. Garbero and G. Baldi, "On the Simulation of Turbulent Precipitation in a Tubular Reactor via Computational Fluid Dynamics (CFD)," *Trans. IChemE*, **79** 998-1004 (2001).

[19]J.C. Tanger, H.C. Helgeson, "Calculation of the Thermodynamic and Transport Properties of Aqueous Species at High Pressures and Temperatures: Revised Equations of State for the Standard Partial Molal Properties of Ions and Electrolytes," *Am. J. Sci.*, **288** 19-98 (1998).

[20]K. Sue, Y. Hakuta, R.L. Smith, Jr., T. Adschiri and K. Arai, "Solubility of Lead (II) Oxide and Copper (II) Oxide in Subcritical and Supercritical Water," *J. Chem. & Engineering Data*, **44** [6], 1422-1426 (1999).

[21]K. Sue, T. Adschiri and K. Arai, "A Predictive Model for Equilibrium Constants of Aqueous Inorganic Species at Subcritical and Supercritical Conditions," *Industrial and Engineering Chem. Res.*, **41** [13] 3298-3306 (2002).

[22]H. Cao, J.Y. Xu, D.Z. Zhang, S.-H. Chang, S.T. Ho, E.W. Seelig, X. Liu and R.P.H. Chang, "Spatial Confinement of Laser Light in Active Random Media," *Phys. Rev. Letters*, **84** [24] 5584-5587 (2000).

[23]M.H. Huang, S. Mao, H. Feick, H. Yan, Y. Wu, H. Kind, E. Weber, R. Russo and P. Yang, "Room-Temperature Ultraviolet Nanowire Nanolasers," *Science*, **292** 1897-1899 (2001).

PRODUCTION OF ORDERED POROUS STRUCTURES WITH CONTROLLED WALL THICKNESS

Yuji Hotta and Koji Watari
National Institute of Advanced
Industrial Science and Technology
Aichi Nagoya 463-8560, Japan

P. C. A. Alberius and L. Bergström
YKI, Institute for Surface Chemistry
Drottning Kristinas väg 45
Stockholm, Sweden

ABSTRACT

Polystyrene (PS) colloidal particles have been used as templates to produce well-ordered porous silica structures. Silica films were grown on the PS particle surfaces using an ethanol solution containing acidic water and tetraethyl orthosilicate. The coated particles were characterized using transmission electron microscopy, scanning electron microscopy, and by measuring zeta potential. The growth rate and the maximum thickness of the surface coating are related to the hydrolysis and condensation rate, respectively. Centrifugal casting of the coated PS particles formed ordered structures. Calcination of the close-packed PS spheres yields a well-ordered inverse opal structure consisting of monodispersed pores. The shrinkage of the silica produced by hydrolyzing TEOS increases as the pH for the hydrolysis increases. The production of the well-defined porous material is influenced by the coating thickness on the PS particles, and by the structure of fine silica particles produced by the hydrolysis of the TEOS.

INTRODUCTION

In macroporous ceramics, pore wall thickness and structural integrity are important because these parameters determine mechanical stability, and influence the physical properties, e., density, thermal conductivity and dielectric permittivity.[1] Macroporous materials[2] can be designed to provide optimal transport properties and efficiency in catalysis and large molecule separation processes, as well as to immobilize and stabilize large guest molecules.[3,4] Many methods have been utilized to produce porous materials, e.g., replication of polymer foam[5], decomposition of foaming agents[6], foaming of sol-gel solutions using surfactants[7-10], and templating of emulsions.[11] However, all of these methods are confronted with issues of pore size distribution, long-range order, pore shape, and control of the wall thickness. In previous work, we demonstrated that a macroporous structure can be produce from silica-coated PS particles.[12] In this study, it is demonstrated how silica coated PS particles can be particulate building blocks to fabricate well-ordered macroporous materials with inverse opal or honeycomb structures. Effects of reaction time, pH, and surface charge on the silica film growth rate and cell wall thickness were investigated.

EXPERIMENTAL

Materials.
Hydrochloric acid (0.1N) and tetraethyl orthosilicate (TEOS) were obtained from Aldrich. Hydrochloric acid (HCl) was diluted to pH 1.5, 2, and 3 with water. Polystyrene (PS) particles with sulfate functional groups on the surface were obtained from Interfacial Dynamics Corporation (Tualatin, OR, USA, diameter: 500 and 1000 nm).

Preparation of coated particles.
TEOS was hydrolyzed by stirring for 2 h at pH 1.5, 2, or 3. Polystyrene particle suspensions were added to the hydrolyzed solution and stirred for 4, 24, 72, and 144 h to coat silica onto the surface of the PS particles. This mixture was centrifuged at 10000 rpm for 1 h, and the solid was washed three times in a 1:1 solution of ethanol and water.

Characterization.
The particle size of the silica coated PS was determined by scanning electron microscopy (SEM, Philips XL30, USA). The interfacial particle surface charge was determined using a zeta potential apparatus (Model 502, Nihon Ruhuto Co. Ltd., Japan). The weight loss of silica produced by hydrolysis of TEOS at various pHs (1.5, 2 and 3), and of the silica-coated PS particles was characterized using thermogravimetric analysis (EXTRA6000Tg/DTA, Seiko Instrument Inc., Japan) at a scan rate of 3°C min^{-1}.

Fabrication.
Centrifugation was used to fabricate macroscopic structures from the coated PS particles. The silica-coated PS particles were dispersed in a solution of ethanol and water, and centrifuged for 1 h. After drying, the compacts were calcined in air at 550°C for 5 h. Scanning electron microscopy (SEM, Philips XL30 and JEOL 820) and transmission electron microscopy (TEM, JEOL 200 FX) measurements were performed to determine the thickness of the silica coating on the PS particles, the cell wall thickness in the calcined samples, and the morphology of the pores.

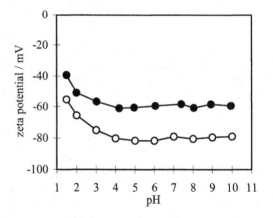

Figure 1. Zeta potential of different size PS spheres: filled-in circle, 500 nm; open circle, 1000 nm.

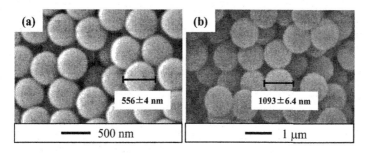

Figure 2.   SEM micrographs of: a) 500 nm; and b) 1000 nm PS particles reacted for 72 h in a hydrolyzed TEOS solution at pH 2.

RESULTS AND DISCUSSION
Silica Film Growth on PS Surface.

Figure 1 shows the zeta potential of the PS particles used in the present study. The PS particles are charged negatively in the pH range investigated, and the isoelectric point (IEP) could not be determined. The zeta potential of the 1000 nm PS particles is about 1.4 times larger than that of the 500 nm PS particles. This is probably related to a higher concentration of the sulfate functional groups on the surface of the larger PS particles.

The increase in particle size with reaction time observed by SEM shows that the PS particles have been coated with a layer of silica (Figure 2). The monodispersity of the coated particles, and the absence of debris indicate that the silica coating is uniform, and that the nucleation and precipitation occur predominantly at the PS particle surface. This is supported by TEM analysis of the coated PS particles. For the 500 nm PS spheres reacted for 72 h in a hydrolyzed TEOS solution at pH 2, the film thickness estimated from TEM is 30 nm. This corresponds well with the thickness obtained from the estimated increase in the particle diameter using SEM.

Figure 3 shows the relation between reaction time and the film thickness estimated from SEM observations. The thickness of the silica film as a function of time shows parabolic growth, with an asymptotic approach to a maximum film thickness. The film thickness increases as the reaction pH decreases.

The growth of the film

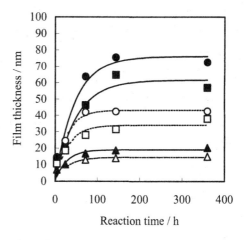

Figure 3.   Relationship between reaction time and film thickness estimated from SEM observations: open symbols, silica coated 500 nm PS spheres; closed symbols, silica coated 1000 nm PS spheres for (●,○) pH 1.5, (□,■) pH 2, and (△, ▲) pH 3.

thickness, δ, at reaction time, t, where δ_M is the maximum film thickness, and κ is the growth rate constant, can be described using equation (1),

$$\delta = \delta_M (1-e^{-\kappa t}) \qquad (1)$$

Figure 4 shows that the simple first-order equation (1) represents the kinetic data well. It is clear that the maximum silica-film thickness increases with decreasing pH, and that the growth rate constant also increases with decreasing pH. The maximum silica-film thickness grown on the 1000 nm PS spheres at pH 1.5, 2, and 3 is 1.6, 1.7, and 1.4 times larger than the film thickness formed on the 500 nm PS spheres, respectively. These results correlate with the observation that the surface charge of the 1000 nm PS spheres is about 1.4 times larger than that of the 500 nm PS spheres. Thus, the growth rate and the maximum film thickness are dependent on reaction time, pH, and the surface charge of the particles, which affords precise control over the final film thickness.

Macroporous Structures.

Figure 5 shows that the silica-coated PS spheres can be fabricated into close-packed structures by centrifugation. The macroporous silica structures are characterized by a well-defined and long-range order. The macroporous structure consists of hexagonally close-packed uniform pores resembling an inverse opal. The average center-to-center distance between the pores of the calcined well-ordered sample is 408 ± 11 and 773 ± 16 nm (Figures 5c and d). The diameters of the silica coated PS particles of 500 and 1000 nm are 556 ± 4 and 1093 ± 6 nm, respectively (Figure 2). The linear shrinkage of the calcined silica-coated 500 and 1000 PS sphere compacts during calcination are 27 and 26 %, respectively.

Figure 6 shows the relationship between linear shrinkage and silica-film thickness. We found that the shrinkage increases as the film thickness increases. The silica coated 500 and 1000 nm PS spheres show similar behavior. The shrinkage is mainly related to the densification of the silica wall by the condensation of the silanol groups, and by the evaporation of the water. The increase in weight loss as the reaction pH increases (Figure 7) suggests that the silica formed at pH 3 contains more hydrolyzed species than the silica formed at lower pH. This correlates well with the shrinkage data in Figure 6.

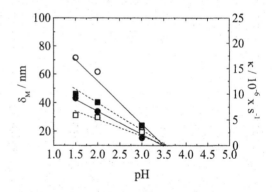

Figure 4. Relationship between reaction pH, maximum film thickness (δ_M) (squares), and growth rate constant (κ) (circles): closed symbol, 500 nm PS; open symbol, 1000 nm PS.

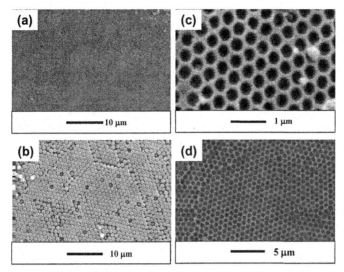

Figure 5. SEM micrographs of close-packed structures formed by centrifugation of silica coated 500 and 1000 nm PS spheres: a) and b) prior to calcination; and c) and d) after calcination.

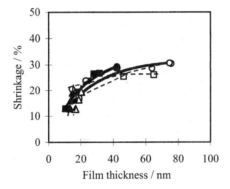

Figure 6. Relationship between shrinkage and film thickness; closed symbols, silica coated on 500 nm PS spheres; open symbol, silica coated on 1000 nm PS spheres for (●,○) pH 1.5, (□,■) pH 2, and (△, ▲) pH 3.

Figure 8 shows that the weight loss from the PS spheres takes place between 80 and 120°C, and above 260°C. This corresponds to the dehydration of adsorbed water, and to the decomposition of PS, respectively. In contrast, the weight decrease in the silica-coated PS particles takes place between 80 and 120°C, between 150 and 265°C, and above 265°C. The weight loss from the silica produced by the hydrolysis of TEOS is related to the dehydration of adsorbed water (120°C), dehydration by the reaction between silanol groups (150-300°C), and dehydration above 300°C due to the reaction of the hydroxyl groups in the internal structure (Figure 7). Hence, the hydrolyzed silica species on the coated particles react between 150 and 265°C to produce the silica film. Therefore, the silica structure forms before the PS decomposes.[13] This is probably related to the structural integrity of the macroporous structure after calcination; the silica film does not rupture or crack because it is flexible during the decomposition of the polystyrene.

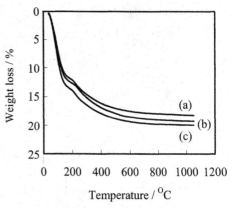

Figure 7. TGA measured weight loss from silica produced by hydrolysis of TEOS at pH: a) 1.5; b) 2; and c) 3.

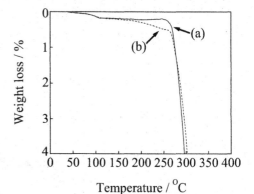

Figure. 8. TGA weight loss from: a) 500 nm PS spheres; and b) silica-coated 500 nm PS spheres.

Macroporous silica with a well-defined pore structure was produced by coating silica on 500 nm PS spheres. The film thickness on PS spheres increases with decreasing reaction pH, and can be controlled. The pore structure produced replicates the hexagonally close-packing of the PS spheres. The silica wall formed becomes thicker as the pH decreases from 3 to 2 to 1.5. Wall thickness also increases as film thickness increases.

CONCLUSIONS

It was shown that the growth rate and the maximum film thickness of silica on PS spheres are dependent on reaction time, pH, and surface charge, allowing for precise control over final film thickness. The structures produced by centrifugation consist of uniform close-packed voids with an inverse or honeycomb structure. To form well-defined porosity after calcination, the film grown on the PS spheres must react before removing the PS sphere.

REFERENCES

[1] T. J. Bandosz, C. Lin and J. A. Ritter, "Porosity and Surface Acidity of SiO_2–Al_2O_3 Xerogels", *J. Colloid. Interface Sci.*, **198**, 347-353 (1998).

[2] P. Langley and J. Hulliger, "Nanoporous and Mesoporous Organic Structures: New Openings for Materials Research", *Chem. Soc. Rev.*, **28**, 279 (1999).

[3] N. A. Melosh, P. Lipic, F. S.Bates, F.Wudl, G. D. Stucky, G. H. Fredrickson and B. F. Chemelka, "Molecular and Mesoscopic Structures of Transparent Block Copolymer-Silica Monoliths", *Macromolecules*, **32**, 4332 (1999).

[4] F. C. Meldrum, and R. Seshadri, "Porous Gold Structures Through Templating by Echinoid Skeletal Plates", *Chem. Commun.*, 29 (2000).

[5] F. F. Lange and K. T. Miller, "Open-Cell, Low-Density Ceramics Fabricated from Reticulated Polymer Substrates", *Adv. Ceram. Mater.*, **2**, 827 (1987).

[6] N. W. Androff, L. F. Francis and B. V. Velamakanni, "Macroporous Ceramics from Ceramic Polymer Dispersion Methods", *AIChE J.*, **43**, 2878 (1997).

[7] M. Wu, T. Fujiu and G. L. Messing, "Synthesis of Cellular Inorganic Materials by Foaming Sol-Gels", *J. Non-Cryst. Solids*, **121**, 407 (1990).

[8] V. Dmitri, V. S. Elena, K. Andreas, G. Nikolai, H. Markus, L. Andrey and W. Horst, "A New Approach to Crystallization of CdSe Nanoparticles into Ordered Three-Dimensional Superlattices", *Adv. Mater.*, **13**, 1868 (2001).

[9] S. W. Sofie and F. Dogan, "Freeze Casting of Aqueous Alumina Slurries with Glycerol", *J. Am. Ceram. Soc.*, **84**, 1459 (2001).

[10] F. Caruso, R. A. Caruso and H. Möhwald, "Nanoengineering of Inorganic and Hybrid Hollow Spheres by Colloidal Templating", *Science*, **282**, 1111 (1998).

[11] A. Imhof and D. J. Pine, "Ordered Macroporous Materials by Emulsion Templating", *Nature*, **389**, 948 (1997).

[12] Y. Hotta, P. C.A. Alberius and L. Bergström, "Coated Polystyrene Particles as Templates for Ordered Macroporous Silica Structure with Controlled Wall Thickness ", *J. Mater. Chem.*, **13**, 496-501 (2003).

[13] P. F. James, "The Gel to Glass Transition: Chemical and Microstructural Evolution", *J. Non-Crystal. Solids*, **100**, 93-114 (1988).

CONTROL OF DISPERSION CHARACTERISTICS OF TiO$_2$ NANO-POWDERS FOR ELECTRONIC PAPER

S.H. Kwon and J.H. Ahn
Department of Material Science &
Engineering, Hanyang University
17 Haengdang-dong, Seongdong-gu
Seoul 133-791, Korea

Y.I. Cho
Korea Photonics Technology Institute
459-3 Bonchon-Dong, Buk-gu
Gwangju 500-210, Korea

W.S. Hong and S.J. Kim[*]
Sejong Advanced Institute of Nano
Technology, Sejong University
98 Gunja-dong, Gwangjin-gu
Seoul 143-747, Korea

ABSTRACT
    The dispersion of rutile (TiO$_2$) nanoparticles electronic paper displays was studied. The dispersions were prepared using a homogeneous precipitation process (HPPLT) in organic media. To use TiO$_2$ nanoparticles as an electronic paper pigment, OLOA1200 dispersant was added to a low dielectric constant organic media containing dispersed TiO$_2$ particles. This produced a polyisobutylene chain in which the OLOA1200 acts as a steric stabilizer and a charge control agent. Then, the dispersed TiO$_2$ particles were coated with a polymeric material to produce a particle with a lower (average) specific gravity. We characterized the dispersion characteristics of TiO$_2$ in organic media, focusing on the surface and electro-optic properties. Particle mobility and particle size were analyzed by measuring the ζ-potential. The optical contrast in an electrophoretic test cell was measured by varying driving frequency and voltage.

INTRODUCTION
    In paper-like display systems, there have been electrophoretic displays, twisting ball displays, in-plane type electrophoretic displays and a cholesteric-type display. Among them, an electrophoretic display using TiO$_2$ particles is the most promising candidate because it offers advantages such as an ink-on-paper appearance, good optical contrast, wide viewing angle, image stability in the off state, and extremely low power consumption.[1-5] The critical technology in electrophoretic displays is the dispersion of the rutile particles in the organic media. Rutile prepared using HPPLT has advantages compared to commercial

---

[*] Corresponding Author: sjkim1@sejong.ac.kr

TiO$_2$ powders.[6] Nano size TiO$_2$ powder with the rutile phase, produced using HPPLT, is expected to improve display resolution and particle mobility, and the rutile phase can also provide high reflectance (n=2.7) and whiteness. Rutile powder has a difference in optical index of refraction of Δ n = 1.3 relative to the surrounding dielectric liquid. This results in a very short scattering length (~1 μm), high reflectivity, and high contrast, similar to that of ink on paper. Though the electrophoretic rutile system has many advantages, the specific gravity of the TiO$_2$ powder is nearly 4.2 g/cm^{-3}, which can result in sedimentation in the dispersing liquid medium; consequently, a polymer coating is necessary to try to produce a lower specific gravity composite particle to match the density of the dielectric suspending liquid medium[6]. In this study, TiO$_2$ was prepared using the HPPLT process, and ζ-potential was measured to characterize the dispersibility of the TiO$_2$ particles in a low dielectric constant organic medium containing a charge control agent to change the TiO$_2$ surface from neutral to negative.

EXPERIMETAL

Transparent titanium tetrachloride (3N; TiCl$_4$, Aldrich Chemical Co., Inc., Milwaukee, WI) was used as a starting material to fabricate TiO$_2$ powder using a homogeneous precipitation method. The procedure used to make TiOCl$_2$ solution from TiCl$_4$ is described in detail elsewhere.[7] The precipitates produced by the HPPLT process were dried at 150°C for 12 h to obtain the final powder.

The recipe used to disperse the TiO$_2$ particles in a low dielectric constant liquid is shown in Table I. To prepare the TiO$_2$ particle dispersions, OLOA1200 was used as the surfactant. In our electrophoretic application, OLOA1200 (Figure 1) is a charge control agent or surfactant that adheres to the TiO$_2$ particle surface, and that has a copolymer type structure such that it can provide steric stabilization with a basic (amine) group.[8]

To optimize the adsorption of the polyisobutene chain onto the TiO$_2$ surface, the TiO$_2$ powder was dispersed in tetrachloroethylene at various levels of pH from 2-12. The pH was measured using a pH Meter (Metter Toledo Co. MP220)

To characterize dispersibility, the change in ζ-potential with increasing surfactant concentration was measured from 0-2 wt%. We added the OLOA1200 in 0.2 wt% increments after dispersing the TiO$_2$ in solution and we characterized the change in dispersibility with increasing surfactant content. ζ-potential measurements were made using a ζ-potential analyzer (Otsuka Co. ELS 8000 electrophoretic light scattering).

Cyclohexane and tetrachloroethylene were mixed in the various ratios (80:20, 70:30 and 60:40), to determine the changes in ζ-potential when two different liquids are present as the dispersing medium.

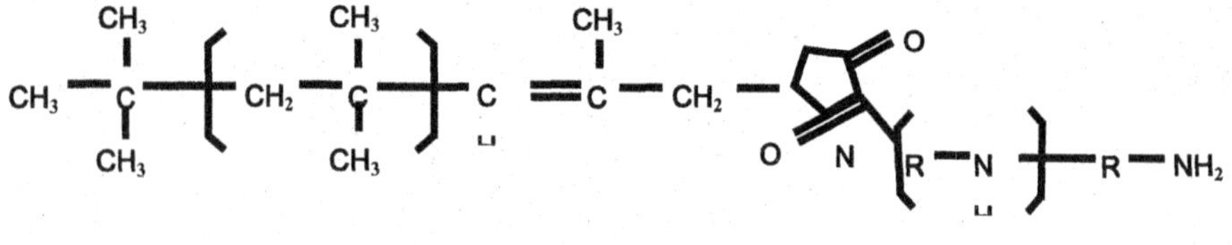

Figure 1. Schematic of chemical structure of the polyisobutene chain.

Table I. The recipe used to disperse TiO_2 in a low dielectric constant liquid.

Component	Concentration (wt%)
Cyclohexane or TCE	97~99
TiO_2	1
OLOA1200	0~2

The TiO_2 particles were coated with polymer (polyethylene, Aldrich Co.) by dispersing in liquid and cooling. Low molecular weight polyethylene was added to the dispersing liquid by stirring at around 100°C. During the dispersing process and cooling, the dissolved polymer adsorbed on to the surface of the TiO_2 particles. FT-IR and SEM were used to characterize the coated TiO_2 surface.

An electrophoretic test kit was fabricated from two ITO-coated glass panels placed face to face and separated by 100~200 µm ribs. The TiO_2 dispersion was injected into the space between the ribs, and the test kit was sealed hermetically. An electrical bias was applied across both sides of the glass to characterize the electrophoretic behavior of the TiO_2 particles. The image stability after cutting off the bias also was tested.

RESULTS AND DISCUSSION

TiO_2 particles for use as the white pigment in an electronic paper display were produced using the HPPLT process. Figure 2 shows a SEM photograph of the TiO_2 particles formed, which have a surface area >150 m^2/g. Figure 3 shows the x-ray diffraction (XRD) pattern for the powder. The rutile phase of TiO_2 forms during the precipitation stage. In contrast, P-25 (commercial TiO_2, Degussa. Co) contains a mixture of the rutile and anatase phase. From this result we know that the HPPLT process can produce TiO_2 with the rutile phase.

SEM SEI 15.0kV ×20,000   1µm WD16mm

Figure 2. A SEM image of the HPPLT TiO_2 powder.

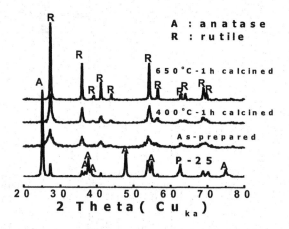

Figure 3.   XRD patterns for the HPPLT TiO$_2$ powder before and after calcination in air, and for the Degussa Co P-25 TiO$_2$ powder.

Figure 4a shows that nanoparticle TiO$_2$ has a maximum ζ-potential of 120 mV when more than 0.6 wt% of OLOA1200 is added to a 1 wt% TiO$_2$ solution. The polyisobutene polymer chain of OLOA1200 is believed to provide steric stabilization of the TiO$_2$ particles. The OLOA1200 helps prevent agglomeration, and helps improve the mobility of the TiO$_2$ particles when an electric field is applied. Surfactant concentrations in excess of 0.6 wt% exceed the Critical Micelle Concentration (C.M.C), resulting in saturation of the ζ-potential.

Figure 4b shows that the ζ-potential of the nanoparticle TiO$_2$ dispersions decreases with increasing pH from 1 to 10. This effect is caused by the chemical structure of the OLOA1200, which has polyisobutene chains that consist of basic main part and an acidic tail. When the particle surface is acidic, the basic main part of the OLOA1200 easily adsorbs onto the TiO$_2$ particle surface. However, for a basic TiO$_2$ surface, adsorption is weak due to the low ζ-potential. To obtain good TiO$_2$ particle dispersibility, it is necessary to control the TiO$_2$ surface condition. The best dispersibility is achieved when the TiO$_2$ particle surface is acidic.

Figure 5a shows a SEM photograph of TiO$_2$ coated with a polymer. Figure 5b reveals that there is no visible characteristic IR peak for TiO$_2$ after coating, which confirms that TiO$_2$ is completely encapsulated by the polymer. The polymer coating produces a composite particle with a lower specific gravity (relative to pure TiO$_2$)

It is very difficult to produce a polymer-coated TiO$_2$ with a low enough specific gravity to match the density of the cyclohexane dispersing solution; therefore tetra-chloroethylene was mixed with cyclohexane to produce a higher density dispersing solution. The tetrachloroethylene mixes well with cyclohexane, and helps control the density of the dispersing solution.

Figure 4a shows that adding tetrachloroethylene in concentrations >20% caused the ζ-potential of the mixture to decrease. Presumably, this occurs because

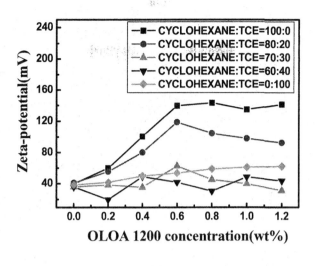

(a)

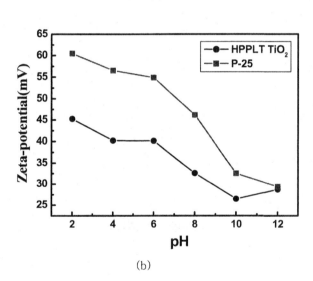

(b)

Figure 4. ζ-potential of nano-sized HPPLT and P-25 TiO₂ powder dispersions as a function of: a) surfactant concentration in different cyclohexane:TCE liquids; and b) pH in TCE.

(a)

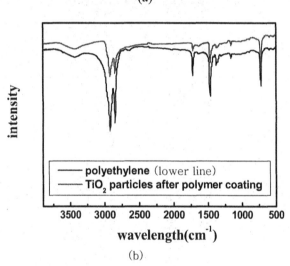

polyethylene (lower line)
TiO₂ particles after polymer coating

wavelength(cm⁻¹)

(b)

Figure 5. An a) SEM image, and b) FT-IR spectrum of polymer-coated TiO₂ particles.

the characteristics of the individual solvents (i.e., viscosity, refractive index, and dielectric constant) work in opposition to each other.

Finally, we fabricated an electrophoretic test kit and verified the electrophoretic phenomenon (Figure 6). Also, after a number of voltage reversal tests, we cut off the bias to confirm the display image was stable. We maintained the image for over one week.

Figure 6. The electrophoretic test kit and an image of the electrophoretic phenomenon (bottom).

SUMMARY

We prepared TiO_2 with the rutile phase, and with an acidic surface using a homogeneous precipitation process (HPPLT). The TiO_2 was dispersed in the various solutions using OLOA1200 as a surfactant. HPPLT processed TiO_2 is a good material for the white pigment in electronic paper because it has a good optical properties, and the ζ-potential can be easily controlled compared with P-25. The best dispersibility is achieved when more than 0.6 wt% of OLOA1200 is added to TiO_2 in 100 % cyclohexane solution.

TiO_2 particles were coated with polymer to produce a composite particle with a lower density for dispersion. Additionally, tetrachloroethylene was added to cyclohexane to produce a higher density liquid dispersion medium to better match that of the polymer-coated TiO_2 particles. Finally, we used an electrophoretic test kit to verify that polymer coated TiO_2 prepared by the HPPLT process has good optical and electrical properties for electronic paper.

ACKNOWLEDGEMENTS

This work was financially supported by the KOSEF(R01-2002-000-00338-0) and Ministry of Commerce, Industry & Energy(10002930).

REFERENCES

[1] J.I. Pankove, "Color Reflection Type Display Panel," Tech. Note No. 535, RCA Lab Princeton, NJ (1962).

[2] I. Ota, J. Honishi, & M. Yoshiyama, "Eletrophoretic Image Display Panel," *Proc. IEEE* **61**, 832-836 (1973).

[3] N.K. Sheridon & M.A. Berkovitz, "The Gyricon-a Twisting Ball Display," *Pro. Soc. Information Display*, **18** (3, 4), 289-293 (1977).

[4] A.L. Dalisa, "Electrophoretic Display Technology," *IEEE trans.* Electron Devices ED-**24** (7), 827-834 (1977).

[5]B. Comiskey, J.D. Albert, H. Yoshizawa, J. Jacoson, "An Electrophoretic Ink for All-Printed Reflective Electronic Display," *Nature*, 394 (1998).

[6]B. Comiskey, J.D. Albert, J.D. Albert, J. Jacobson., "Elecrophoretic Ink: a Printable Display Material," *In Digest of Tech. Papers* 75-76 (1997).

[7]S.J. Kim, S.D. Park, Y.H. Jeong, S. Park, "Homogeneous Precipitation of TiO_2 Ultrafine Powders from Aqueous $TiOCl_2$ Solution," *J. Am. Ceram. Soc.*, **82** (4) 927-32 (1999).

[8]R.J. Pugh, T. Matsunaga, and F.M. Fowkes. "The Dispersibility and Stability of Carbon Black in Media of Low Dielectric Constant," *Colloids and Surfaces*, **7**, 183-207(1983).

DIRECT FORCE MEASUREMENTS OF CERAMIC NANOPARTICLES IN LIQUID MEDIA

Jeong-Min Cho and Wolfgang M. Sigmund
Department of Materials Science and Engineering
University of Florida
225 Rhines Hall
Gainesville, FL 32611

ABSTRACT
Surface forces for ceramic particles in slurries have recently been of interest in both theory and practice. As particle size decreases from the micron size down to the nano size, surface forces become increasingly important. The colloid probe technique based on atomic force microscopy (AFM) that is well established for micron size particles can be extended to nanosize particles (10 nm) by using carbon nanotubes as proximal probes. Nanotubes, with their high aspect ratio, avoid contributions from cone shapes as happens with standard AFM tips. Interaction forces of micron- and nano-sized colloidal probes will be presented. The difference in the particle size has an influence on surface forces for sterically dispersed colloidal systems. AFM measurements clearly demonstrate that surface forces for micron size particles are affected by polymer concentration, whereas surface forces for nanosize objects are independent of polymer concentration. Additionally, while polymeric force of nanoparticles is inversely proportional to the third power of separation distance at long interaction range, its dependence on the inverse of one-third power of separation distance for short interaction range was observed.

INTRODUCTION
Surface forces for ceramic particles play a considerable role in many particulate systems; including ink and paint pigments, adhesion, coatings, chemical mechanical planarization (CMP) slurries and biological applications.[1-4] Theoretical and experimental approaches have been developed to elucidate surface forces for particles of sizes in the range of millimeters to micrometers. However, the extension of these theories down to the nanometer size range is questionable because the premises that were required to develop the current theories directly exclude the nanometer scale. For instance, the Derjarguin approximation, which translates the interaction energy between two flat surfaces to other shapes, like sphere-sphere and sphere-plane, is only valid when the radius of the particle is much greater than the separation distance between the particles.[5] The Poisson-Boltzmann equation is based on point charges, which is not applicable for a very short separation distance or small particles, because

the contribution to the increase of effective volume by ion size is not negligible for very small particles.

Chemical additives such as surfactants, polymers, or polyelectrolytes are normally utilized to efficiently control the stability of a colloidal system.[6] The presence of chemical additives can change the interaction behavior of colloidal particles. Polymer induced surface forces are affected by the nature of the polymer (i.e., structure of polymer, ionizability, conformation on surfaces) and the physicochemical condition of the system (i.e., solvent quality, pH, and ionic strength). In addition to these parameters, the relative size of polymer molecules to the particles also must be considered for nanoparticles. For example, the theories developed for the colloidal behavior of micron size particles with polymers or polyelectrolytes have simply assumed the particle surface to be flat because relative sizes difference between particles and polymer molecules are pretty large. However, when the particle size decreases to a few nanometers, this assumption must be reconsidered. The surface chemical structure can change due to a large radius of curvature of a particle as shown by Kamiya et al.[7] They utilized a Si_3N_4 AFM tip over one primary silica particle, after imaging the silica compacts in water, to study the effect of the particle size on the formation of a hydration layer around the surface. They observed with FT-IR and force measurement, that no hydration force was found for nanoparticles with a diameter less than about 30 nm. They suggested this could result from the different silanol structure on the surface of SiO_2 depending on the particle size. In addition, the relative size of the polymer molecule and the colloidal particle may affect the conformation of the polymer to the particle surface. According to molecular dynamics (MD) simulations for the interaction of polyethylene oxide (PEO) with colloidal silica,[8] polymers adopt a flat conformation on the surface when the polymer molecules are much larger than the particles. In contrast, when particle size increases, the polymer adopts an extended conformation consisting of loops and tails.

Therefore, the fundamental understanding and more precise control of the surface chemistry of nanometer-sized colloidal particles is especially important. As an approach to the above purpose, the extension of colloidal probe technique to the nanometer size scale using nanotubes will be demonstrated, and surface force measurements between polymer coated layers will be compared and discussed depending on the probe size

EXPERIMENTAL

A single glassy carbon sphere with a radius of about 25 μm was used as a micron size probe. It was attached to the Si cantilever (ESP, Digital Instrument, CA) with a spring constant of 0.05 N/m, using a low melting temperature resin (Epon R 1004, Shell Chemicals Co.) following the Ducker method [9]. For a nanosize probe, multi-walled carbon nanotubes (MWNT) were extracted from the uncrushed carbon soot that was produced by the arc-discharge method (Alfa Aesar, MA). This carbon soot was purified by heat-treating in a thermo gravimetric analyzer (TGA) at 650°C in air to oxidize the carbonaceous materials such as amorphous carbon until the residual mass was less than 10 wt% of the original mass. No apparent defects such as an open cap or partially oxidized nanotubes were observed in SEM and TEM images. Many bundles of MWNT were aligned on the edge of a sharply folded SEM carbon tape by lightly touching the purified carbon soot. One MWNT bundle was then glued to the AFM tip with a small amount of epoxy glue in a dark field mode of an inverted optical microscope (Carl Zeiss, Germany) using a method similar to that developed by H. Dai et al.[10] After a long nanotube bundle was attached to the end of the AFM tip, the assembly was heated to 200°C for 5 min. to harden the epoxy glue. Then it was air-cooled to room temperature. Next, the AFM tip was weakly flushed with deionized (DI) water, leaving a strongly attached MWNT bundle with a single protruding nanotube. The length of the attached MWNT is usually about 1 µm, and the diameter is about 20 nm. A plasma enhanced chemical vapor deposited (PE-CVD) 2 µm thick SiO_2 coating of 0.3 nm RMS roughness on a silicon wafer supplied by the Motorola Company was used as a substrate. A branched polyethyleneimine (PEI, Mw=70,000 g/mol, 30 w/v%, Alfa Aeser, MA) was diluted to the targeted concentration with ultrapure water for the steric force measurements. PEI has a high affinity for the silica and the nanotube at high pH. Good adsorption of PEI on the CNT in aqueous solution has been reported by Seeger et al.[11]

Force measurements were conducted in a fused silica liquid cell with a commercial multimode AFM (Nanoscope III, Digital Instrument, CA). The working space between the tip and the substrate in the cell was secured with silicone O-ring. Prior to making measurements, the silica substrate was cleaned by ultrasonicating in detergent, acetone, and then ethanol for 30 minutes, followed by rinsing with DI water. The probe was cleaned with ethanol and water right before the measurement. After the cantilever and substrate were installed in the AFM, 20 ml of the PEI solution was injected through the silicone tubing, and PEI molecules were allowed to adsorb onto the silica surfaces for 15 minutes. Force measurements were conducted at three different positions on the silica substrate. At least five measurements were taken at each position. A Z-scan rate of 0.5 Hz, and a scan size of 600-1000 nm were used. A force-separation plot was converted from the deflection signal – piezo scanner movement data, and averaged with data analysis software.

RESULTS AND DISCUSSION

The use of a nanotube as an alternative to a single nanoparticle to investigate interaction forces for nanoparticles can be validated with a theoretical calculation as shown in Figure 1. A standard AFM tip has of pyramidal shape with an irregular apex, which may influence the direct force measurement between the tip and the substrate. A standard AFM tip can be treated as a cone with an angle of 2θ, attached by a half sphere with radius R. The van der Waals force between the cone-shaped AFM tip and the flat substrate then can be described as[12]

$$F(D)_{cone-plate} = -\frac{A}{6}\frac{D^2(R+D)\sin^2\theta + D^2R\sin\theta + R^2(R+D)}{D^2(R+D)^2} \tag{1}$$

where A is the Hamaker constant, R is the radius, and D is the separation distance.

The force calculated with a conventional AFM tip deviates significantly from the force for a nanoparticle. On the other hand, using a nanotube tip, which has a high aspect ratio, sidewall effects are minimized, and the interaction forces between nanoparticles can be more accurately measured.

Figure 2 shows the normalized force-separation distance plot measured using a glassy carbon sphere with different amounts of PEI. It was observed that the surface forces measured using a micron size colloidal probe are dependent on the

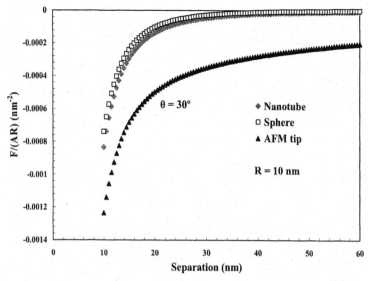

Figure 1. The calculated van der Waals interaction force (from Eq. 1) between a colloidal probe and a flat plate depending on the probe tip geometry. A is the Hamaker constant and R is the radius.

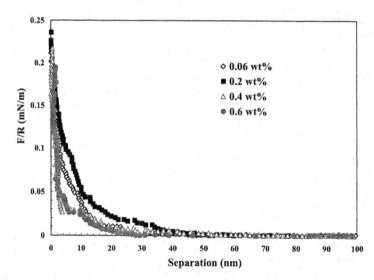

Figure 2. Force vs. separation plot between a glassy carbon sphere and a SiO_2 substrate. The force dependence on the PEI concentration was determined using a micron size particle (pH 10, 0.01 M NaCl).

polymer concentration. With 0.06 wt% PEI, there appears to be incomplete coverage of the surface by the polymer layer, considering the highest repulsive force occurs at a separation of 50 nm with 0.2 wt% of PEI. This repulsive barrier thickness is reasonable considering that the radius of gyration (R_g) for a linear PEI molecule is 15.9 nm. As the concentration of polymer increases over 0.2 wt%, the interaction distance and repulsive forces decreases. One feasible interpretation is that free, non-adsorbed polymers are present. When a large amount of polymer is not adsorbed on the surface and the separation between surfaces is reduced to a critical distance, attractive forces may occur by the depletion due to the induced osmotic pressure. Additionally, it is possible that polymer entanglement may reduce the repulsive forces with increasing polymer concentration.

The force-separation distance profile for the MWNT probe with the SiO$_2$ substrate is shown in Figure 3. In contrast to the micron size particle, the same magnitudes of repulsive forces are observed, independent of the polymer concentration. The interaction distance with the MWNT tip is significantly shorter than for that with the micron size glassy carbon tip. This behavior can be explained by the difference in probe size. The glassy carbon probe is much larger than the polymer, which allows it to interact with both tail and loop structures of the adsorbed PEI. Consequently the glassy carbon probe measures a thicker PEI layer. However, the nanosize MWNT is smaller than the dimensions of the polymer chain (i.e., radius of gyration of PEI 70000 $\cong$ 15.9 nm), and the polymer tails easily diffuse out of way when the nanosize tip approaches. Additionally, depletion effects cannot play a significant role, because the nanotube tip size is similar to that of the polymer molecule. Thus, once the surface of the nanoparticle

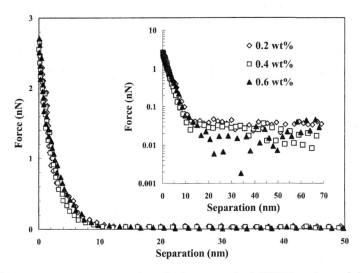

Figure 3.   Force vs. separation plot between the MWNT probe and the PEI coated SiO$_2$ substrate.   The measured forces are independent of PEI concentration (pH 10, 0.01 M NaCl). The inset is the logarithmic plot to show clearly no electric double layer force.

is covered with a monolayer of the polymer, the particle interaction forces are less affected by the presence of free polymer that has not been adsorbed.

The interaction force between two physically adsorbed polymer layers follows an inverse of a power law relationship with separation distance by mean field and scaling theory.[13] The interaction force with a nanotube tip was assumed to have a similar dependence. Therefore, the fitting equation for the nanotube tip was assumed to be

$$F(H) = \frac{A}{H^n} \tag{2}$$

where H is the separation distance, A is the coefficient, and n is the exponent.

As shown in Figure 4, the fit of equation (2) to the measured interaction force of nanotube tip with 0.2 wt% PEI was made in two regions because scaling theory identifies the conformation transition of an adsorbed polymer from a confined state to an escaped state when polymer chains are compressed by a finite-sized object. Interaction forces measured with a nanotube tip for other concentrations of PEI 70,000 solution were also fitted by equation (2), and the coefficients and exponents are shown in Table I. Above 5 nm (long range), the average force is inversely proportional to the separation distance to the power of three. Below 5 nm (short range) it is inversely proportional to the separation distance to the power of one third. This is a significant difference compared to the micron size particle, whose force is inversely proportional to separation distance squared at short range. This again implies that nanoparticles with adsorbed polymers may interact only with the crowded PEI layers (i.e., loop-tail or loop-loop interaction)

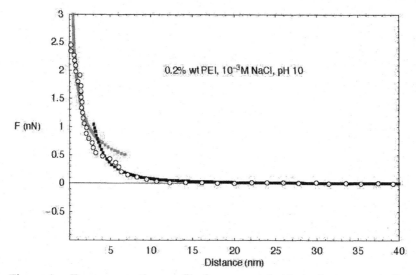

Figure 4.   Force-separation profile for a nanotube tip in the presence of PEI 70,000 fitted using the inverse power law (from Eq. 1). The open circles denote the measured force data.

Table I. Fitting equations for the interaction force with a MWNT tip at a short (< 5 nm) and long (> 5 nm) separation distances.

Concentration	Short distance	Long distance
0.2 wt%	$1.73 \dfrac{1}{H^{0.56}}$	$155.6 \dfrac{1}{H^{3.53}}$
0.4 wt%	$1.59 \dfrac{1}{H^{0.27}}$	$44.8 \dfrac{1}{H^{3.24}}$
0.6 wt%	$1.68 \dfrac{1}{H^{0.19}}$	$48.0 \dfrac{1}{H^{2.85}}$
Average	$\dfrac{1}{H^{0.34\pm0.2}} \left(\sim \dfrac{1}{H^{1/3}} \right)$	$\dfrac{1}{H^{3.20\pm0.34}} \left(\sim \dfrac{1}{H^{3}} \right)$

rather than with loose PEI layers (i.e. tail-tail interaction). The other possible interpretation is that a flatter conformation of PEI may be built on nanoparticles due to the large radius of curvature.

CONCLUSION

Use of a MWNT as a nanometer-size colloidal probe in AFM opens the possibility to directly measure the interaction forces of true nanosize particles in various colloidal conditions by minimizing the contribution from the side of the tip. A micron size glassy carbon sphere shows a larger interaction distance, and that interaction forces are influenced by the concentration of polymer. In contrast, a nano size MWNT probe clearly shows that the interaction force is independent of polymer concentration. Effects of the free, non-adsorbed polymer are not observed with the MWNT probe because of the similarity in size of the probe and the polymer molecule. A shorter interaction range and different polymeric force dependence with separation distance was measured with the nanoprobe.

REFERENCES

[1]C. A. Johnson and A. M. Lenhoff, "Adsorption of Charged Latex Particles on Mica Studied by Atomic Force Microscopy", *J. Colloid and Interface Sci.*, **179** [2] 587 (1996).

[2]M. Kappl and H. J. Butt, "The Colloidal Probe Technique and Its Application to Adhesion Force Measurements", *Particle & Particle Systems Characterization*, **19** [3] 129 (2002).

[3]R. G. Horn, "Surface Forces and Their Action in Ceramic Materials", *J. Am. Ceram. Soc.*, **73** [5] 1117 (1990).

[4]J. N. Israelachvili, Intermolecular and Surface Forces. 2nd ed., San Diego: Academic Press, 1992.

[5]B. V. Derjaguin, "Friction and Adhesion. IV: The Theory of Adhesion of Small Particles", *Kolloid Zeits*, **69** 155 (1934).

[6]D. H. Napper, Polymeric Stabilisation of Colloidal Dispersions, London: Academic Press, 1983.

[7]H. Kamiya, M. Mitsui, H. Takano, and S. Miyazawa, "Influence of Particle Diameter on Surface Silanol Structure, Hydration Forces, and Aggregation Behavior of Alkoxide-Derived Silica Particles", *J. Am. Ceram. Soc.*, **83** [2] 287 (2000).

[8]T. Cosgrove, P. C. Griffiths, and P. M. Lloyd, "Polymer Adsorption - The Effect of the Relative Sizes of Polymer and Particle", *Langmuir*, **11** [5] 1457 (1995).

[9]W. A. Ducker, T. J. Senden, and R. M. Pashley, "Direct Measurement of Colloidal Forces Using an Atomic Force Microscope", *Nature*, **353** [6341] 239 (1991).

[10]H. J. Dai, J. H. Hafner, A. G. Rinzler, D. T. Colbert, and R. E. Smalley, "Nanotubes as Nanoprobes in Scanning Probe Microscopy", *Nature*, **384** [6605] 147 (1996).

[11]T. Seeger, P. Redlich, N. Grobert, M. Terrones, D. R. M. Walton, H. W. Kroto, and M. Ruhle, "Siox-Coating of Carbon Nanotubes at Room Temperature", *Chemical Physics Letters*, **339** [1-2] 41 (2001).

[12]H. G. Pedersen, Particle Interactions: An AFM Study of Colloidal Systems, in *Department of Chemistry*. 1998, Technical University of Denmark.

[13]M. C. Guffond, D. R. M. Williams, and E. M. Sevick, "End-Tethered Polymer Chains under AFM Tips: Compression and Escape in Theta Solvents", *Langmuir*, **13** [21] 5691 (1997).

DIRECT FORCE MEASUREMENTS BETWEEN ZIRCONIA SURFACES: INFLUENCE OF THE CONCENTRATION OF POLYACRYLIC ACID, pH, AND MOLECULAR WEIGHT

Jing Sun and Lian Gao
State Key Lab
Shanghai Institute of Ceramics
Shanghai 200050, P. R. China

Lennart Bergstrom
Institute for Surface Chemistry
Box 5607, S-114 86
Stockholm, Sweden

Mikio Iwasa
National Institute of Advanced
Industrial Science and Technology
AIST Kansai, Midorigaoka 1-8-31
Ikeda, Osaka 563-8577, Japan

ABSTRACT

The adsorption of three different molecular weights polyacrylic acid (PAA, MW 1800, 50000, 90000) on zirconia has been studied as function of pH, and the concentration of polymer. Atomic force microscopy was used to directly monitor the changes in interaction force. Regardless of the concentration, the interaction force is always attractive with a low molecular weight (1800) polymer when the pH is below the pKa of PAA. When the polymer molecular weight is as high as 50000, and the concentration is 300 ppm, the total interaction force is repulsion. For PAA of MW 90000, the interaction force is always repulsive, regardless of the concentration and pH.

INTRODUCTION

Surface forces play an important role in wetting and adsorption, and in the formation of emulsions and suspensions for ceramic powder processing.[1,2] Fundamental understanding of the nature and origin of surface forces will allow greater control of the rheological properties of suspensions for more controlled processing. Acidic, acrylic-based polymers are commonly used as dispersants during wet powder processing.[3,4] Polymer adsorption onto a surface is highly dependent on pH, polymer concentration, and molecular weight. When a polymer is adsorbed or grafted onto the surface of a particle, it can lead to a range of possible interaction forces.[5] These may include a steric barrier between the particles, as well as bridging and depletion forces.[6] Recently, atomic force microscopy (AFM) has been used to measure interparticle forces for rutile[7], zirconia[8,9] and alumina[10,11], as well as metals[12] and polymeric latex materials.[13] The present work describes the use of the AFM to study polymer interactions between colloidal-sized objects. In particular, the influence of polymer molecular weight and concentration on the interaction force between zirconia is presented and discussed.

EXPERIMENTS

All AFM force measurements were performed using a spherical zirconia colloidal probe against a flat zirconia plate. The AFM colloidal probe was made from spray dried and sintering 3T-TZP powder (Tosoh, Japan). ZrO_2 spheres 10-25 μm in diameter were mounted on the AFM cantilever using a technique adapted from Ducker.[14] The zirconia plate was made from the same 3Y-TZP ceramic powder.

Analytical grade NaCl, HCl, and NaOH were used to adjust pH. Millipore Milli Q water was used throughout, which had a conductivity of less than 18.2 MΩ•cm at 20°C. Polyacrylic acid (PAA) with molecular weight of 1800, 50000, or 90000 was studied (Polysciences, Inc., USA). A Nanoscope III atomic force microscope (Santa Barbara, CA) was used for the force measuring experiments. The spring constants of the AFM cantilevers were measured following the method of Cleveland.[15]

RESULTS

Force Measurement in 10^{-3} M NaCl Solutions

Figure1 shows force-distance profiles between the zirconia sphere and the zirconia plate at different values of pH in 10^{-3} M NaCl without PAA. Except at pH 6.0, a repulsive force is always observed. Figure 2 shows the zeta potential of 3Y-TZP measured using the electrophoretic method with and without PAA (MW 1800). Without PAA, the isoelectric point is around pH 8.4. It moves to pH 3 after the addition of PAA. The concentration of PAA has no obvious influence on the zeta potential.

From force measurements, the apparent isoelectric point (iep) of the zirconia powder in NaCl solution is pH 6.6. This is significantly different from the plate and probe measurement. The difference is possibly due to less hydrolysis on the plate and the probe. The dehydration of an oxide surface tends to lower the iep.[10] High-temperature sintering of the probe and plate likely reduces the degree of surface hydroxylation.

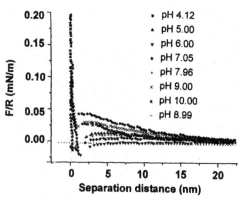

Figure1. Force-distance profiles between zirconia surfaces in 10^{-3} M NaCl (without PAA).

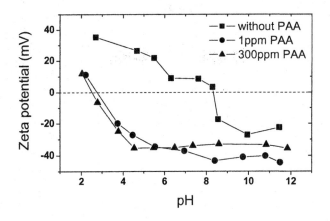

Figure 2.   Zeta potential of 3Y-TZP powder as a function of pH, with and without PAA.

AFM Force Measurement in the Presence of PAA

All of the AFM measurements were performed at relatively high ionic strength (I=0.01M) in order to screen the electrostatic repulsion. The isoelectric point of 3Y-TZP is around pH 8.4. The pKa of polyacrylic acid is around 4.5.[16] Figure 3 shows how the force-distance curves change with pH in 300 ppm PAA (MW 1800) solution.   At pH 3.57, the interaction force is attractive. The interaction force turns repulsive at pH 5.03. Figure 3b, which

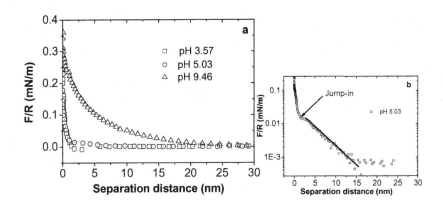

Figure 3.   Force-distance profiles as a function of pH for zirconia with 300 ppm of MW 1800 PAA. (F/R = Force/Radius data)

Figure 3b, which shows the data on a semi-log scale, clearly shows two distinct repulsive regimes at pH 5.03. The long-range part is exponential in form, and is due to the electrical double-layer overlap between the surfaces. Within 2.5 nm of separation, a second steeper repulsive region is also observed. This region is attributed to the steric repulsion between the surfaces as the adsorbed polymer layers overlap.

Force curves measured with 1 ppm PAA (MW 50000) are shown in Figure 4. At pH 4.05, no repulsive or attractive force can be seen. At pH 5.20 and 9.46, there is a repulsive force up to a distance of around 15 nm. The polymer can be totally dissociated in the basic range such that the steric force provided by the polymer is more prominent.

Figure 5 shows the force curves measured with 1 ppm of MW 90000 PAA. Increasing the molecular weight increases the distance of the repulsive force. Regardless of concentration, only repulsive forces are observed with increasing PAA molecular weight at pH ranging from 4.06 to 9.42 (Figure not shown).

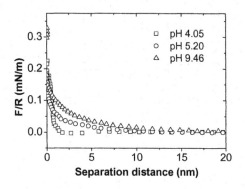

Figure 4.   Force-distance profiles between zirconia surfaces with 1 ppm of MW 50000 PAA.

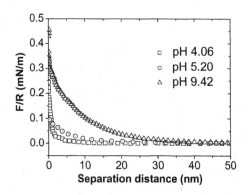

Figure 5.   Force-distance profiles between zirconia surfaces with 1 ppm of MW 90000 PAA.

DISCUSSION

Three factors: pH; polymer concentration; and polymer molecular weight all play an important role in determining the interaction force between the zirconia surfaces. pH determines how the PAA conforms to the 3Y-TZP surface.[16] The pH relative to the pKa determines the form of the polymer (e.g., coiled or flat). The concentration and molecular weight of the polymer determine the number of segments available to be adsorbed.[17] It is well known that, as the molecular weight of the polymer increase, the number of segments available for adsorption increases.[18] For a given system, a high molecular weight polymer generally provide better stabilization than a low molecular-weight polymer of the same composition.[18] In essence, the adsorption of a high molecular weight polymer is irreversible. Although the adsorption of each polymer segment may be reversible, because many segments of a long chain high molecular weight polymer are adsorbed at any given instant, the probability of all of the adsorbed segments of the chain being desorbed simultaneously is extremely small. This is not the case for low molecular weight polymers, in which there are only a few points of attachment. This is the reason why high molecular weight PAA (MW 90000) at pH 4.06 shows a repulsive interaction force. By providing more segments to be adsorbed, concentration also can be a factor. This is the case for low molecular weight polymer PAA (MW 50000) at pH 3.44. For 1 ppm PAA (MW 50000) at pH 4.05, the interaction force is attractive, but by increasing the PAA concentration, more segments can be adsorbed onto the surface, and the final interaction is repulsive.

Figure 6 compares the force profiles for the different molecular weights of polymer at pH 9.5. It is interesting to notice that, at low concentration, the MW 1800 and MW 50000 PAA are very similar. The steric force range is less than 10 nm, and much shorter than for MW 90000 PAA, which induces the steric force up to 30 nm. In contrast, the force curves for MW 50000 and MW 90000 PAA are more similar at high concentration. They overlap each other down to 8 nm, and the repulsive range is more than 40 nm long. For PAA with a molecular weight of 1800, the repulsive force is weaker and the repulsive range is short, only 25 nm.

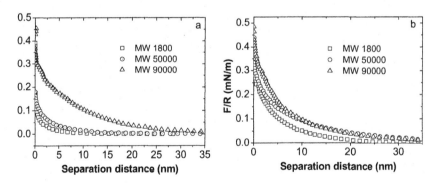

Figure 6.   Force-distance profile between zirconia surfaces at pH 9.5 with:
a) 1 ppm; and b) 300 ppm of PAA.

CONCLUSION

Based on this investigation, the following conclusions can be drawn. pH influences the surface interaction forces between 3Y-TZP. For low molecular weight PAA, the interaction is attractive at pH 5.2, which is higher than the pKa. With increasing PAA concentration from 1 ppm to 300 ppm, when the pH is higher than the pKa, the interaction is repulsive. For higher molecular weight PAA, MW 50000 and MW 90000, when pH value is higher than the pKa, the interaction is always repulsive. At pH 9.5, where the polymer is totally dissociated, the force-distance behavior of MW 50000 PAA at 1 ppm is very similar to that of MW 1800 concentration. At 300 ppm, the force-distance behavior is more similar to that of MW 90000 PAA.

REFERENCES

[1] J.N. Israelachvili, "Intermolecular and Surface Forces," Academic Press, New York, 1995.

[2] R.J. Pugh and L. Bergstrom, "Surface and Colloid Chemistry in Advanced Ceramic Processing," Published by Marcel Dekker Inc., 1994

[3] J. Cesarano III, I.A. Aksay and A. Bleier, "Stability of Aqueous α-Al$_2$O$_3$ Suspensions with Poly(methacrylic acid) Polyelectrolyte," *J. Am. Ceram. Soc.,* **71** [4], 250-55 (1998).

[4] J. Cesarano III, I.A. Aksay and A. Bleier, "Processing of Highly Concentrated Aqueous α-Alumina Suspensions Stabilized with Polyelectrolytes," *J. Am. Ceram. Soc.,* **71** [12], 1062-1067 (1998).

[5] K. F. Tjipangandjara and P. Somasundaran, "Effect of Changes in Adsorbed Polyacrylic Acid Conformation on Alumina Flocculation," *Colloids and Surf.,* **55**, 245-255 (1991)

[6] S. Biggs, "Steric and Bridging Forces between Surfaces Bearing Adsorbed Polymer: An Atomic Force Microscopy Study," *Langmuir,* **11**, 156-162 (1995).

[7] I. Larson, C.J. Drumnond, D.Y.C. Chan and F. Grieser, "Direct Force Measurements Between Titanium Dioxide Surfaces," *J. Am. Chem. Soc.,* **115**, 11885-11890 (1993).

[8] M.S. Hool, P.G. Hartley and P.J. Thistthwaite, "Fabrication and Characterization of Spherical Zirconia Particles for Direct Force Measurement Using the Atomic Force Microscope," *Langmuir,* **15**, 6220-6225 (1999).

[9] G.P. Henrik and L. Bergstrom, "Forces Measured Between Zirconia Surfaces in Poly (Acrylic Acid) Solutions," *J. Am. Ceram. Soc.,* **82** [5], 1137- 45 (1999).

[10] G.P. Henrik, "Alumium Oxide Probes for AFM Forces Measurements: Preparation, Characterization and Measurements," *Langmuir,* **15**, 3015-17 (1999).

[11] I. Larson, C.J. Drummord, D.Y.C. Chan and F. Grieser, "Direct Force Measurements between Silica and Alumina," *Langmuir*, **13**, 2109-2112 (1997).

[12] S. Biggs, P. Mulvaney, C.F. Zukoski and F. Grieser, "Study of Anion Adsorption at the Gold-Aqueous Solution Interface by Atomic Force Microscopy," *J. Am. Chem. Soc.,* **116**, 9150- 9157 (1994).

[13] W.R. Mark and T. J. Senden, "Adsorption of the Poly(oxyethylene) Nonionic Surfactant C12E5 to Silica: A Study Using Atomic Force Microscopy," *Langmuir*, **9**, 412-418 (1993).

[14] W.A. Ducker, T.J. Senden and R.M. Pashley, "Direct Measurement of Colloidal Forces Using An Atomic Force Microscope, " *Nature*, **353** [6341], 239-241 (1991).

[15] J.P. Cleveland, S.J. Manne, D. Bocek and P.K. Hansma, " A Non-destructive Method for Determining the Spring Constant of Cantilevers for Scanning Force

Microscopy," *Rev. Sci. Instrum.*, **64**, 403-405 (1993).

[16]V.A. Hackley, "Colloidal Processing of Silicon Nitride with Poly(Acrylic Acid): I, Adsorption and Electrostatic Interactions," *J. Am. Ceram. Soc.*, **80** [9], 2315-25 (1997).

[17]D.H. Napper, "Stablilization of Colloids by Free Polymer," *pp.199-203* in The Effect of Polymers on Dispersion Properties. Edited by Th. F. Tadros. Academic Press, London, U.K., 1982.

[18]Th.F. Tadros, "Polymer Adsorption and Dispersion Stability," *see Ref.17, pp.1-35.*

EFFECTS OF POLYMER DISPERSANT MOLECULAR STRUCTURE ON NONAQUEOUS CERAMIC SUSPENSIONS

Toshio Kakui
Chemicals Research Laboratories
Lion Corporation
13-12, Hirai 7-chome, Edogawa-ku
Tokyo 132-0035, Japan

Hidehiro Kamiya
Graduate School of Bio-Application &
Systems Engineering, BASE
Tokyo University of Agriculture &
Technology, Koganei
Tokyo184-8588, Japan

ABSTRACT

The dispersion and aggregation behavior of Al_2O_3, Si_3N_4, and SiC in ethanol suspensions were investigated using three types of novel polymer dispersants with multi-amino functional groups, branched and linear polyethyleneimine (PEI), and comb-graft copolymers of PEI with a polypropylene glycol (PPG) side chain. When various branched PEIs with different molecular weights (Mw) ranging from 300 to 30,000 g/mol are used, the optimum molecular weight to produce a highly fluid SiC suspension is in the range from 1,800 to 10,000. The viscosity of the suspension increases significantly with an increase in the molecular weight from 10,000 to 30,000. For Al_2O_3 and Si_3N_4, ethanol suspensions with high fluidity are produced with the addition of PEIs having a Mw over 1,800. The optimum molecular weight of the polymer dispersant for high fluidity depends on the material. Linear PEIs without a branched chain are only adsorbed onto the SiC, and do not act to reduce the viscosity of a 20 vol.% SiC ethanol suspension. When using comb-graft copolymers with PPGs introduced into a branched PEI with a Mw of 300, the SiC suspension has a higher fluidity than with the original PEI. The steric repulsion from the branched chains of the PEI promotes the dispersability of the ceramic particles in an ethanol suspension.

INTRODUCTION

To produce homogeneous suspensions with high fluidity for ceramic processing, the dispersion and aggregation behavior of particles can be engineered using a polymer dispersant.[1] However, the dispersants and their use are often selected on the basis of empirical data, and the role of molecular structure and molecular weight are poorly understood. Previous work conducted using a colloidal probe AFM reported the effect of the molecular structure of a polymer dispersant in a concentrated aqueous suspension.[2] Although there is much information available, and various types of dispersants are known for aqueous suspensions, little is known about the dispersion of ceramic particles in nonaqueous suspensions. Nonaqueous suspensions are preferred when the presence of water causes a deleterious effect, or when a complicated ceramic forming process is required such as tape casting.[3]

A simple suspension consists of three elements: particles, a dispersant, and a dispersing media. The dispersion of particles is achieved by electrostatic repulsion and/or steric repulsion. The aggregation and dispersion of ceramic particles in nonaqueous suspensions using electrostatic repulsion has been addressed in a previous paper,[4] which discussed the effects of active sites on particle surfaces, the physical-chemical properties of solvent and dispersants, and the proportions of mixed solvents.

The dispersion of alumina in a nonaqueous suspension like ethanol is difficult because ethanol is a weak acid and has strong hydrogen bonding.[5] For nonaqueous suspensions, dispersants and surfactants,[6] polymer dispersants,[7] and coupling agents that react chemically[8] have been described. However, only a few polymer dispersants have been reported.

This work focuses on the effect of the molecular structure and molecular weight of polymer dispersants in concentrated ethanol suspensions with different ceramic powders; SiC, Si_3N_4, and Al_2O_3. Branched and linear polyethyleneimines (PEIs) and comb-graft copolymers of PEI with polypropylene glycol (PPG) side chains have been examined as novel polymer dispersants with multi-amine functional groups. Moreover, the mechanism for dispersing particles in a nonaqueous system is discussed based on the adsorption behavior of the polymer on particle surfaces.

MATERIALS AND METHODS
Polymer Dispersants, Ceramic Powders, and Organic Solvent
Branched PEIs and linear PEIs (Nippon Syokubai Co., Ltd., Japan) and the comb-graft copolymers of PEI with PPG side chains were used in this study as the polymer dispersants with multi-amine functional groups (Figure 1). Seven kinds of branched PEIs with molecular weights ranging from 300 to 30,000, and two kinds of linear PEIs with a Mw of 600 and 1,800 were used. The comb-graft copolymers were prepared by combining propylene oxide into the branched PEI with a Mw of

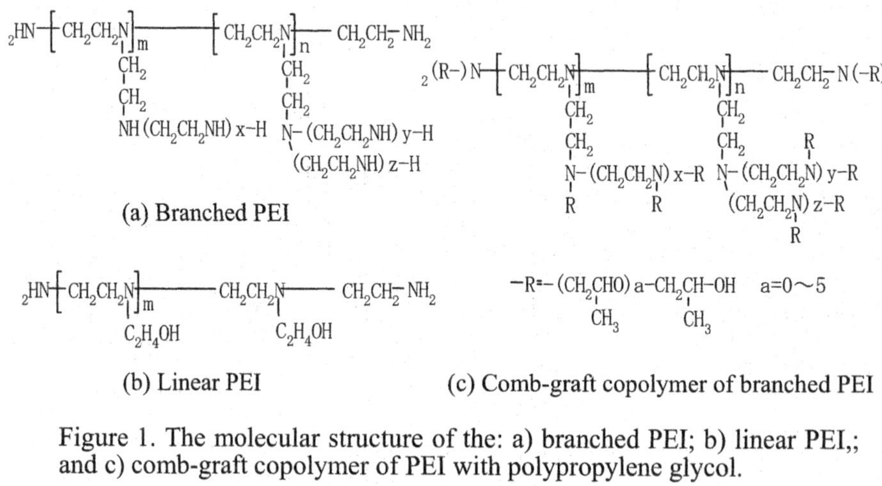

(a) Branched PEI

(b) Linear PEI   (c) Comb-graft copolymer of branched PEI

Figure 1. The molecular structure of the: a) branched PEI; b) linear PEI,; and c) comb-graft copolymer of PEI with polypropylene glycol.

Table I. Representative characteristics of the fine ceramics powders.

	Average particle size	Specific surface area	True specific gravity	Isoelectric point pH
SiC	0.4 μm	14.0 m^2/g	3.20 g/cm^3	3-5
Si$_3$N$_4$	1.5	8.5	3.20	5-6
Al$_2$O$_3$	0.6	7.0	3.93	8.5-9.5

300 using KOH catalysis.[9] The molecular weights of the PPG modified PEI were 580 and 3,050.

Three different ceramic powders were studied: α-Al$_2$O$_3$ (AL-160SG from Nippon Light Metal Company, Ltd., Japan); and SiC and Si$_3$N$_4$ (A-1 and NU-1, respectively, from Showa Denko K.K., Japan). Table I shows the representative characteristics of each ceramic powder. Ethanol (HPLC grade, Junsei Chemical Co., Ltd., Japan) was used as received, without further purification as the suspension medium.

Preparation and Evaluation of Concentrated Ceramic Ethanol Suspensions

Each ceramic powder and the polymer dispersant were mixed in ethanol and then ball milled for 24 hrs. The apparent viscosity of each ceramic suspension was determined at shear rates of 0.63, 1.29, 3.14, and 6.28s^{-1} using a BL-type viscometer (Tokimec Co., Ltd., Japan). The amount of polymer dispersant adsorbed onto the ceramic particles in each 20 vol.% solids suspension (i.e., the g of polymer per g of ceramic) was determined from the difference in the concentration of polymer in the original polymer/ethanol mixture and the concentration in the supernatant after adsorption, and normalized to the specific surface area of the ceramic (m^2/g). The supernatants containing the free polymer were obtained from each suspension using centrifugal sedimentation at 5,000 rpm for 30 min. The amount of free polymer was determined by removing the ethanol from the supernatant through distillation and drying in a vacuum dryer.

RESULTS

Dispersion Behavior of Suspensions with Branched PEI

By using branched PEIs with a molecular weight ranging from 300 to 30,000, the effect of the molecular weight on the viscosity of the ethanol suspension was investigated. The dispersant concentration was fixed at 1.0 wt%, which was a sufficient amount for good dispersability. As shown in Figure 2, the viscosity of the ethanol suspensions with SiC and Si$_3$N$_4$ decreases with increasing molecular weight up to 1,800. The optimum molecular weight of the branched PEI for high fluidity in a SiC suspension was determined to be in the range from 1,800 to 10,000. The viscosity of the suspension increases with increasing molecular weight up to 30,000. The viscosity of the Si$_3$N$_4$ ethanol suspension is independent of molecular weight at greater than 1,800. The viscosity of the Al$_2$O$_3$ ethanol suspension gradually decreases with increasing molecular weight up to 30,000. The optimum molecular weight of the branched PEI to obtain a high-fluidity suspension depends on the powder in the ethanol system.

The effect of polymer dispersant concentration on viscosity was investigated using the branched PEI with a Mw of 5,000. This Mw PEI has good dispersability

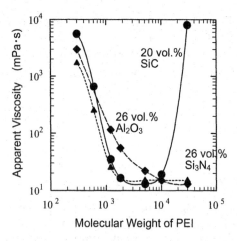

Figure 2. The effect of the molecular weight of the branched PEI on the apparent viscosity of ethanol suspensions with SiC, Si_3N_4, and Al_2O_3.

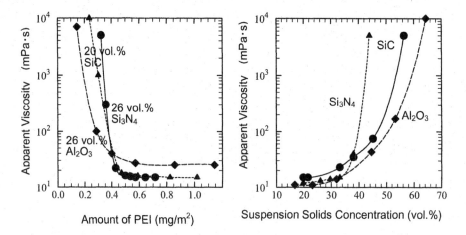

Figure 3. The apparent viscosity of ethanol suspensions as a function of the amount of the branched PEI.

Figure 4. The apparent viscosity of ethanol suspensions with different solid volume fractions.

in all of the suspensions. As shown in Figure 3, the viscosity of the Al_2O_3 suspensions gradually decreases with increasing PEI concentration from 0.14 to 0.42 mg/m^2. The viscosity of the suspensions with SiC and Si_3N_4 drastically decrease at about 0.4 mg/m^2 PEI. The lowest concentration of PEI required to produce a highly fluid suspension is about 0.5 mg/m^2.

To determine the differences between the different ceramic powders, the effect

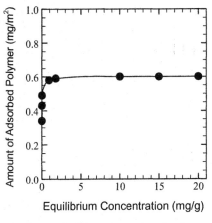

Figure 5. The adsorption isotherm of branched PEI (Mw = 1,800) on SiC in ethanol.

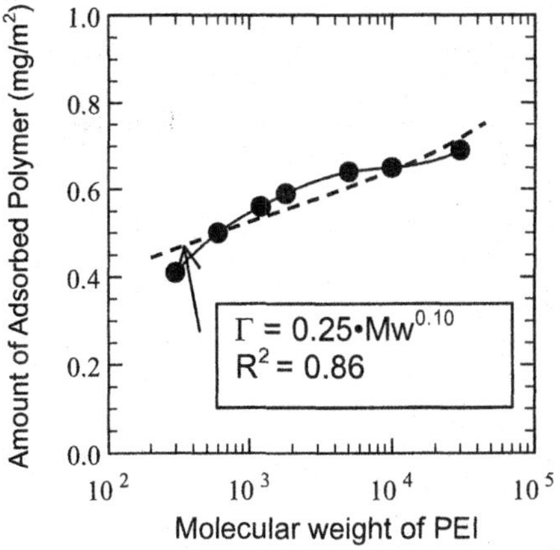

Figure 6. The amount of adsorption of branched PEI on SiC particles in a 20 vol.% solids ethanol suspensions.

of the solids concentration on each suspension was examined with 1.0 wt% PEI. As shown in Figure 4, all viscosities rise quickly above 35 vol.%. However, the change in suspension viscosity depends on the specific material being dispersed. Some potential reasons for these differences include differences in particle size, and differences in the number of active sites on particle surfaces.

Adsorption Properties and the Configuration of Branched PEI onto SiC.
To investigate the affinity of the branched PEI for the ceramic particles, the adsorption properties of the branched PEIs on SiC particles were examined. These

Table II. The amount of polymer adsorption on the SiC particles, and the viscosity of a 20 vol.% suspension with branched and linear PEIs, and a comb-graft copolymer.

	Mw of PEI	Mw of PPG	Amount adsorbed	Apparent viscosity
	300	-----	0.41 mg/m^2	5,600 mPa·s
Branched PEI	600	-----	0.50	750
	1,800	-----	0.59	15
Linear PEI	600	-----	0.47	4,500
	1,800	-----	0.57	4,100
Comb-graft PEI with PPG	300	580	0.29	510
	300	3,050	0.21	440

experimental limitations were set because the viscosity of the ethanol suspension changes depending on the molecular weight of the branched PEIs.

As shown in Figure 5, the adsorption isotherm of the 1,800 Mw branched PEI on SiC particles in a 20 vol.% ethanol suspension shows Langmuir adsorption characteristics and a high affinity of the PEI for the SiC. The branched PEI shows an isotherm with substantial adsorption up to 0.5 mg/m^2 at a low equilibrium concentration, and a plateau value of 0.6 mg/m^2 at a high equilibrium concentration. The saturated adsorption concentration is approximately the same as for the smallest amount (0.5 mg/m^2) of dispersant required to produce a highly fluid suspension. The amount of the adsorption as a function of the molecular weight of the PEI was investigated. As shown in Figure 6, the amount of adsorption does not significantly change with molecular weight, though it gradually increases from 0.4 to 0.7 mg/m^2 with an increase in the molecular weight from 300 to 30,000.

Effect of Polymer Dispersant Molecular Structure

To evaluate the effect of the molecular structure of the branched chain PEI on a ceramic suspension, the adsorption of liner PEI (without a branched chain) on SiC, and the dispersability of SiC were investigated. As shown in Table II, the amount of adsorption of the linear PEIs with Mw of 600 and 1,800 is almost equal to that of the branched PEIs. However, the 20 vol.% SiC ethanol suspensions with linear PEIs do not have a low viscosity.

A new comb-graft copolymer was designed by introducing PPG side chains into a branched PEI of Mw 300. The comb-graft copolymers with a Mw of 580 and 3,050 reduced the viscosities of the SiC suspensions. The amount of adsorption of the comb-graft copolymer decreases with increasing molecular weight of the PPG side chains, and is about half of the 0.41 mg/m^2 measured for the original branched PEI with a Mw of 300. The comb-graft copolymers are effective in dispersing the SiC particles in an ethanol suspension.

DISCUSSION

Based on the results presented, it appears that the multi-amino functional groups of the PEIs selectively adsorb onto the acidic sites of the SiC surface with an isoelectric point of pH 3-5. A model for the adsorption of polymer onto a particle surface can be described by an equation from R. Ullman et al. [10], which relates the amount of adsorbed polymer (Γ) to molecular weight (Mw).

$$\Gamma = K \cdot Mw^{\alpha} \qquad (1)$$

where, K is a constant, and α is the adsorption constant. The adsorption constant has been defined relative to three models in Figure 7. Values of K and α were determined from the fitting curve (dashed line) in Figure 6 to be about 0.25 and 0.1, respectively.

The adsorption of the branched PEI onto the SiC surface follows a head-tail model (II). A polymer dispersant with an effective steric repulsion will favor model (II) versus model (III). However, equation (1) does not consider the effect of a branched chain polymer. The viscosity of SiC ethanol suspensions significantly decreases with a branched chain polymer having a molecular weight ranging from 10,000 to 1,800. A linear PEI without a side chain only adsorbs onto particle surfaces, and does not reduce the viscosity of a SiC suspension. This result suggests that steric repulsion by the branched chain of the PEI contributes to the dispersability of SiC in this molecular weight range. The branched PEI with a Mw

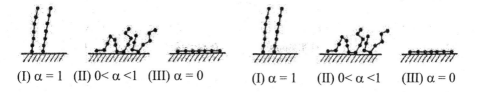

(I) α = 1 (II) 0< α <1 (III) α = 0   (I) α = 1 (II) 0< α <1 (III) α = 0

Figure 7. A model for the adsorption structure of a polymer on a particle surface, including the α value in the equation (1). I) bonding at the head or tail of the polymer chain only, II) bonding intermittently along the length of the polymer chain, and III) bonding along the total length of the polymer chain.

of 30,000 might increase the viscosity as a result of bridging aggregation among the particles due to the high affinity of the PEI for the SiC surface. In the case of the comb-graft copolymer, the PEI will adsorb onto the particle surface, and the PPG side chains will work as a steric repulsive chain. The molecular structure of the polymer dispersant can be engineered to control the dispersion and aggregation of particles in a concentrated nonaqueous suspension by controlling the adsorption properties and the molecular structure of the polymer.

CONCLUSION
Branched PEI is an excellent polymer dispersant for concentrated ethanol suspensions with Al_2O_3, Si_3N_4, and SiC. The optimum molecular weight and amount of polymer necessary to produce a highly fluid suspension depends on the ceramic powder. The adsorption isotherm for a branched PEI on the acidic SiC particle surface in an ethanol suspension displays Langmuir adsorption characteristics. There is a high affinity of the PEI for the ceramic surface, and the adsorption properties are almost independent of the molecular weight of the branched PEIs. Based on the viscosity of the SiC suspension and the amount of the adsorbed polymer. The branched chain polymer significantly affects the fluidity of a SiC ethanol suspension.

REFERENCES
[1]G. Cordoba and R. Moreno, "Evaluation of the Effectiveness of a Phosphate Ester in Al2O3 Tape Casting Slips," *Ceramic Transactions*, **51**, 379-383 (1995).
[2]H. Kamiya, Y. Fukuda, Y. Suzuki, M. Tsukada, T. Kakui and M. Naito, "Effect of Polymer Dispersant Structure on Electrosteric Interaction and Dense Alumina Suspension Behavior," *J. Am. Ceram. Soc.,* **82**, 3407-3412 (1999).
[3]L. Wang, W. Sigmund and F. Aldinger, "A Novel Class of Dispersants for Colloidal Processing of Si_3N_4 in Non-Aqueous Media," *Materials Letters*, **40**, 14-17 (1999).
[4]B. Siffert, A. Jada, and J. Eleli-Letsango, "Location of the Shear Plane in the Electric Double Layer in an Organic Medium," *J. of Colloid and Int. Science*, **167**, 327-333 (1994).
[5]B.I. Lee and U. Paik, "Dispersion of Alumina and Silica Powders in Non-aqueous Media: Mixed – Solvent Effects, " *Ceramics International*, **19**, 241-250 (1993).

[6]J. H. H. ter Maat, "Critical Factors Determining the Activity of Oligomeric Dispersants," *Ceramic Transactions,* **51**, 355-359 (1995).

[7]W. M. Sigmund, G. Wegner and F. Aldinger, "Interaction of Organic Additives with Alumina Surfaces in a Ceramic Slurry," *Mat. Res. Soc. Symp. Proc.* **Vol.407**, 313-318 (1996).

[8]T. Kramer and F. F. Lange, "Rheology and Particle Packing of Chem- and Phys- Adsorbed, Alkylated Silicon Nitride Powders," *J. Am. Ceram. Soc.*, **77**, 922-928 (1996).

[9]A. Naka, Y. Nishida, H. Sugiyama and T. Sugiyama, "Ability of N-[Poly(oxyethylene)]-Polyethyleneimine to Increase the Coal Content of Coal-Water Slurry," *J. Chem. Soc. Jpn.*, **13**, 227-230 (1986).

[10]R. Perkel and R. Ullmann, " The Adsorption of Polydimethylsiloxanes from Solution," *J. Polymer Sci.*, **54**, 127-148 (1961).

INFLUENCE OF THE MOLECULAR STRUCTURE OF A POLYMER DISPERSANT ON CONCENTRATED SiC AQUEOUS SUSPENSIONS

Kimitoshi Sato, Madoka Hasegawa, Toshio Kakui, Mayumi Tsukada, Shinzo Omi, and Hidehiro Kamiya
Graduate School of Bio-Applications and Systems Engineering, BASE
Tokyo University of Agriculture and Technology, Koganei,
Tokyo 184-8588, Japan

ABSTRACT

To analyze the effect of the molecular structure of a polymer dispersant on aggregation and the dispersion behavior of silicon carbide particles in a concentrated aqueous suspension, polymer dispersants with different hydrophilic to hydrophobic ratios, m:n = 30:70, 50:50, 75:25, were added silicon carbide suspensions. The dispersant polymers were synthesized from an ammonium salt of acrylic acid and styrene, or from an acrylic methyl monomer. The average molecular weight of each polymer was fixed at 10,000 g/mol. For the copolymer of acrylic acid and styrene, the optimum ratio to minimize suspension viscosity was determined at m:n = 50:50. With this polymer dispersant, the suspension viscosity was lower than with a commercial dispersant for silicon carbide. For the copolymer of acrylic acid and acrylic methyl, the suspension viscosity decreased with increasing hydrophobic group ratio. Based on these results, the adsorption structure and mechanism of dispersion for each polymer dispersant are discussed.

INTRODUCTION

To control the aggregation and dispersion behavior of fine particle size silicon carbide powder in a concentrated aqueous suspension, various kinds of polymer dispersants have been used. The effects of polymer molecular weight and structure have been studied in alumina and silicon nitride powder.[1-4] However, in the case of silicon carbide ceramics, the dispersion mechanism has not been determined. This is because, in an aqueous suspension, the surface of silicon carbide has a complicated structure with a hydrophobic group and a partially oxidized hydrophilic group.[5] Therefore, it is important to analyze the effect of the molecular structure of polymer dispersants on the aggregation behavior of silicon carbide particles and on suspension viscosity. In our previous work[5-7], the effect of the hydrophilic to hydrophobic ratio of a copolymer of acrylic acid and an acrylic methyl monomer on alumina and silicon nitride suspension viscosity was measured and analyzed using a colloidal probe atomic force microscope.

In the case of silicon carbide, the effect of the molecular structure of the polymer dispersant on the aggregation and dispersion behavior has been studied[5,8,9] using the same copolymer of acrylic acid and acrylic methyl used for alumina and silicon nitride powders.[6,7] However, the viscosity of a suspension with this copolymer is higher than with a commercial polymer dispersant. Previous work[5,8,9] showed that suspension viscosity decreases with increasing hydrophobic group ratio; consequently, an increase in the hydrophobicity of the hydrophobic group of the polymer dispersant is needed to improve the dispersion stability of silicon carbide particles.

Because it appears that a phenyl group has a higher hydrophobicity than a methyl group, polymer dispersants with different hydrophilic to hydrophobic ratio were synthesized from acrylic acid and styrene. With these polymer dispersants, we tried to determine the optimum molecular structure and dispersion mechanism to obtain the minimum viscosity in a concentrated silicon carbide suspension.

EXPERIMENTAL PROCEDURE

Fine silicon carbide powder (GMF UH-4, Pacific Rundum Co., Ltd) having a mean diameter of 0.69 μm, and a specific surface area of 16.4 m²/g, was mixed in water with a polymer dispersant, and ball-milled for 24 h with 3 mm diameter silicon carbide balls. The solid volume fraction and the concentration of polymer dispersant in solution were fixed at 35 vol% and 0.48 mg/m², respectively, which is the amount of polymer added per surface area of powder. Suspension viscosity was determined as a function of shear rate ranging from 1 to 300[1/s] at 25°C using a concentric cylinder viscometer (VT550, HAAKE, Germany). The polymer dispersants used in this study were copolymers of styrene and acrylic acid monomer with different hydrophilic to hydrophobic group ratios, m:n. The hydrophilic to hydrophobic group ratio of each polymer dispersant was engineered to fall within the range of 30:70 to 75:25, as shown in Table. The molecular structure of dispersants is shown in Figure 1. A commercial polymer dispersant having a molecular weight of 10,000 (Rejit SM101, Sanyo Kasei Co., Ltd) also was examined and compared to the model polymer dispersants.

RESULTS

To determine the effect of polymer dispersant molecular structure on suspension viscosity, the relationship between the shear stress and shear rate of SiC suspensions with dispersants was determined (Figure 2). The optimum hydrophobic to hydrophilic group ratio (m:n) that produced the minimum suspension viscosity was 50:50. The suspension viscosity with this dispersant was the same or slightly lower than with the commercial dispersant. When the hydrophilic to hydrophobic ratio used was 30:70, the suspension viscosity increased. The maximum viscosity was observed with m:n = 75:25.

The effect of polymer molecular weight on silicon carbide aqueous suspension

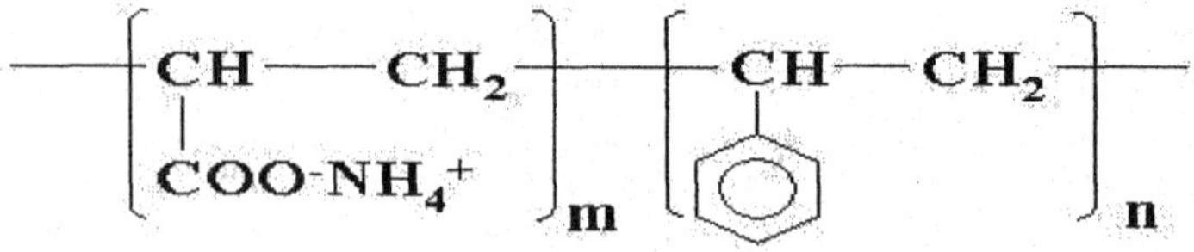

Figure 1. The molecular structure of the acrylic acid/styrene copolymer ammonium salt dispersant.

Table I . The hydrophilic to hydrophobic group ratio and the molecular

	monomer ratio	molecular weight (avg.)	
		Lot.1	Lot.2
AA-St-30	30:70	1.15×10^4	2.43×10^4
AA-St-50	50:50	1.60×10^2	2.38×10^4
AA-St-75	75:25	-	1.31×10^4

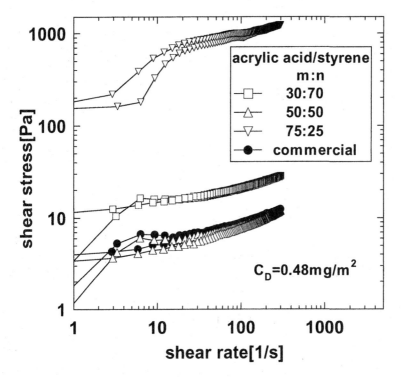

Figure 2. Relationship between shear rate and shear stress for silicon a carbide suspension with different polymer dispersants (pH=9).

viscosity was studied using two kinds of polymer dispersant with the same hydrophilic to hydrophobic ratio, 50:50, but different molecular weight Mw; 160 and 23,800 (Figure 3). The suspension viscosity with the high molecular weight (Mw = 23,800) polymer was the same or slightly lower than that with low molecular weight polymer dispersant (Mw = 160).

The effect of the hydrophilic to hydrophobic ratio (m:n) on the apparent

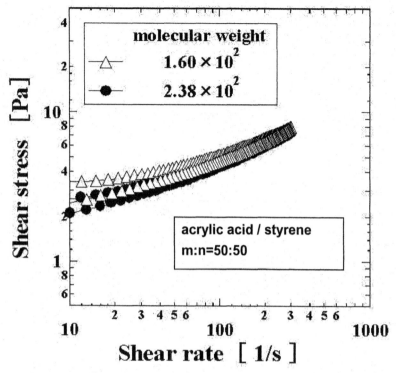

Figure 3. Effect of polymer dispersant molecular weight on the shear stress - shear rate response of a silicon carbide suspension (pH=9).

viscosity of suspensions with two kinds of dispersants is shown in Figure 4. The figure shows the apparent viscosity of suspensions with polymer dispersants synthesized from acrylic acid and methyl acrylate monomer (AA-MA) in a previous paper[3]. In Figure 4, the apparent viscosity (η) of each suspension is given as the ratio of the apparent viscosity of the model dispersant suspension (η_m) relative to the apparent viscosity of the suspension made with the commercial polymer dispersant (η_c). For the acrylic acid and styrene copolymer, an optimum ratio of m:n = 50:50 produced the minimum suspension viscosity. In contrast, for the acrylic acid and methyl acrylate copolymer, the suspension viscosity decreases with an increase of the hydrophobic group ratio in the polymer dispersant. At a ratio m:n = 75:25, the viscosity of the suspension with acrylic acid and methyl acrylate copolymer is lower than with the acrylic acid and styrene copolymer.

DISCUSSION

For the acrylic acid and styrene copolymer dispersants, the molecular ratio of hydrophilic to hydrophobic groups, m:n, necessary to produce the minimum apparent viscosity is 50:50. This produces an apparent viscosity that is lower than

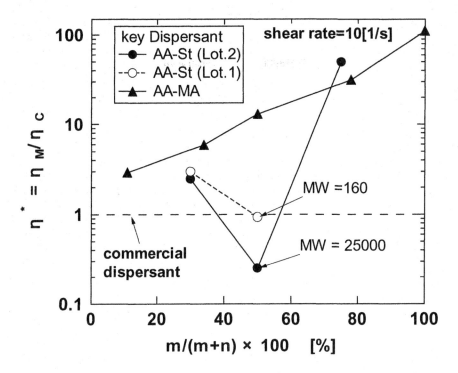

Figure 4. Effect of the hydrophilic group ratio of a polymer dispersant on the apparent viscosity of silicon carbide suspensions (compared to a commercial dispersant).

is achieved with a commercial dispersant. These results suggest that the structure of the adsorbed polymer dispersant on the surface of the silicon carbide particles depends on the molecular structure of polymer dispersants.on silicon carbide powders are charged negatively under at a relatively high pH that is higher than the iso-electric point of silicon carbide (pH = 5-6)[5], the hydrophilic group in the polymer dispersant cannot adsorb onto the site of the negatively charged hydroxyl group on silicon carbide. Since the negatively charged hydrophilic group cannot absorb on to the negatively charged hydroxyl group on silicon carbide, it is hypothesized that the hydrophobic group in the polymer dispersant will have to adsorb onto the hydrophobic parts of the silicon carbide powder surface.

It is hypothesized that phenyl groups in the dispersants can promote hydrophobic adsorption between silicon carbide and polymer dispersant, and the negatively charged hydrophilic groups, COO⁻, in the polymer dispersant will be repelled by the negatively charged hydroxyl group on the silicon carbide. At a ratio, m:n = 50:50, it is believed that a loop-and-train polymer structure is adsorbed onto the silicon carbide surface. Since these loop-train structures give rise to steric repulsion between particles[2], the silicon carbide particles are dispersed, and the apparent viscosity of the suspension is reduced. For a polymer dispersant with a high hydrophilic ratio, m:n=75:25, it seems that the density of adsorption sites on the silicon carbide surface is very low. Consequently, the dispersant cannot form a loop-train structure on the silicon carbide surface, and the suspension viscosity increases. With an excess hydrophobic group ratio, m:n=30:70, it is believed that a flat structure is adsorbed onto the silicon carbide surface that increases the suspension viscosity.

CONCLUSIONS

To determine the effect of polymer dispersant molecular structure on the aggregation, dispersion behavior, and viscosity of a SiC aqueous suspension, polymer dispersants, with different hydrophilic to hydrophobic group ratios m:n=70:30,50:50,75:25 were characterized. An optimum hydrophilic to hydrophobic group ratio of m:n=50:50 produced the minimum suspension viscosity. The suspension viscosity is the same or slightly lower than that obtained using a commercial polymer dispersant. The proposed adsorption mechanism for the copolymers of acrylic acid and styrene monomer is hydrophobic adsorption between the phenyl groups of the polymer and the hydrophobic structure on the silicon carbide.

ACKNOWLEDGEMENT

This work is supported by a Grant-in-Aid for Scientific Research (B), Japan and the Structurization of Material Technology Knowledge Project in the Nano-Technology Program by METI Japan.

REFERENCES

[1] H. Okamoto, M. Hayashiba, Y. Nurishi, and K. Hiramatsu, "Fluidity and Dispersion of Alumina Suspension at the Limit of Thickening by Ammonium Poly-acrylates," *J. Mater. Sci.*, **26**, 383-387 (1991).

[2] J. Cesarano III and I.A. Aksay, "Processing of Highly Concentrated Aqueous Alumina Suspension Stabilized with Polyelectrolytes," *J. Am. Ceram. Soc.*, **71**, 1262-1267 (1988).

[3] V.A. Hackley, "Colloidal Processing of Silicon Nitride with Poly(acrylic acid):I, Adsorption and Electrostatic Interactions," *J. Am. Ceram. Soc.*, **80** [9] 2315-25 (1997).

[4] K. Wada and H. Abe, "Viscosities of Highly Concentrated Alumina Suspensions with Polyacrylic Ammonium," *J. Ceram. Soc. Japan*, **103**, 979-982 (1995).

[5] M. Nojiri, S. Matsui, H. Hasegawa, T. Ono, Y. Fukuda, M. Tsukada, and H. Kamiya, "Analysis of Anionic Polymer Dispersant Behavior in Dense Silicon Nitride and Carbide Suspensions Using an AFM" *J. Nanoparticle Res.*, **3**, 237-244 (2001).

[6] H. Kamiya, Y. Fukuda, Y. Suzuki, M. Tsukada, T. Kakui and M. Naito, "Effect of Polymer Dispersant Structure on Electrosteric Interaction and Dense Alumina Suspension Behavior," *J. Am. Ceram. Soc.*, **82** [12] 3407-12 (1999).

[7] H. Kamiya, S. Matsui and T. Kakui, "Analysis of Action Mechanism of Anionic Polymer Dispersant with Different Molecular Structure in Dense Silicon Nitride Suspension by Using Colloidal Probe AFM," *Ceram. Trans.*, in press.

[8] M. Nojiri, H. Hasegawa, T. Ono, T. Kakui, H. Kamiya, "Influence of Molecular Structure of Anionic Polymer Dispersants on Dense Silicon Carbide Suspension Behavior and Microstructures of Green Bodies Prepared by Slip Casting," *J. Ceram. Soc. Japan*, **111** [5], 327-332 (2003).

[9]M. Nojiri, H. Hasegawa, T. Ono, and H. Kamiya, "Action Mechanism of Anionic Polymer Dispersant on SiC Particles and Suspension Characteristics," Ceram. Trans., **Vol. 112**, 191-196 (2001).

SYNTHESIS OF TiO$_2$ NANO-POWDERS FROM AQUEOUS SOLUTIONS WITH VARIOUS CATION AND ANION SPECIES

Yong-Ick Cho
Korea Photonics Technology Institute
459-3, Bonchon-dong, Buk-gu
Gwangju 500-210, Korea

Sang-Chul Jung
Sunchon National University
315, Maegok, Sunchon
Jeonnam 540-742, Korea

Doo-Sun Hwang and Sun-Jae Kim
SAINT, Sejong University
98, Gunja-dong, Gwangjin-gu
Seoul 143-747, Korea
sjkim1@sejong.ac.kr

Naito Makio and Kiyoshi Nogi
Osaka University
11-1, Mihogaoka, Ibaraki
Osaka 567-0047, Japan

Keizo Uematsu
Nagaoka University of Technology
1603-1, Kamitomioka-cho
Nagaoka 940-2188, Japan

ABSTRACT

TiO$_2$ nano-powders were synthesized from aqueous TiOCl$_2$ solutions containing different cation and anion additives using the homogeneous precipitation process at low temperatures (HPPLT). Nano-structured rutile powder is directly synthesized by homogeneous precipitation during the hydrolysis of TiOCl$_2$ in a highly acidic aqueous solution. In aqueous TiOCl$_2$ solutions with SO$_4^{2-}$ ions, spherical anatase particles form; however, the efficiency of preparing TiO$_2$ powder is greatly suppressed by the addition of SO$_4^{2-}$ ions.

Doped TiO$_2$ powders are completely crystallized, and have the rutile structure when Cu^{2+} or Fe^{3+} ions are added to aqueous TiOCl$_2$ solutions as metal-chlorides, and when there are no additives at all. The anatase phase is produced when Zr^{4+} or Al^{3+} ions are added to aqueous TiOCl$_2$ solutions as metal-chlorides. The rutile and anatase phases precipitate as acicular and spherical primary particles, respectively.

INTRODUCTION

Recently, nanocrystalline TiO$_2$ has attracted increasing attention because of its wide application in cosmetics, optics, electronics, and solar cells. Both the anatase and rutile phases are used in the TiO$_2$ industry. Generally, in TiO$_2$, the anatase phase is stable at room temperature, and rutile is stable at high temperature. A high temperature treatment is typically required to transform anatase to rutile. However, Kim and co-workers[1,2] have reported that the rutile phase can be produced near room temperature from a heated aqueous TiOCl$_2$ solution using the homogeneous precipitation process at low temperatures (HPPLT). Yang et al.[3] and Okada et al.[4] have reported that the anatase to rutile phase transition is retarded by Zr^{4+} and Si^{4+}.

We are interested in developing ways to easily and directly produce ultrafine rutile or anatase TiO_2 powder by means of the HPPLT. This includes determining the effects of various cation and anion additions to the aqueous Ti solutions during the HPPLT, as well as elucidating the mechanism of TiO_2 crystal formation. In this paper, the synthesis of nano-structured TiO_2 powders during the HPPLT was investigated. The effects of adding cations to $TiOCl_2$ solution were examined, including the effect of ionic radii and valence state. Additionally, the effects of the anion species and the concentration of SO_4^{2-} in $TiOCl_2$ solutions were examined.

EXPERIMENTAL

Transparent $TiCl_4$ was used as a starting material to fabricate TiO_2 powder by the HPPLT. The procedure to prepare the $TiOCl_2$ solution from $TiCl_4$ has been described in detail elsewhere.[1,2] The aqueous $TiOCl_2$ solutions were adjusted to contain 1.5 mol% of Zr^{4+}, Ni^{2+}, Cu^{2+}, Fe^{3+}, Al^{3+}, or Nb^{5+}. Aqueous $TiOCl_2$ solutions were adjusted to 0.03-0.08M by adding the appropriate amount of water. Similarly, aqueous $TiO(NO_3)_2$ and $TiOSO_4$ solutions were adjusted to 0.03-0.08M by adding the appropriate amount of water. On heating at 100°C for 4h, TiO_2 particles were spontaneously precipitated from these aqueous solutions in a closed Teflon reactor. Membrane filters with pores ranging from 0.1 μm to 1 μm were used to separate the Cl^- and SO_4^{2-} ions from the TiO_2 precipitate after the reaction, followed by washing with purified water. Crystalline TiO_2 powder was obtained after drying at 105°C for 24h in air. Crystallographic examination of the TiO_2 powder was performed using X-ray diffraction (XRD) with CuK_α radiation, and Raman spectroscopy. The morphology of the primary and secondary TiO_2 particles was examined using scanning electron microscopy (SEM) and transmission electron microscopy (TEM), respectively. A quantitative chemical analysis of the doping elements in the TiO_2 precipitates was conducted using inductively coupled plasma emission spectroscopy (ICP).

RESULTS AND DISCUSSION

Effect of Cations on the Phase Transition in TiO_2 Nano-powders.

Figure 1 shows the XRD patterns for the doped TiO_2 powders produced from the $TiOCl_2$ solutions with the various additives. The TiO_2 powder prepared using $ZrOCl_2$ shows only anatase peaks. In contrast, the TiO_2 powders produced by adding $CuCl_2$, $FeCl_3$, or $NbCl_5$ show only rutile peaks. The powders made with $AlCl_3$ have both (anatase + rutile) phases. Those prepared with $NiCl_2$ show mainly rutile peaks, and some small anatase peaks. The differences in the XRD patterns for the doped TiO_2 powders may be due to the ionic radius difference relative to Ti^{4+} (r=68pm). The powders produced by adding ions having ionic radii similar in size to Ti^{4+} have the rutile phase. These include Cu^{2+} (r=72pm), Fe^{3+} (r=64pm) and Nb^{5+} (r=69pm). On the other hand, the powders made with ions with an ionic radius different than Ti^{4+} have the anatase phase, or a mixture of the anatase and rutile phases. These include Zr^{4+} (r=79pm) and Al^{+3} (r=51pm).

Figure 2 shows the morphology of the TiO_2 powders prepared from $TiOCl_2$ solutions with Zr, Ni, Cu, Fe, Al or Nb. The doped TiO_2 powders produced with ions having a similar ionic radius to Ti^{4+} have a spherical shape and uniform size (Figures 2c, d, and f). However, the powders prepared with ions with radii that are different from Ti^{4+} have irregular shapes and are larger in size (Figures 2a and e). Additionally, the powders that contain a small amount of anatase have a relatively

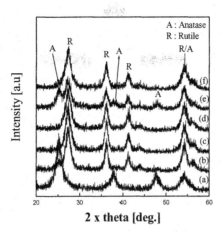

Figure 1. XRD patterns for TiO$_2$ powder synthesized from TiOCl$_2$ solution with: a) ZrOCl$_2$; b) NiCl$_2$; c) CuCl$_2$; d) FeCl$_3$; e) AlCl$_3$; and f) NbCl$_5$.

Figure 2. Morphology of the TiO$_2$ powder synthesized from TiOCl$_2$ solutions with: a) ZrOCl$_2$; b) NiCl$_2$; c) CuCl$_2$; d) FeCl$_3$; e) AlCl$_3$; and f) NbCl$_5$.

uniform particle size, but a rougher particle morphology. Considering both the XRD and the SEM results, it appears that the shape of the powder is spherical when only rutile is present, and it is irregular when the anatase phase is present.

Figure 3 shows TEM photographs of the TiO$_2$ powders synthesized from TiOCl$_2$ solutions with Zr, Ni, Cu, Fe, Al, or Nb. The TEM shows that the powder is actually comprised of larger secondary particles that are clusters of finer size

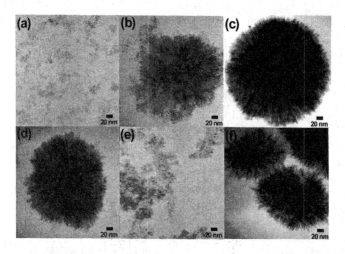

Figure 3. TEM photos of doped TiO_2 powder produced from a $TiOCl_2$ solution with: a) $ZrOCl_2$: b) $NiCl_2$; c) $CuCl_2$; d) $FeCl_3$; e) $AlCl_3$; and f) $NbCl_5$.

primary particles. In combination with the XRD results, TEM determined that the pure rutile phase of the TiO_2 powder has a needle shape, with a length of about 200 nm (Figures 3c, d and f). The anatase phase forms spherical primary particleTiO_2 powder of ~20 nm in diameter, as shown in Figures 3a and e. When small anatase peaks are observed with rutile peaks by XRD, some spherical primary particles exist around acicular primary particles, as shown in Figure 3b.

The present investigation demonstrates that either rutile or anatase can be produced by the HPPLT, depending on the additive and its ionic radius relative to Ti^{4+}. From the results presented, it can be anticipated that the addition of a cation with a size that differs significantly from that of Ti^{4+} will impede the formation of rutile due to a strain effect that favors the formation of a spherical precipitate having the anatase phase.

Ions having a different valence state, such as Ni^{2+} and Cu^{2+}, also determine the crystalline TiO_2 structure and morphology. However, it was not possible to isolate the effect of valence state from the effect of ionic radius. When Ni^{2+} is added, which has a similar ionic radius to Ti^{4+}, the powder produced is mostly rutile with some anatase phase. In contrast, the addition of Cu^{2+}, which has the same valence state as Ni^{2+}, and a similar ionic radius to Ti^{4+} produces only the rutile phase. All of the XRD, TEM, and ICP results for this study are summarized in Table I.

Effect of Anions on the Phase Transition of TiO_2 Nano-powders.

Figure 4 shows the Raman spectrum for a commercial TiO_2 powder that contains both the anatase and rutile phases. This standard was compared to the spectra for TiO_2 powders produced by the HPPLT at 50°C in aqueous $TiOCl_2$ solution with various Ti^{4+} contents. All of the powers were dried at 150°C for 1 h prior to the analysis. The peak at 241 cm^{-1} is from phonon scattering, and has no effect on the TiO_2 crystal structure. The $E_{1g}(R)$ and $A_{1g}(R)$ peaks, which are the characteristic peaks for the rutile phase, appear in the dried TiO_2 powder produced

Table I. A summary of the chemical analysis and crystalline phase of the TiO_2 powders homogeneously precipitated in aqueous $TiOCl_2$ solution with various additives.

	Major/ (Minor) Phase	Valence State and Ionic Radius (pm)				Dopant (1.5 mole%)	Crystalline Morphology	ICP Measured Content (mol%)
		+2	+3	+4	+5			
Ti	R			61, 68			Acicular	-
Zr	A			72, 79		$ZrOCl_2$	Spherical	1.41±0.06
Ni	R/(A)	69				$NiCl_2$	Acicular	0.10±0.02
Cu	R	72				$CuCl_2$	Acicular	0.05±0.02
Fe	R		64			$FeCl_3$	Acicular	1.44±0.11
Al	A/(R)		51, 53			$AlCl_3$	Spherical	0.13±0.04
Nb	R				64, 69	$NbCl_5$	Acicular	1.27±1.21

* R=rutile, A=Anatase

by the HPPLT. The reaction to form rutile depends on the amount of Ti^{4+} ions in the solution, so three different Ti^{4+} concentrations were studied. Generally, stable hydroxides such as $Ti(OH)_4$ will form first and then transform from an amorphous phase to anatase. As the heat-treatment temperature increases, the anatase phase transforms to the rutile phase. However, the dried powder produced by the HPPLT does not show any transformation when heated from 400 to 650°C, which is the temperature range in which the anatase phase is stable. Therefore, it can be assumed that the dried TiO_2 powder produced by HPPLT does not contain anatase, amorphous TiO_2 or Ti hydroxide.

Figure 5 shows the Raman spectra for $TiCl_4$ and aqueous $TiOCl_2$ solutions with 5M and 0.5M Ti^{4+} concentrations. The characteristics of the $TiCl_4$ spectrum are quite different from those of $TiOCl_2$ in water. After reacting for 4 h at 50°C, precipitation occurred in the $TiOCl_2$ solution, and the spectrum shows peaks

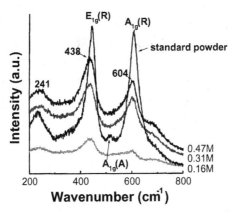

Figure 4. Raman spectra of crystalline TiO_2 powder with different Ti^{4+} contents.

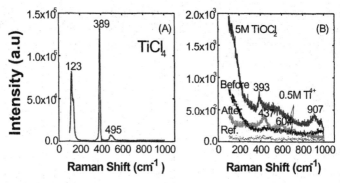

Figure 5. Raman spectra for: A) TiCl$_4$; and B) aqueous TiOCl$_2$ solutions with 0.5M and 5M Ti^{4+}.

characteristic of the rutile phase. Therefore, it can be concluded that the ultrafine rutile powder prepared from the aqueous TiOCl$_2$ solution by the HPPLT, is crystalline after the precipitation process.

To help understanding what determines the crystallinity and the shape of the precipitates, a precipitation reaction (HPPLT) in aqueous TiOCl$_2$ solution was carried out. TiOSO$_4$ is supposed to from the anatase phase. Only rutile precipitates from TiOCl$_2$ at temperatures below 70°C as our work has determined. The crystalline phase of the TiO$_2$ precipitate produced depends on the type of cation present in the aqueous Ti solution. The type of anion in the Ti compound solution will affect the O-Ti-O bonding structure, which will determine whether the anatase phase or the rutile phase precipitates. The crystallinity of the precipitate produced by reacting aqueous Ti solution with different cations is summarized in Table II. The XRD and TEM results for dried TiO$_2$ powder produced from TiOSO$_4$, TiOCl$_2$, and TiO(NO$_3$)$_2$ solutions produced from TiCl$_4$ are shown. Acicular primary particles of rutile always precipitate in an aqueous TiOCl$_2$ solution without additives, and in an aqueous TiOCl$_2$ solution with Cl$^-$ and NO^{3-}.

Table II. Effects of anion type additives on the crystallinity of TiO$_2$ powder synthesized from Ti aqueous solutions by the homogeneous precipitation process at low temperature.

Solutions	Additives	Crystallinity	Shapes
TiOCl$_2$	Not added	Rutile	Acicular
TiOCl$_2$	HCl	Rutile	Acicular
TiOCl$_2$	H$_2$SO$_4$	Anatase	Spherical
TiOCl$_2$	HNO$_3$	Rutile	Acicular
TiOCl$_2$	TiOSO$_4$	Anatase	Spherical
TiOCl$_2$	TiO(NO$_3$)$_2$	Rutile	Acicular
TiOCl$_2$	FeSO$_4$	Anatase	Spherical
TiOCl$_2$	CuSO$_4$	Anatase	Spherical
TiOSO$_4$	Not added	Anatase	Spherical
TiO(NO$_3$)$_2$	Not added	Rutile	Acicular
TiO(NO$_3$)$_2$	HCl	Rutile	Acicular
TiO(NO$_3$)$_2$	TiOSO$_4$	Anatase	Spherical

Spherical anatase precipitates in an aqueous $TiOCl_2$ solution with SO_4^{2-}. The efficiency of preparing TiO_2 powder is greatly suppressed by the addition of SO_4^{2-} ions. The same phenomena is observed in aqueous $TiO(NO_3)_2$ solution. Only spherical anatase forms in aqueous $TiOSO_4$ solutions, regardless of the anion additives.

When the additives are added appropriately, a mixture of anatase and rutile is produced (Figures 6 and 7). No precipitation of Fe_xO_y or Cu_xO, or doping of TiO_2 is observed with the addition of $FeSO_4$ and $CuSO_4$ under the experimental conditions examined. A transition from the precipitate of rutile to anatase occurred in aqueous $TiOCl_2$ solutions with increasing concentration of SO_4^{2-} ions (Figure 6). Table II shows that anatase is formed in the aqueous $TiOCl_2$ solution regardless of the anion, e.g., H_2SO_4, $FeSO_4$, or $CuSO_4$. It is concluded that anatase phase formation is primarily affected by the presence of the SO_4^{2-} versus the cation species. It is concluded that the addition of SO_4^{2-} to the aqueous $TiOCl_2$ solution promotes anatase precipitation while suppressing the precipitation of the rutile phase. Furthermore, SO_4^{2-} ions lead to the formation of a spherical precipitate, while suppressing directional growth. According to Park *et al.*, a high

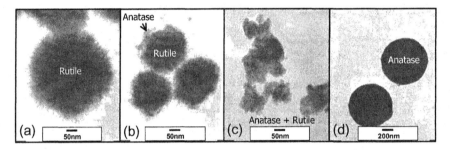

Figure 6. TEM micrographs of the TiO_2 precipitates produced from $TiOCl_2$ solutions containing different amounts of SO_4^{2-}: a) 0M; b) 0.03M; c) 0.05M; and d) 0.08M.

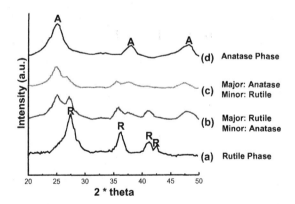

Figure 7. XRD patterns for TiO_2 precipitated from an aqueous $TiOCl_2$ solution with various amounts of SO_4^{2-} as shown in Figure 6.

vapor pressure of H_2O during the formation of TiO_2 results in the formation of the rutile phase.[2] The capillary pressure within the acicular primary particles may also be a factor. The SO_4^{2-} ions may act to decrease the capillary pressure in the acicular primary particles.[2]

CONCLUSIONS

TiO_2 nano-powders have been produced from aqueous $TiOCl_2$ solutions with various cation and anion additives by the HPPLT. Cations with ionic radii that are different from Ti^{4+} produce a spherical shape anatase particle. Increasing the amount of SO_4^{2-} ions in the aqueous $TiOCl_2$ solution promotes the formation of spherical anatase particles. It is, therefore, concluded that the cation radii and the concentration of SO_4^{2-} control the crystallinity and morphology of the nano-structured TiO_2 precipitates that form during the HPPLT.

REFERENCES

[1]S.J. Kim, S.D. Park, Y.H. Jeong and S. Park, "Homogeneous Precipitation of TiO_2 Ultrafine Powders from Aqueous $TiOCl_2$ Solution," *J. Am. Ceram. Soc.*, **82** [4] 927-32 (1999).

[2]S.D. Park, Y.H. Cho, W.W. Kim and S.J. Kim, "Understanding of Homogeneous Spontaneous Precipitation for Monodispersed TiO_2 Ultrafine Powders with Rutile Phase around Room Temperature," *J. Solid State Chemistry*, **146** [1] 230-38 (1999).

[3]J. Yang and J.M.F. Ferreira, "On the Titania Phase Transformation by Zirconia Additive in a Sol-Gel Derived Powder," *Mater. Res. Bull.*, **33** 389 (1998).

[4]K. Okada, N. Yamamoto, Y. Kameshima, A. Yasumori and K J. D. Mackenzie, "Effect of Silica Additive on the Anatase-to-Rutile Phase Transition", *J. Am. Ceram. Soc.*, **84** [7] 1591-96 (2001).

CHARACTERIZATION OF NANO-PARTICLE DISPERSION IN A SILICA SLURRY

Chika Takai, Masayoshi Fuji, and Minoru Takahashi
Ceramics Research Laboratory, Nagoya Institute of Technology
Tajimi 507-0071, Japan

ABSTRACT

TEM observation of an *in situ* solidified slurry was used to characterize the dispersion and structure of nano-particle silica in a slurry. Direct observations were made by TEM to visualize the stereoscopic structure of the silica nano-particle structure in a slurry. These observations were then compared with the measured zeta potential and slurry viscosity. Flocculation is observed at pH 1-2, where the zeta potential is low, and the viscosity is high. The nano-particles form a three-dimensional network at pH 6.35, and the network structure disappears at pH 7.02. The silica nano-particles re-agglomerate at pH 8.43-10.43. Particle dispersion/aggregation changes significantly with pH although slurry viscosity is relatively constant in the range of pH 6-10. By comparing TEM observations to zeta-potential and viscosity measurements, we verified that that the observed structure reflects the actual structure in the slurry.

INTRODUCTION

To develop a fabrication process for nano-ceramics, it is necessary to prepare well-dispersed slurry. Rheological analysis and electrokinetic measurements are generally used to characterize ceramic slurries.[1,2] However, these methods provide only indirect information on particle dispersion. Direct visualization of particle dispersion is required to obtain more specific information. A new method has been developed that allows direct observation of particle dispersions using *in situ* solidification.[3,4] As illustrated in Figure 1, *in situ* solidification is accomplished using gel casting,[5] which is a method to fabricate ceramic green bodies by means of polymerization of organic monomers. The organic monomers hardly influence the particle dispersion and structure in slurries.[3] In gel casting, a gel structure is created that holds the ceramic particles together. Owing to the good machinability of the solidified green bodies, thinned samples for transparent observation (e.g. by transmission electron microscopy, TEM) can be easily prepared from gel cast bodies.

The dispersion of sub-micron size alumina particles in a slurry has been evaluated[3,4] using the direct observation method. A polymer dispersant was used to systematically change the particle dispersion in the slurry. Additionally analysis on thinned solidified bodies using a transmission optical microscopy provided information on dispersion behavior that cannot be explained by a rheological measurement. These structures verified that *in situ* solidification has prospects for direct visualization of particle dispersion

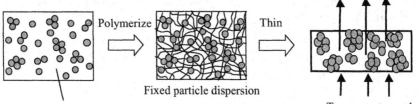

Polymerize

Thin

Fixed particle dispersion

Slurry + Organic Monomer

Transparent sample

Figure 1. A schematic illustrating the *in situ* solidification technique used to produce transparent samples to directly characterize particle dispersion. The structure of the particle dispersion is "fixed" by *in situ* solidification, and thinning produces a transparent specimen.

in slurries.

Particles <100 nm in diameter have great potential for a wide variety of applications in electronic, magnetic, and engineering ceramics.[6,7] Unfortunately, nano particles aggregate easily in comparison with sub-micron particle.[8] So far, the dispersion of nano-particles in a slurry has been examined only by rheological and electrokinetic measurements. In this study, we characterize silica nano-particle dispersions at different pH using transmission electron microscopy (TEM), and compare these results with zeta potential,[9,10] and viscosity measurements.

EXPERIMENT

A premix solution was prepared by dissolving an organic monomer, methacrylamide (MAM, Wako Chemical, Osaka, Japan) and a cross-linker, N,N'-methylenebisacrylamide (MBAM, Wako Chemical, Osaka, Japan) in distilled water. The concentration of organics in the premix solution was   18 wt%, and the mass ratio of MAM:MBAM was 10:1. Preliminary examination determined that this composition produces a gelled body that is strong enough not to deform during drying.[3]

Silica powder (AEROSIL200, Nippon Aerosil Corp., Tokyo, Japan, primary particle diameter 12 nm) was suspended in the premix solution. The solids loading was fixed at 2.5 wt%. To vary the degree of particle dispersion, the pH of the slurry was changed using HNO_3 and NaOH. After adjusting pH from 1 to 11, the slurry was ball-milled for 24 h. Maintaining the slurry temperature at 25°C, the polymerization of the solvent was initiated by adding ammonium persulfate $(NH_4)_2S_2O_8$ (Wako Chemical, Osaka, Japan). Immediately after the addition of the initiator, the slurry was cast into a plastic mold. The reaction was accelerated with a catalyst: N, N, N', N'- tetramethyl ethylenediamine (TEMED, Wako Chemical, Osaka, Japan). The amounts of initiator and catalyst were 0.02 and 0.05 wt%, respectively. These concentrations ware used to prevent phase separation in the ceramic slurries. After polymerization, the solidified green body was de-molded and dried in a controlled humidity chamber (Model FX206P, ETAC, Tokyo, Japan).

The dried bodies were ground using an agate pestle and mortar to obtain fragments ~1 μm in length. A specimen for TEM analysis was prepared by dispersing some fragments on a micro-grid after an ultrasonic cleaning treatment in ethanol. TEM images were obtained using a JEOL JEM-2000EX II operated at 160 kV.

The viscosity of the slurry was measured just after ball milling using a

Brookfield viscometer (Model BL, Toki Industries, Tokyo, Japan) and a shear rate of 15.24 s^{-1}. Zeta potential measurements on the slurry were performed using an acoustic and electroacoustic spectrometer (DT-1200, Dispersion Technology, Inc, N.Y., USA).

RESULTS
TEM Observation of the Silica Structure
TEM photographs of *in situ* solidified nano silica slurries prepared at different pH are shown in Figures 2a-f. The dark areas are silica nano-particles. The structure of the particle dispersion changes significantly with pH. At pH 1.92, large flocculation structures are observed (Figure 2a). The floc size decreases as pH increases (Figure 2b). The nano-particles form a three-dimensional network structure at pH 6.35 (Figure 2c). The network disappears and particle flocculation occurs at pH 7.02 (Figure 2d). Further increases in pH increase the size of the flocculate units (Figure 2e-f). Figure 3 is a schematic of the TEM results shown in Figure 2. Figure 3a illustrates the flocculation at low pH. Figure 3b illustrates the 3D network structure that forms as one approaches a neutral pH (as shown in Figure 2c). Small particle flocs of about 50 nm form at neutral pH (Figure 3c) where the slurry is assumed to be well dispersion. The floc size increases with increasing pH (Figure 3d).

Zeta Potential Measurement
The zeta potential of nano-particle silica slurries as a function of pH are shown in Figure 4. The potential of the silica strongly depends on the pH. The isoelectric point (iep) of the silica surface was found to be in the range of pH = 1-2. The surface charge increases in a negative direction until pH 7.41 where it shows a maximum charge of -45.13 mv. Above pH 7.41, the absolute value of the surface charge decreases.

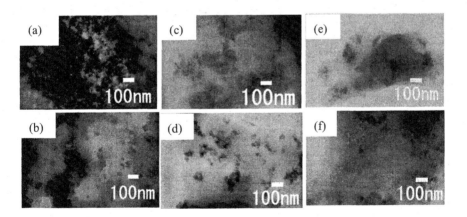

Figure 2. TEM images of thin sections of *in situ* solidified silica slurries prepared at different pH: a) strong acid (pH 1.92); b) acid (pH 2.48); c) neutral (pH 6.35); d) neutral (pH 7.02); e) base (pH 8.04); and f) strong base (pH 10.43).

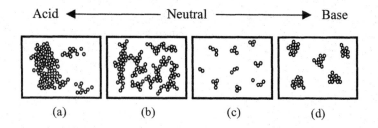

Acid ◄──────── Neutral ────────► Base

(a)   (b)   (c)   (d)

Figure 3. An illustration of the changes in structure in a silica particle dispersion with increasing in pH: a) acidic (pH 1.92); b) neutral (pH 6.35); c) neutral (pH 7.02); and d) basic (pH 10.43).

Slurry Viscosity Measurement

Figure 5 shows the viscosity of nano-particle silica slurries as a function of pH. The viscosity decreases with pH up to pH = 6. The viscosity is substantially constant at pH 6 and higher.

DISCUSSION

The results of the zeta potential measurements suggest that, in the acidic region near the iep of silica, the silica nano-particles strongly attract each other to form a large floc structure. With increasing pH, the repulsive electrostatic force between the particles increases, and the floc size becomes smaller. Then, the smaller flocs form a three-dimensional network. At pH 7.41 where the silica surface has a negative maximum charge, the strong repulsive electrostatic force between particles stabilizes the slurry dispersion. Further increasing pH into the basic region decreases the zeta potential toward zero again, which promotes flocculation in the slurry.

The viscosity measurements indicate that the slurries prepared in the acidic region are flocculated. However, the degree of flocculation decreases with increasing pH. A minimum and nearly constant viscosity is observed at pH>6, which suggests that the floc structure is soft and easily broken down by shear force.

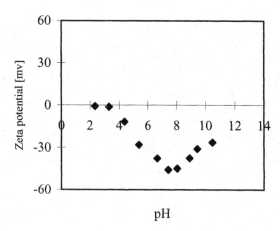

Figure 4. Zeta potential of a nano-particle silica slurry as a function of pH.

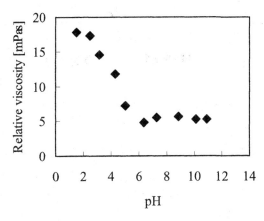

Figure 5. The viscosity of a nano-particle silica slurry as a function of pH (shear rate: 15.24 s^{-1}).

TEM observations of the solidified slurries are consistence with the zeta potential results. The large flocs observed in the acidic slurry can be explained by the iep of the silica as shown in Figure 4. At around pH 7, the slurry has small flocs, although the zeta potential shows a maximum repulsive force between particles. These may be small flocs of nano-particles that were not broken down to primary particles during ball milling.

The silica particles in the slurry prepared at pH>8 are possibly dissolved and re-crystallized silica. An increase of ionic strength, owing to the dissolution of silica in the basic region, might bring about the decrease of zeta potential observed[2] and could cause the re-flocculation observed.

It should be emphasized that the viscosity and zeta potential do not provide any direct information on the silica structure in the particle dispersion. Results obtained using these more traditional rheological characterization methods can be correctly interpreted by directly imaging the particle packing structure in the slurry.

CONCLUSION

An *in situ* solidification technique was used to "freeze in" the structure of the silica nano-particles in a slurry. TEM examination of the solidified bodies provided clear images of the particle structure in the dispersion, which changed significantly with slurry pH. Direct observations of the nano-particle silica structure are consistent with results from viscosity and zeta potential measurements. This new technique also has prospects for direct visualization of nano-particle dispersion in other slurries.

REFERENCES

[1]T. Fengqiu, H. Xiaoxian, Z. Yufeng, G. Jingkun, "Effect of Dispersants on Surface Chemical Properties of Nano-zirconia Suspensions," *Ceramics International,* **26** (2000) 93-97.

[2]R.R. Rao, H.N. Roopa, T.S. Kannan, "Effect of pH on the Dispersability of Silicon Carbide Powders in Aqueous Media," *Ceramics International,* **25** (1999) 223-230.

[3]M. Oya and M. Takahashi, "Transparent Observation of Particle Dispersion in Alumina Slurry Using *in situ* Solidification Technique,"

Ceramics Tran., **133,** (2002) 47-52.

[4]M. Takahashi, M. Oya, and M. Fuji, "Transparent Observation of Particle Dispersion in Alumina Slurry Using *in situ* Solidification Technique,"Advanced Powder Technology, in press.

[5]A.C. Young, O. Ogbemi, M.A. Janney, and P.A. Menchhofer, "Gelcasting of Alumina," *J. Am. Ceram. Soc.,* **74** (3), (1991) 612-18.

[6]M.D. Mukadam, S.M. Yusuf, P. Sharma and S.K. Kulshreshtha, "Particle Size-Dependent Magnetic Properties of γ -Fe$_2$O$_3$ Nanoparticles," *J. Magnetism and Magnetic Materials,* in press

[7]O. Cintora-Gonzalez, C. Estournes, M. Richard-Plouet, J.L. Guille, "Nickel Nano-Particles in Silica Gel Monoliths: Control of the Size and Magnetic Properties," *Materials Science and Engineering C,* **15** 2001 179–182

[8]Z. Li, Y. Zhu, "Surface-Modification of SiO$_2$ Nanoparticles with Oleic Acid," *Applied Surface Science,* **211** (2003) 315–320

[9]R.A. Hayes, "The Electrokinetic Behavior of Surfaces Modified by Particle Adsorption," *Colloids and Surfaces A: Physicochemical and Engineering Aspects,* **146,** (1999) 89-94.

[10]M. Ishimori, "Chemical Properties of Pigment Surface and Dispersion Stability," DIC Technical Review No.5 (1999).

EFFECTS OF SLURRY PREPARATION CONDITIONS ON GRANULE PROPERTIES AND THE STRENGTH OF ALUMINA CERAMICS

Makoto Furukawa and Tadashi Hotta
Japan Fine Ceramics Center
2-4-1, Mutsumo, Atsuta-ku,
Nagoya, 456-8587, Japan

Kenji Okamoto
Nippon Denko Co., Ltd.
28-1, Kouno, Tatibana-Cho,
Anan, 774-0028, Japan

Hiroya Abe and Makio Naito
Joining and Welding Research Institute,
Osaka University
11-1, Mihogaoka,
Ibaraki, 567-0047, Japan

ABSTRACT

Effects of slurry preparation conditions on the properties of spray-dried granules, green compact microstructure, and sintered ceramic strength were investigated. Two different slurries of low soda alumina powder were prepared and studied. One was prepared by ball milling with a dispersant of ammonium polycarboxylate; the other was prepared by ball milling and adjusting pH with HNO_3 to obtain the same apparent viscosity as that of the slurry prepared with the dispersant. The slurry prepared by adjusting the pH made hard and brittle granules that produce a uniform microstructure green body. The brittle granules fracture during compaction to produce primary particles that fill the void spaces between the granules. The granules produced from the slurry prepared by adding the dispersant produce a less homogeneous green microstructure. Pressed compacts retain the interfaces between the original granules, as well as internal voids. These microstructure heterogeneities lower the fracture strength of the sintered body.

INTRODUCTION

The strength of ceramics strongly depends on the processing conditions before firing.[1] In the powder compaction process, granule properties play an important role in producing high reliability ceramics. A recent study demonstrated that heating alumina granules can effectively reduce the size and population of potential flaws in the green bodies during compaction.[2] The hard and brittle character of heat-treated granules contributes to a uniform packing structure in the green body, and the spaces between granules were efficiently filled with primary powder particles caused by granule fracture during pressing. Granule characteristics are also strongly affected by the slurry properties.[3] Therefore, the slurry preparation conditions are key to achieving high quality granules and ceramics.

The objective of this paper is to examine the effect of slurry preparation conditions on the compressive strength and compaction behavior of granules, and on the structure and properties of green and sintered bodies. Two typical slurry preparation conditions were used to prepare granules for pressing. The fracture behavior of the two different granules is examined, and the differences in compaction behavior are evaluated. Variations in the pore structure and in the properties of the green and sintered bodies are discussed relative to the granule characteristics.

EXPERIMENTAL PROCEDURE

Figure 1 shows the processes used to produce the alumina ceramics. For this study, low soda alumina powder (AL-160SG-4, Showadenko, Japan) was used. One slurry was prepared by ball milling and adjusting the pH with HNO_3 to control viscosity. In this case, the pH of the processed slurry was 4. Another slurry was prepared by ball milling with 0.4 wt % of ammonium polycarboxylate. (CERUNA D-305, Chukyo Yushi Co., Japan). The solids content of the slurries was 35 vol.%. After ball milling, organic processing aids were added to each slurry. For this experiment, 0.4 wt% binder (CELUNA WF-804+WF-610, Chukyo Yushi Co., Japan) and 0.14 wt% lubricant (Serosol 920, Chukyo Yushi Co., Japan) were added, relative to the weight of alumina powder, and the slurries were stirred for 1 h. In both cases, the apparent viscosity of the slurry was adjusted 50 MPa · s at a shear rate of $300 \ s^{-1}$.

Each slurry was spray-dried with inlet air temperature 105°C, outlet air temperature 70°C, 8000 rpm disk rotation speed, and 100 g/min slurry feed rate. The spray-dried granules were screened through a 150 μm sieve to remove any

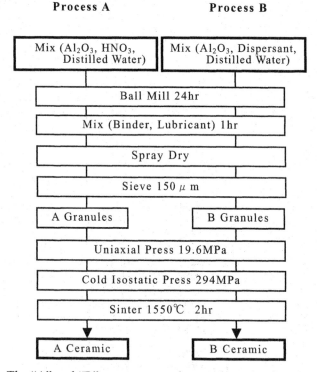

Figure 1. The "A" and "B" processes used to produce alumina ceramics.

large agglomerates and foreign objects. The granules were uniaxially pressed in a rectangular die (60×50 mm) at 19.6 MPa, and then cold isostatically pressed (CIP) at 294 MPa. The green bodies were sintered in an electric furnace and at 1550°C for 2 h in air. The densities of the sintered samples were measured using the Archimedes method, with water as the immersion medium. Test samples measuring 3 mm×4 mm×40 mm were cut from the sintered bodies to measure fracture strength and toughness. The surface of the samples was finished with a No. 800 diamond grinding wheel. Four-point bending strength was measured according to JIS R 1601 using a universal testing machine with a crosshead speed of 0.5 mm/min. The single-edged-precracked-beam method was used to measure fracture toughness according to JIS R 1607.

The liquid immersion method [4] was used to examine the structure of the granules produced by spray drying. Samples were characterized in the transmission mode using an optical microscope. The fracture behavior of single granules was analyzed using a diametral micro-compression testing machine (PCT-200, Shimadzu Corp., Kyoto, Japan). Mercury porosimetry (AutoPore III 9420, Shimadzu Corp. , Kyoto, Japan) was used to determine the size and the size distribution of the pore channels in the pressed green compacts. The internal structure of pressed green bodies was assessed from fracture surfaces characterized using SEM. The internal structure of thinned samples of sintered bodies was examined in the transmission mode with no immersion liquid.[5]

RESULTS

Figure 2 shows an optical micrograph of a typical B granules characterized using the liquid-immersion method.[4] All of the granules contain dimples. There are no clear differences in the internal structure of the A and B granules, and both have the same average size of 55 μm. The fracture behavior of the A and B granules, however, is different. Strength was determined from the Hiramatsu's

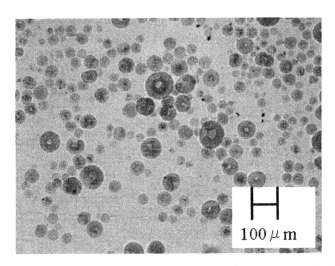

Figure 2. Optical micrograph of typical B granules taken using the liquid immersion method.

Table I. Density of green and sintered bodies

	Density of green bodies ($\times 10^3$kg/m^3)		Density of sintered body ($\times 10^3$ kg/m^3)
	19.6 MPa	294 MPa	
Process A	2.14	2.37	3.95
Process B	2.12	2.38	3.95

equation.[6]   An average granule tensile strength of 0.36 MPa and 0.24 MPa was measured for the A and B granules, respectively. The relationship between compressive load and strain for single a granule reveals that the A granules are more brittle than the B granules. In this case, the displacement needed to break a granule was 1.53 μm and 1.98 μm for the A and B granules, respectively.

Table 1 shows the apparent density of green compacts after uniaxial pressing and after CIPing. Both A and B granules compact to almost the same apparent density.

Figure 3 shows the pore-size distribution curves of A and B powder compacts made by CIPing at 294 MPa. No significant difference is observed between the green A and B ceramic bodies.

Figure 4 shows SEM micrographs of fracture surfaces in a green A and B ceramic compact formed by uniaxial pressing at 19.6 MPa. A marked difference is

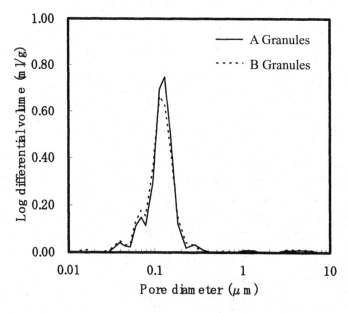

Figure 3. Pore size distribution curves of green compacts made from the A and B granules after CIP.

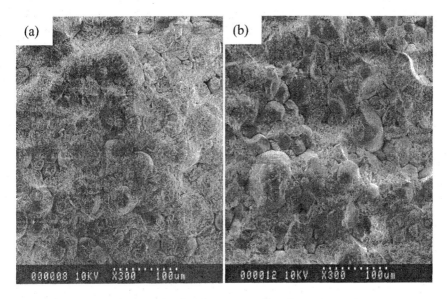

Figure 4. SEM micrograph of a fracture surface in a green power compact formed by uniaxial pressing at 19.6 MPa: a) "A" granules; and b) "B" granules.

evident in the fracture of individual A and B granules. For the B granules prepared with the dispersant, even though most of the granules undergo substantial plastic deformation. during pressing, remnants of the original granules are clearly visible in the pressed powder compact. Only some of the granules appear to have been crushed. When the pressure is increased to 294 MPa, more granules are fractured compared to Figure 4; however, unfractured granules are still present in both the A and B green compacts. Characterization of the green compact microstructure using the liquid immersion method [4] revealed the same results as shown in Figure 4.

Table 1 also shows the density of the sintered A and B ceramics. Both sintered bodies show almost the same density. The fracture toughness of both sintered bodies is 3.6 MPa $\cdot$ m$^{1/2}$. Figure 5 shows the Weibull modules and the distribution of fracture strength measured for the sintered A and B ceramics. The fracture strength of the sintered A ceramic body (average strength = 510 MPa) is clearly higher than that of the sintered B ceramic body (average strength = 475 MPa). The Weibull modulus is almost the same for both.

Figure 6 shows photomicrographs, taken in the transmission mode, of thinned samples of sintered A and B ceramics. Many large pores are present in both samples, even after sintering. The size and number of the pores are clearly smaller in the sintered A ceramics made from the slurry prepared by adjusting the pH.

DISCUSSION

As shown in Figure 6, the sintered A ceramic has large pores that are smaller in size than the sintered B ceramic. This is most likely the reason why the sintered A ceramic body has a higher strength. The large pores are evidence of the

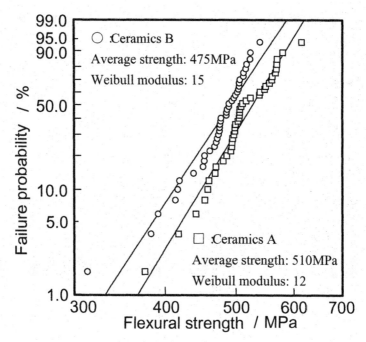

Figure 5. Probability of failure, Weibull modules, and fracture strength measured for sintered A and B ceramics.

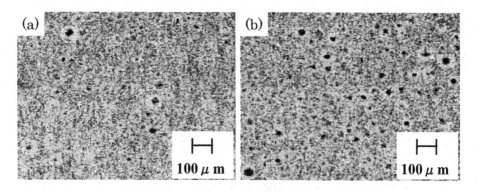

Figure 6. Photomicrographs taken in the transmission mode on thinned specimens for the: a) sintered A ceramics; and b) sintered B ceramics.

incomplete deformation and fracture of the granules in the green compacts during powder pressing.

As illustrated in Figure 4, the A ceramic has a more uniform green microstructure than the B ceramic. This reflects a difference in compaction behavior of the A and B granules. The A granules show characteristics of hard and brittle granules. These characteristics contribute to more uniform packing of the granules during die filling, and a more uniform compact microstructure after compaction. During power pressing, the spaces between the original A granules are efficiently filled with the smaller, primary particles produced by the granule fracture process. In contrast, the B ceramics retain more of the original granule structure, and more of the large pore structure between the original granules. The result is a sintered B ceramic that has larger pores, and a lower fracture strength. A similar trend has also been observed in a comparative study on as-sprayed granules and heat-treated granules [2].

The dispersant used in this study functions as a solid bridge on the surfaces of granules and between primary particles. This may favor granule deformation over granule fracture during compaction. Further work is needed to determine the mechanism that changes the fracture behavior of granules prepared from different slurries.

CONCLUSIONS

An alumina slurry prepared by adjusting pH made hard and brittle granules. Brittle granules favor granule fracture during powder compaction to produce smaller, primary particles that efficiently fill the void spaces between granules. Less brittle granules made from a slurry prepared with a dispersant retain more of the original granule structure, and the large inter-granular porosity after compaction. This results in a sintered body with a lower fracture strength.

REFERENCES

[1]G.Y. Onoda and L.L. Hench, "Ceramic Processing Before Firing," John Wiley & Sons, Inc., New York, 1978.
[2]N. Shinohara, S. Katori, M. Okumiya, T. Hotta, K. Nakahira, M. Naito, Y-I Cho, and K. Uematsu, "Effect of Heat Treatment of Alumina Granules on the Compaction Behavior and Properties of Green and Sintered Bodies," J. Euro. Ceram. Soc., 22, 2841-2848 (2002).
[3]M. Naito, Y. Fukuda, N. Yoshikawa, H. Kamiya, and J. Tsubaki, "Optimization of Suspension Characteristics for Shaping Processes," J. Euro. Ceram. Soc., 17, 251-257 (1997).
[4]K. Uematsu, "Immersion Microscopy for Detailed Characterization of Defects in Ceramic Powders and Green Bodies," Powder Technology, 88, 291-298 (1996).
[5]T. Hotta, K. Nakahira, M. Naito, N. Shinohara, M. Okumiya, and K. Uematsu, "Origin of Strength Change in Ceramics Associated with the Alteration of Spray Dryer," J. Mater. Res., 14, 2974-2979 (1999).
[6]Y. Hiramatsu, Y. Oka, and H. Kiyama, "Determination of the Tensile Strength of Rock by a Compression Test of an Irregular Test Piece," J. Min. Eng., Japan, 81, 1024-1030 (1965).
[7]W.J. Walker, Jr., J.S. Reed, and S.K. Verma, "Influence of Slurry Parameters on Granule Characteristics," J. Am. Ceram. Soc., 82, 1711-1719 (1999).

High Temperature Interfaces

EFFECTS OF TITANIUM ON WETTABILITY AND INTERFACES IN ALUMINUM/CERAMIC SYSTEMS

Natalia Sobczak
Foundry Research Institute, 73 Zakopianska St., 30-418 Cracow, Poland

ABSTRACT
An overview of recent research on the effects of titanium on wetting and on the mechanical properties, structure, and chemistry of Al/ceramic interfaces is presented. This paper summarizes work completed over a wide temperature range on non-reactive (Al/AlN and Al/Al$_2$O$_3$) and reactive (Al/SiO$_2$, Al/mullite, Al/Si$_3$N$_4$, Al/(Si$_3$N$_4$+Y$_2$O$_3$), Al/TiO$_2$, Al/C, Al/SiC) systems.

INTRODUCTION
The addition of Ti to noble metals (e.g., Cu, Sn, Ag) has beneficial effects on wettability and bonding. This has been demonstrated for many metal/ceramic systems, and has been successfully utilized in industrial applications to manufacture composites and to join dissimilar materials.[1-3] For Ti additions to Al, the real effect of Ti can be masked by molten Al reactions with ceramic materials. Moreover, to achieve the advantageous effect of Ti, a larger amount of Ti must be added to Al (e.g., Al/C system [4]). However, Ti drastically increases the liquidus temperature of the Al-Ti alloy. Additionally, most of the Ti is in the solid Al$_3$Ti phase, which requires time to melt during processing; thus, the Al-Ti alloy/ceramic interaction might depend on kinetic factors. In this study, the effects of Ti additions on wettability and interfaces of several Al/ceramic couples are summarized.

EXPERIMENTAL PROCEDURE
Contact angle (θ) measurements were completed under vacuum (2-5x10^{-4} Pa) using the sessile drop method and equipment described elsewhere.[4-6] Figure 1 illustrates the four different procedures used, including: CH- contact heating of a metal/ceramic couple to the test temperature; CP- capillary purification, assuring separate heating and removal of the oxide film from the Al drop; CP$_c$- a modified CP method, in which the Al drop is strained between the substrate and a capillary by the fixed position of the piston; and CP$_r$- modified CP$_c$, which involves drawing part of the Al drop into the capillary (by raising the piston) at the end of the test, followed by fast cooling.
For selected couples, the interfacial strength (τ) was measured directly on solidified sessile drop samples using an improved push off shear test.[7]
The substrates used include SiO$_2$ (fused quartz), SiC (0.5%B, sintered in argon), and Al$_2$O$_3$ (99.9%), TiO$_2$ (99.95%), AlN (99.3%), Si$_3$N$_4$ (99.9%) and

mullite (99.95%) all hot pressed in argon.

Pure Al (99.9999%) and AlTi6 alloy (5.9 wt.%Ti) were used in the study. As-extruded AlTi6 alloy was selected as a model system to demonstrate the role of kinetic factors in reactive systems; it melts over a wide temperature range (937-1395 K), and it has a specific structure that allows one to easily distinguish the liquid and semi-liquid regions in the drop from the morphology of Al_3Ti phase (Figure 2). Additional tests were completed with AlSi11 alloy (11 wt.% Si) to determine the effect of alloying on chemical reactions in the system. For simplification, the Al/ceramic systems studied are divided into two groups that take into account only possible displacement reactions between Al and the ceramic: 1) non-reactive systems (Al/Al_2O_3, Al/AlN), and 2) reactive systems (Al/TiO_2, Al/SiO_2, $Al/mullite$, Al/Si_3N_4, Al/C).

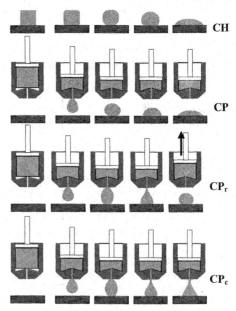

Figure 1. The sessile drop testing procedures used (see explanations in the text).

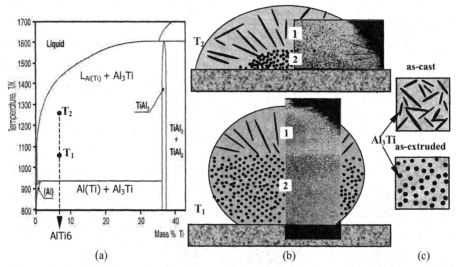

Figure 2. a) Al-Ti phase diagram, b) a schematic of the microstructure of a solidified AlTi6 drop after wettability test at different temperatures where $T_1<T_2$ (where, 1- liquid state and 2- semi-liquid state, and c) a comparison of as-cast and as-extruded AlTi6 alloy).

RESULTS AND DISCUSSION

Both pure Al and AlTi6 alloy do not form any bulk reaction products with Al_2O_3 and AlN. Under the conditions of this study, high temperature wettability in the Al/Al_2O_3 system (Table I) is affected by the dissolution of the substrates in the Al drop, and the formation of fine Al_2O_3 crystals directly at the drop-side interface.[8] This results in two effects, secondary roughness and interface reinforcing, that together with improved wettability, contribute to an increase in interfacial strength with increasing temperature (Figure 3).

The addition of 6 wt.% Ti to Al results to poorer wetting and bonding properties with Al_2O_3 (Tables I and II, and Figure 3). After removing the metal drop by dissolution in an acid, characterization of the substrate surface using a profilometer shows that a much smoother interface forms with AlTi6 compared to pure Al. This suggests that there is less dissolution and a weaker interaction after alloying with Ti. A Similar wetting-interface structure-bonding relationship is observed for Al and the AlTi6 alloy on oxygen-free AlN (1) substrates. Better wetting and stronger interfaces are obtained with oxygen containing AlN (2) substrates (Table 1, Figure 3).

Compared to pure Al, the AlSi11 alloy has poorer wetting on Al_2O_3, and better wetting of AlN (2). Among the couples examined, the $AlSi11/Al_2O_3$ and AlSi11/AlN samples show the highest shear strength. Since Si and Ti in the Al alloy do not react with Al_2O_3 or AlN, the improved strength measured may be due to a beneficial interface structure formed during solidification. Si nucleates at the interface, and a large amount of Si precipitates in the AlSi11 alloy, which may

Table I. Contact angle, θ (CH), at different temperatures for non-reactive couples

T [K]	Al_2O_3			AlN(1) [9]		AlN(2) [9]		
	Al [7]	AlTi6	AlSi11	Al	AlTi6	Al	AlTi6	AlSi11
953	126							
1023	121							
1123	96	114	118					
1173				102	112	108	77	88
1223	79	90	117	59	85	57	57	48
1323	74	89	85					

Table II. Effects of testing procedure, processing time, and temperature on θ and τ in Al/Al_2O_3 couples

Metal	Conditions	θ [°]	τ [MPa]
Al	1223 K, 30 min,CH	79	46.6
	973 K, 15 min,CP	94	42.1
AlTi6	1223 K, 30 min,CH	90	20.2
	1073 K, 30 min,CP	91	5
AlSi11	1123 K, 30 min,CH	118	68
	973 K, 15 min,CP	98	66.5

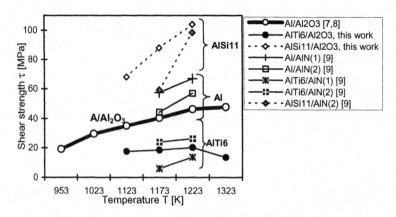

Figure 3. Effect of processing temperature and alloying on the shear strength of Al/Al$_2$O$_3$ and Al/AlN couples (CH).

reinforce the interface.

A comparison of wetting using the CH and CP procedures shows that the removal of the oxide film from the metal drop (CP) improves low temperature wetting and bonding in Al/Al$_2$O$_3$ and AlSi11/Al$_2$O$_3$ couples, but not in AlTi6/Al$_2$O$_3$ couples (Table II).

For the Al/TiO$_2$ system, thermodynamic calculations suggest that molten Al should react with TiO$_2$ to form Al$_2$O$_3$ and Ti as follows [10]:

$$4Al + 3TiO_2 = 2Al_2O_3 + 6Ti \tag{1}$$

An SEM study [11] has shown that, at 1173 K, TiO$_2$ is dissolved in molten Al, and Al$_2$O$_3$ crystals precipitate. This results in a similar interface strengthening effect (Table III) as in the Al/Al$_2$O$_3$ system. Higher temperature favors better wetting, which is accompanied by reactive metal infiltration (RMI) of the substrate. However, after solidification, samples with a thick RMI zone (1373 K) have numerous cracks at the substrate/RMI interface, which lowers the shear strength. Alloying Al with 6 wt.% Ti suppresses both TiO$_2$ dissolution and the displacement reaction, resulting in poorer wetting and a weaker interface (Table III).

Table III. θ-τ relationships for Al/TiO$_2$ couples

T [K]	Al, 120 min		AlTi6, 120 min	
	θ [°] [10]	τ [MPa]	θ [°]	τ [MPa]
1173	96	45		
1273	80	35	90	18
1373	64	25		

Similar interactions are observed in Al/SiO$_2$, Al/mullite and Al/Si$_3$N$_4$ couples.[11-14] This is due to the formation of the same Si metal by-product, and the same reaction products (Al$_2$O$_3$ or AlN) that have similar interactions with Al and AlTi6, i.e.,

$$4Al + 3SiO_2 = 2Al_2O_3 + 3Si \qquad (2)$$
$$8Al + 3Al_6Si_2O_{13} = 13Al_2O_3 + 6Si \qquad (3)$$
$$4Al + Si_3N_4 = 4AlN + 3Si \qquad (4)$$
$$Ti + 2Si = TiSi_2 \qquad (5)$$

All these systems show strong reactivity, as evidenced by the development of a thick reaction product region (RPR). This is true even at low temperatures (e.g., Figure 4) when molten Al and AlTi6 do not wet the ceramics (Table IV). For Al/SiO$_2$, there are two hypotheses to explain the contradiction between a high reactivity and a lack of low temperature wettability: 1) it is due to an oxide film on the Al drop [15]; and 2) it is due to pinning of the triple line by surface cavities caused by the volume mismatch between SiO$_2$ and freshly formed Al$_2$O$_3$ [16]. The first hypothesis has been proven to be true in a CP test that demonstrated low temperature wettability in an Al/SiO$_2$ couple after the oxide film on the Al drop was removed (Table IV). The oxide film might affect θ in other Al/ceramic systems, particularly those measured in a gaseous atmosphere (e.g. for Al/mullite in the work [18]). The oxide film can be eliminated by *in situ* cleaning of the Al drop in vacuum at T>1173 K [1] by the following reaction:

$$4Al + Al_2O_3 = 3Al_2O\uparrow \qquad (6)$$

In contrast to pure Al, the AlTi6 alloy forms a higher θ in the CP tests than in the CH test on both SiO$_2$ and mullite (Table IV). On the other hand, $\theta^{Al/SiO2}=\theta^{Al/mullite}$ and $\theta^{AlTi6/SiO2}=\theta^{AlTi6/mullite}$ for each testing procedure at 1273 K and under vacuum. These results suggest kinetic factors may play an essential role. Kinetics may affect possible reactions with the AlTi6 alloy, and different phenomena during the wettability tests such as the rates of dissolution and diffusion of the Si by-product into the Al drop, and the formation of the TiSi$_2$ phase following the dissolution of the Al$_3$Ti precipitates in the metal drop. To demonstrate this, a special test was completed with a AlTi6/SiO$_2$ couple at 973 K using the modified CP$_r$ procedure whereby after 60 min., the sessile drop was drawn back into the capillary and quenched. A detailed SEM analysis provided a few interesting observations (Figure 5), including: 1) in the drop near the substrate, there is a 0.4 mm thick layer almost completely free of Si, Ti, and any

Table IV. Contact angles of Al and Al alloys on SiO$_2$ and mullite [11-13]

T [K]	SiO$_2$, 120 min			Mullite, 120 min		
	Al	AlTi6	AlSi11	Al	AlTi6	AlSi11
973	59*		84*			
1073	112	82		130		
		87*				
1173	88	78	77	94		102
1273	53	48	60	52	47	123
		69*			70*	

*CP

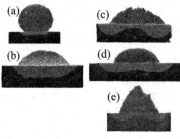

Figure 4. Cross-sections of sessile drop samples of Al/mullite at: a) 1073 K, CH; and b) 1273 K, CH; and of AlTi6/mullite at 1273 K: c) CH; d) CP; and e) CP_c, where for the CP and CP_c, the drop was produced at 1073 K.

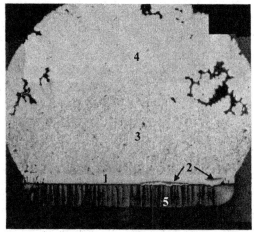

Figure 5. Cross-section of AlTi6/SiO_2 (973 K, 60 min., CP_r) showing the different structure and chemistry explained in the text as numbered.

precipitates; 2) there is a layer that contains RPR pieces formed due to the longitudinal cracking of the RPR and that detached due to metal movement during the test; 3) the central part of the drop shows the initial semi-liquid state (L_{Al}+Al_3Ti) composed of $TiSi_2$, a small amount of non-reacted spherical Al_3Ti particles, and new precipitates of the $Al_3Ti(Si)$ phase containing Si; 4) the upper part of the drop that contains a large amount of (Al+Si) eutectic without the needle-like Al_3Ti precipitates, typical of the liquid state area in the AlTi6 alloy; and 5) the thick RPR that contains negligible amounts of Si and Ti.

The above observations suggest: 1) a dominant role of reaction (2) between Al and SiO_2; 2) a secondary role of reaction (5), which takes place between the Ti from the alloy and the Si from the former reaction; and 3) a high rate of diffusion of freshly formed Si into the AlTi6 drop, compared to the rate of dissolution of the Al_3Ti solid particles in the molten Al.

In the AlTi6/mullite (Figure 4c) and AlTi6/Si_3N_4 [14] couples, the RPR produced in the CH test has an unusual shape characterized by a much thinner region near the center of the sessile drop. Again, kinetic factors, and not the oxide film on the drop, affect different phenomena that take place during heating and testing of the couple, as shown schematically in Figure 6.

An important role of the AlTi6/substrate interaction during heating has been demonstrated in a modified CP_r test. An oxide-free AlTi6 drop was produced at 1073 K and kept strained between the substrate and a capillary during heating to and holding at 1273 K to suppress spreading. As expected, and contrary to the CP test (Figure 4d), both the CH and CP_r tests (Figures 4c and 4e, respectively) produce a similar, unusual shape RPR, characterized by non-uniform thickness. For CP_r this effect is less evident because heating from T_s=937 K to 1073 K was instantaneous.

Considering the high reliability of the experimental data obtained with deoxidized drops (CH, 1223-1273 K, vacuum), it may be concluded that the

Table V. θ-τ relationship for Al/Si$_3$N$_4$ couples (CH, 120 min).[14]

T [K]	Si$_3$N$_4$				Si$_3$N$_4$+2%Y$_2$O$_3$				
	Al	AlTi6		AlSi11	Al		AlTi6		AlSi11
	θ [°]	θ [°]	τ [MPa]	θ [°]	θ [°]	τ [MPa]	θ [°]	τ [MPa]	θ [°]
1123	88		*		123	23.7			
1173	56	71	*		70	47.5	76	61.53	
1223	46	43	86.8	56	58	59.14	60	90	78

*numerous interfacial cracks

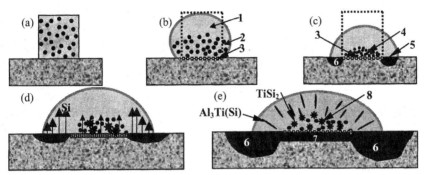

Figure 6. A schematic of the phase transformation in a AlTi6/mullite couple during a wettability test: a) initial homogeneous distribution of spherical Al$_3$Ti precipitates in the as-extruded alloy; b) the settling of heavier Al$_3$Ti precipitates during heating, and the separation of the residual liquid L$_{Al(Ti)}$ (1) from the semi-solid region (L$_{Al(Ti)}$+Al$_3$Ti) (2), which take place along with the formation of reaction product barrier layer (3); c) spreading and interaction while holding at temperature that is accompanied by the formation of the Ti-rich (4) and Ti-free (5) regions in the Al-Ti drop that produce thick barrier-free RPR area (6) and of thin RPR layer (7), respectively; d) Si diffusion into the molten drop and the formation of a new semi-liquid state composed of L$_{Al(Ti)}$, Al$_3$Ti, and reactively produced TiSi$_2$ (8); and e) the final structure after cooling, in which some Al$_3$Ti phase solidifies to form long needles containing up to 10%Si.

contact angle relationships between Al and AlTi6 on SiO$_2$ or mullite (Table IV) and on Si$_3$N$_4$ (Table V) are comparable to those on Al$_2$O$_3$ and AlN (1), respectively. The similarity is attributed to the co-continuous Al(Si)/Al$_2$O$_3$ and Al(Si)/AlN structures produced. Compared to pure Si$_3$N$_4$, the reactivity of (Si$_3$N$_4$+2%Y$_2$O$_3$) substrates is lower because of the formation of a dense and continuous barrier layer of reaction products (i.e., AlN interspersed with Al$_2$O$_3$ particles [14]) that degrade wettability. In spite of the lack of wettability, an improvement is observed after alloying Al with 6 wt.% Ti, as these couples show higher shear strength than those with pure Al. Also, the reaction between Si and Ti contributes to a beneficial interface structure, i.e., a dense, continuous and crack-free composite layer composed of dispersed AlN and TiSi$_2$ phases [14].

In contrast to Ti, the addition of Si to Al suppresses reactions (2-4). Therefore, relative to pure Al, the AlSi11 alloy shows poorer wetting on SiO_2, Si_3N_4, and mullite (Tables IV and V).

In the Al/SiC system, alloying Al with 6% Ti improves wetting (e.g., at 1327 K, $\theta^{Al} = 73°$ and $\theta^{AlTi6} = 58°$ [5]), but, similar to Al/C system [4,19], the main benefit is an improvement of compatibility in the system by the formation of the Ti-rich products instead of the undesirable Al_4C_3 phase. Additionally, in the AlTi6/SiC couples the Ti creates a beneficial graded structure at the interface composed of dense and continuous reaction product layers ($TiC/TiSi_2/Ti_3AlSi_2$) that effectively prevent the formation of the Al_4C_3 phase.

CONCLUSIONS

Alloying aluminum with 6 wt.% Ti degrades wetting and the bonding properties of non-reactive Al/Al_2O_3 and Al/AlN couples. A similar effect is observed in the reactive Al/TiO_2 couple because the Ti addition suppresses dissolution and oxidation-reduction reactions.

In reactive Al/ceramic couples the interface structure and properties have a strong dependence on kinetic factors that affect chemical reactions, dissolution, and diffusion processes in the system. Therefore, the real effect of Ti addition can be masked by the reaction between the Al metal and the ceramic substrate. In the extremely reactive Al/SiO_2, Al/mullite, and Al/Si_3N_4 systems, displacement reactions are accompanied by a reaction between the titanium from the Al alloy and the Si metal by-product to form $TiSi_2$. Kinetic factors contribute to the formation of unusual shape of RPR. For the SiC and carbon materials, a positive effect of Ti is obtained through improved chemical compatibility by the preventing the formation of the undesirable Al_4C_3 phase. In contrast to the Al/C couples, titanium promotes the development of a dense interfacial graded structure in the Al/SiC couples.

ACKNOWLEDGEMENTS

The work was supported by the State Committee for Scientific Research of Poland under the project No 7 T08B 003 20.

REFERENCES

[1] N. Eustathopoulos, M.G. Nicholas, and B. Drevet, *Wettability at High Temperatures*, Pergamon, 1999.

[2] M.M. Schwartz, *Ceramic Joining*, ASM International, USA, 1990.

[3] In Situ Reactions for Synthesis of Composites, Ceramics, and Intermetallics, edited by E.V. Barrera, S.G. Fishman, F.D.S. Marquis, N.N. Thadhani, W.E. Frazier, Z.A. Manir, TMS publications, Pennsylvania, USA, 1995.

[4] R. Asthana, *Solidification Processing of Reinforced Metals*, Trans Tech Publications, USA, 1998.

[5] N. Sobczak, Z. Gorny, M. Ksiazek, W. Radziwill, and P. Rohatgi, "Interaction between Porous Graphite Substrate and Liquid or Semi-Liquid Aluminium Alloys Containing Titanium," *Materials Science Forum*, **217-222** [1] 153-158 (1996).

[6] N. Sobczak, M. Ksiazek, W. Radziwill, J. Morgiel and L. Stobierski, "Effect of Titanium on Wettability and Interfaces in the Al/SiC System"; pp. 138-144 in *Reviewed Proc. Second Int. Conf. „High Temperature Capillarity,"* edited by N. Eustathopoulos and N. Sobczak, Foundry Research Institute, Poland, 1998.

[7] N. Sobczak, R. Asthana, M. Ksiazek, W. Radziwill, B. Mikulowski and I. Surowiak, "Influence of Wettability on the Interfacial Shear Strength in the Al/alumina System"; pp. 129-142 in *State of Art in Cast Metal Matrix Composites in the Next Millennium*, edited by P.K. Rohatgi, TMS Publications, Pennsylvania, USA, 2000.

[8] N. Sobczak, "Wettability, Structure and Properties of Al/Al$_2$O$_3$ Interfaces," *Kompozyty*, **3** [7], 301-312 (2003).

[9] N. Sobczak, M. Ksiazek, W. Radziwill, L. Stobierski, and B. Mikulowski, "Wetting-Bond Strength Relationship in Al-AlN System," *Transactions of Joining and Welding Research Institute*, **30**, 125-130 (2001).

[10] N. Sobczak, "Wettability and Reactivity between Molten Aluminum and Selected Oxides", be published in *Proc. E-MRS Fall Meeting, Symposium G: Bulk and Graded Materials*, Warsaw, Poland (2003).

[11] N. Sobczak, L. Stobierski, W. Radziwill, M. Ksiazek and M. Warmuzek, "Wettability and Interfacial Reactions in Al/TiO$_2$," to be published in *Polish Ceramic Bulletin* (2003).

[12] N. Sobczak, M. Ksiazek, W. Radziwill, M. Warmuzek, R. Nowak, and A. Kudyba, *Odlewnictwo – Nauka i Praktyka*, 3, 3-14 (2003).

[13] N. Sobczak, L. Stobierski, M. Ksiazek, W. Radziwill, R. Nowak, and A. Kudyba, be published in *Polish Ceramic Bulletin* (2003).

[14] N. Sobczak, L. Stobierski, M. Ksiazek, W. Radziwill, J. Morgiel, and B. Mikulowski, "Factors Affecting Wettability, Structure and Chemistry of Reaction Products in Al/Si$_3$N$_4$ System," *Transactions of Joining and Welding Research Institute*, **30**, 39-48 (2001).

[15] V. Laurent, D. Chatain, and N. Eustathopolos, "Wettability of SiO$_2$ and Oxidized SiC by Aluminium," *Materials Science and Engineering*, **A135**, 89-94 (1991).

[16] X.B. Zhou, Th.M. De Hosson, "Reactive Wetting of Liquid Metals on Ceramic Substrates," *Acta Materialia*, **44** [2] 421-426 (1996).

[17] M.C. Breslin, J. Ringnalda, J. Seeger, A.L. Marasco, G.S. Daehn and H.L. Fraser, "Alumina/Aluminum Co-Continuous Ceramic Composite (C^4) Materials Produced by Solid/Liquid Displacement Reactions: Processing Kinetics and Microstructures," *Ceramic Engineering and Science Proceedings*, **15** [4] 104-112 (1994).

[18] R.E. Loehman, K. Ewsuk, and A.P. Tomsia, "Synthesis of Al$_2$O$_3$-Al Composites by Reactive Metal Penetration," *J. Am. Ceram. Soc.,* **79** [1] 27-32 (1996).

[19] N. Sobczak, J. Sobczak, S. Seal, and J. Morgiel, "TEM Examination of the Effect of Titanium on the Al/C Interface Structure," *Materials Chemistry and Physics*, 81, 319-322 (2003).

INTERFACE PHENOMENA AND WETTABILITY IN THE B_4C/(Me-Si) SYSTEMS (Me =Cu, Au, Sn)

N. Froumin, M. Aizenshtein, N. Frage and M. P. Dariel
Materials Engineering Department,
Ben-Gurion University of the Negev, 84105 Beer-Sheva, Israel

ABSTRACT
    The solubility of boron in the molten alloy and the interaction of silicon dissolved in the melt with the boron carbide substrate determine the interface structure and the wetting behavior in the B_4C/(Me-Si) systems. The equilibrium contact angle at the B_4C/(Cu-Si) interface is affected by the composition of the near-surface layer of boron carbide that has shifted to a higher boron content. Wetting in the B_4C/(Au-Si) and B_4C/(Sn-Si) systems reflect the formation of a SiC interlayer, and the wetting behavior of the SiC/(Au-Si) and SiC/(Sn-Si) systems.

INTRODUCTION
    The wetting behavior of a wide variety of molten metals and alloys on non-oxide ceramic substrates has been the subject of numerous studies that have revealed a range of behavior patterns.[1-4] Subtle changes in the solubility of the substrate components in the liquid medium, or in the concentration of additives may significantly affect wetting behavior via the formation of intermediate surface layers and various triple line configurations. Boron carbide exists over a relatively wide range of boron/carbon ratios.[5] Deviations from the stoichiometric composition are known to have a significant affect on the wetting behavior, as has been observed on titanium carbide and titanium nitride. In previous studies [6,7], non-wetting behavior was established for the so-called non-reactive metals (Cu, Ag and Sn) on B_4C ceramic substrates. In our recent investigation of the B_4C/Cu system [8], it was shown that boron carbide decomposes in contact with molten copper alloys, resulting in the dissolution of boron in the melt, and the release of free carbon to form graphite precipitates. The addition of boron to molten Cu increases the near-surface layer concentration of boron in the boron carbide substrate to improve wetting. The boron addition also

prevents boron carbide decomposition, and produces a flat metal-ceramic interface. The addition of elements with high affinity for C to the melt may lead to an interaction that reduces the carbon content of the boron carbide phase, and improves conditions for wetting. Moreover, these interactions may lead to the formation of new phases, and alter the nature of the metal-ceramic interface.

In the present work, the effect of Si additions to molten metals (Cu, Au, Sn) on wetting and the interface structure of B_4C were studied. Thermodynamics are used to explain the experimental results.

EXPERIMENTAL PROCEDURES

Hot pressing was used to produce B_4C ceramic samples of near-theoretical density.[8] The B_4C substrate surface was polished down to a 0.25 μm finish for the wetting experiments. A surface roughness of Ra = 0.12±0.02 μm was measured using an atomic force microscope (AFM). Wetting experiments were performed by means of the sessile drop method at 1423 K in a vacuum furnace (10^{-3} Pa) [9]. Me-Si alloys were prepared *in-situ* by co-melting appropriate amounts of the metals (99.999% pure). Contact angles were measured directly from the magnified profile images of the molten metal drop (drop mass was about 0.2 g) at 5 min. time intervals up to 90 min. The structure of the metal-ceramic interface was studied by SEM analysis of the cross-sections of solidified drops. The interface composition was determined using energy dispersive spectrometry (EDS) and wavelength dispersive spectrometry (WDS) analysis.

EXPERIMENTAL RESULTS

Wetting of B_4C by Me-Si (Me=Cu, Au, Sn) alloys

As was shown in our previous work [8], a high contact angle close to 120° was observed for Cu on the B_4C substrate at 1423 K after 30 min. Under the same conditions, high contact angles of 134° and 128°, also were obtained in the B_4C/Au and B_4C/Sn systems, respectively. The time dependences of the contact angle measured for Me-Si alloys at 1423 K, are shown in Figures 1a-c. The kinetics of wetting in the B_4C/(Cu-Si) system is sluggish, and depends on the concentration of silicon in the melt. For example, for the Cu alloy containing 50 at% Si, the equilibrium value of the contact angle was reached after 20 min. For the Cu-20 at% Si alloy, 90 min. was required to achieve the equilibrium contact angle. For the B_4C/(Au-Si) and B_4C/(Sn-Si) systems, over the whole concentration range, a rather short time of about 15 min. was needed to reach the equilibrium angle.

The variation in the equilibrium values of the contact angle in the B_4C/(Me-Si) alloys system vs. the Si content is shown in Figure 1d. Alloying Cu with Si leads to enhanced wetting of B_4C, and wetting is observed in alloys containing more than 13 at% Si. Similar results are obtained in the B_4C/(Au-Si) system, in which wetting is achieved with alloys containing >12 at% Si. In the B_4C/(Sn-Si) system, even a minute addition of Si (<1 at %) provides good wetting.

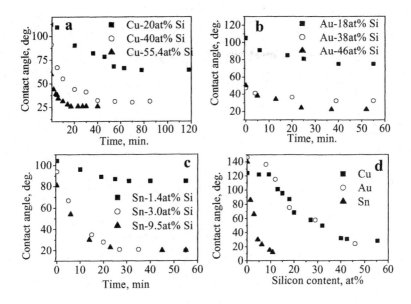

Figure1. Time dependence of the alloy contact angle on a B$_4$C substrate: a) Cu-Si alloys; b) Au-Si alloys; and c) Sn-Si alloys; and d) the concentration dependence of the equilibrium contact angle in the B$_4$C/(Me-Si) systems at 1423 K.

Interface structure in the B$_4$C/(Me-Si) systems

The interface structure in the B$_4$C/Cu system is shown in Figure 2a. The interaction of molten Cu with the B$_4$C substrate releases carbon that forms a very thin surface layer located at the initial substrate-metal interface. Below this layer, a crater forms, and is filled by carbon agglomerates. The interface in the B$_4$C/Au and B$_4$C/Sn systems is flat, and no evidence of boron carbide dissociation is observed in the liquid metal (Figures 2b and c).

The interface structures in the B$_4$C/(Cu-Si) system with various Si contents are shown in Figures 3a-c. For alloys containing less than 13 at% Si, the presence of a small crater with a thin discontinuous layer on its top was observed (Figure 3a). Based on X-ray mapping of B and C, this layer consists of graphite particles dispersed in the metallic melt. In the B$_4$C/(Cu-15 at% Si) system, a composite SiC-metal layer forms (Figure 3b). A different interface structure is observed close to the triple line after the equilibrium contact angle is reached (Figure 3d). The silicon carbide becomes discontinuous, and the interval between the SiC particles increases until the layer eventually disappears. Ultimately, the carbide-metal interface becomes flat without any evidence of crater formation. In the

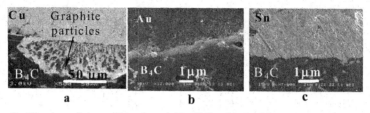

a   b   c

Figure 2. The interface region in: a) B_4C/Cu; b) B_4C/Au; and c) B_4C/Sn after 30 min. at 1423 K.

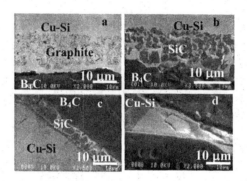

Figure 3. SEM images of the interface in $B_4C/(Cu\text{-}Si)$ systems after 90 min. at 1423 K: a) Cu-9 at% Si; b) Cu-15 at% Si; and c) Cu-50 at% Si under the drop; and d) Cu-50 at% Si near the triple point.

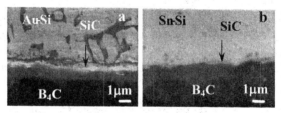

Figure 4. SEM images of the interface in $B_4C/Me\text{-}Si$ systems after 90 min. at 1423 K: a) Au-38 at% Si; and b) Sn-8 at% Si.

$B_4C/(Au\text{-}Si)$ system, starting with >20 at% Si, a very thin (~1 μm) and compact interface layer are formed. This layer consists of the SiC phase. The same interface structure is observed in the $B_4C/(Sn\text{-}Si)$ system over the whole concentration range (Figure 4b).

DISCUSSION

Thermodynamics can be used to explain the interface reactions and the enhanced wetting in the $B_4C/(Me-Si)$ systems. First, according to the binary Me-B phase diagrams, the boron solubility in liquid Cu at T = 1423 K is relatively high (up to 25 at%). By comparison, the B solubility is only 3 at.% in the Au-B system, and only several ppm in the Sn-B system [10, 11]. The solubility of carbon in molten Cu, Au, and Sn is extremely limited, and of the order of a few ppm.[4, 10] Thus, in the B_4C-Cu system, chemical interaction leads to boron dissolution in the melt, and the formation of graphite precipitates that become dispersed in the liquid. In the B_4C-Au and B_4C-Sn systems, a flat interface region is present, and there is no evidence of boron carbide dissociation. Alloying the basic metals with silicon provides another option for the metal-ceramic interaction. Boron carbide exists over a wide range of carbon content, and its interaction with an element with a high affinity to carbon (Si for instance) leads to the formation of a new carbide phase. In turn, this leads to a shift in the boron carbide composition to a higher boron content. This interaction depends on the activity of Si in the molten solution, and on the carbon activity in the boron carbide phase at the various compositions. According to reported data [12-14], and using the Redlich and Kister equation, the partial excess Gibbs energy of Si in the binary solutions may be expressed as:

$$G_{Si}^E = x_{Me}^2 \left(L_{Me-Si}^0 + 3L_{Me-Si}^1 \right) - 4L_{Me-Si}^1 x_{Me}^3 \qquad (1)$$

where Me denotes Cu, Au or Sn, L^0 and L^1 are the interaction parameters presented in Table I.

For the binary systems Cu-Si and Au-Si, the liquid solutions display a strong negative departure from ideality. The Sn-Si [14] liquid solution has a strong positive departure. The activity of carbon in the boron carbide phase as a function of composition has been reported elsewhere.[15]

A thermodynamic analysis of the interaction between boron carbide and the Cu-Si melt was previously reported.[16] This analysis was based on the extension of the boron carbide phase over a wide composition range from B_4C to $B_{10}C$, and on the reaction of boron carbide with a Cu-Si molten solution. By considering the boron carbide phase as a solid solution of B and C with a variable composition [17], and by considering the corresponding activity of Si in Cu, the reaction for the formation of silicon carbide may be rewritten as

$$(Si) + [C] = SiC \qquad (2)$$

where, the parentheses denote silicon dissolved in metallic liquid solution, and the brackets denote carbon in the solid B-C solution. The resulting equilibrium relationships for a certain boron carbide composition (X_C) in equilibrium with the

Table I. Thermodynamic parameters of the Me-Si binary solutions.

System	L^0, J/mole	L^1	Reference
Cu-Si	-46860	16640	[12]
Au-Si	-39300	21210	[13]
Sn-Si	23960	-1950	[14]

corresponding Cu-Si solution at a definite composition (x_{Si}) were calculated. Similar calculations were completed for the Au-Si and Sn-Si alloys, taking into account the thermodynamic properties of the metallic solutions reported elsewhere.[13, 14] The phase equilibria in the B_4C/Me-Si system are shown in Figures 5a-c. Each point on the curves corresponds to a certain boron carbide composition (X_C) in equilibrium with the liquid Me-Si solution at a definite composition (x_{Si}). According to these diagrams the Me-Si alloys that are in equilibrium with B_4C and SiC contain 12 at.% Si in the B_4C/Cu-Si system and 10 at.% Si in the B_4C/Au-Si system. In contrast, even a very small addition (<0.1 at.%) of Si to liquid Sn leads to SiC formation from the carbon that originates from the boron carbide phase.

On the basis of this analysis and the experimental observations [16], the following scenarios of the metal-ceramic interaction and wetting are suggested. In the B_4C/(Cu-Si) system with silicon contents higher than 14 at.%, the stages of the interaction evolution are shown on Figures 6a-d. The first stage (Figure 6a) consists of the dissolution of boron in the melt, the release of carbon into the melt

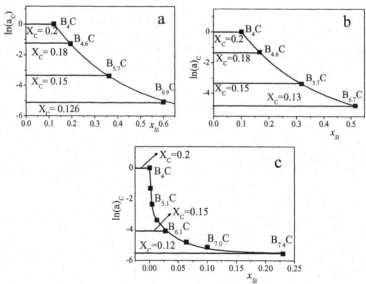

Figure 5. Phase equilibria in boron carbide-Me-Si melt systems at 1423 K: a) Cu-Si; b) Au-Si; and c) Sn-Si.

and the formation of a crater containing graphite agglomerates. In the second stage (Figure 6b), the silicon dissolved in the melt reacts with the graphite, and silicon carbide particles form. However, the metal drop does not yet wet the substrate (dashed line in Figures 6a-b). In the third stage, the substrate composition changes, the boron concentration in the melt increases, and boron diffuses to the drop's periphery and interacts with the substrate surface to generate favorable conditions for wetting in the vicinity of the triple line (region A, Figure 6d). Finally, the metal drop spreads over a sub-stoichiometric boron carbide.

For the $B_4C/(Au-Si)$ and $B_4C/(Sn-Si)$ systems, on account of the very limited boron solubility in the melts, boron carbide does not dissociate, and a flat interface is observed. A direct interaction between the Si dissolved in the metal and the boron carbide takes place. This results in the formation of a continuous thin SiC layer that covers the substrate surface (Figure 7). The measured contact angles reflect the characteristic of (Au-Si) and (Sn-Si) melts on silicon carbide. The equilibrium contact angles obtained in this study are in a good agreement with the experimental results for the $SiC/(Au-Si)$ and $SiC/(Sn-Si)$ systems reported elsewhere.[18, 19]

CONCLUSIONS

Interface structure and wetting behavior in the $B_4C/(Me-Si)$ systems is affected by the solubility of boron in the melt, and the interaction between silicon

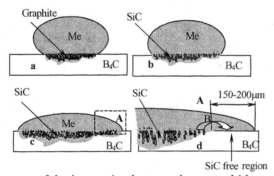

Figure 6. The stages of the interaction between boron carbide and Cu-Si.

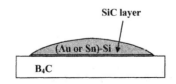

Figure 7. Schematic of the $B_4C/(Au-Si)$ and $B_4C/(Sn-Si)$ interfaces.

dissolved in the melts with the boron carbide substrate. The equilibrium contact angle in the B_4C/(Cu-Si) is affected by composition of the near-surface layer of the boron carbide that has a higher boron content. Wetting in the B_4C/(Au-Si) and B_4C/(Sn-Si) systems reflect the formation of a SiC interlayer, and the wetting behavior of the SiC/(Au-Si) and SiC/(Sn-Si) systems.

REFERENCES

[1]G.G. Gnesin, and Y.V. Naydich, "Contact Interaction of Silicon Carbide with Molten Copper", *Poroshkovaya Metallurgiya*, **74** 57-61 (1969).

[2]K. Suganuma and K. Nogi, "Interface Structure Formed by Characteristic Reaction Between alpha-SiC Single Crystal and Liquid Cu", *Journal of the Japan Institute of Metals,* **59** [12] 1292-1298 (1995).

[3]S. Takahashi and O. Kuboi, "Study on Contact Angles of Au, Ag, Cu, Sn, Al and Al Alloys to SiC", *Journal of Material Science,* **31** [7] 1797-1802 (1996).

[4]C. Rado, B. Drevet and N. Eustathopoulos, "The Role of Compound Formation in Reactive Wetting: the Cu/SiC System", *Acta Materialia* **48** 4483-4491 (2000).

[5]F. Thevenot, "Boron Carbide-A Comprehensive Review", *J. European Ceram. Soc.,* **6** [4] 205-225 (1990).

[6]Ju.V. Naidich, " The Wettability of Solids by Liquid Metals"; pp.353-484 in *Progress in Surface and Membrane Science*, **14** Edited by Cadenhead DA, et al. Academic Press., New York (1981).

[7]A.R. Kennedy, J.D. Wood and B.M. Weager, "The Wetting and Spontaneous Infiltration of Ceramics by Molten Copper", *J. Mater. Sci.,* **35** 2909-2912 (2000).

[8]N. Froumin, N. Frage, M. Aizenshtein and M.P. Dariel, "Ceramic-Metal Interaction and Wetting Pphenomena in the B_4C/Cu System", *J. European Ceram. Soc.,* **23** [15] 2821-2828 (2003).

[9]N. Froumin, N. Frage, M. Polak and M. P. Dariel, "Wettability and Phase Formation in the TiC_x/Al Systems", *Scripta Materialia,* **37** [8] 1263-1266 (1997).

[10]T.B. Massalski, et al., Binary Alloy Phase Diagrams. 2 ed. ASM Int.: Materials Park, OH 1990.

[11]F. Wald and R.W. Stormont, "Investigations on the Constitution of Certain Binary Boron-Metal Systems", *J. Less-Common Metals*, **9** 432-433 (1965).

[12]O. Kubashevski and C.B. Alcock, Metallurgical Thermochemistry, 5th edition, International Series on Materials Science and Technology, **Vol. 24.** Editors: G. Raynor, M. Phil. Pergamon Press 1979.

[13]R. Castanet, R. Chastel and C. Bergman, "Thermodynamic study of the Gold-Silicon System", *Materials Science and Engineering,* **32** [1] 93-98 (1978).

[14]M.C. Heuzey and A.D. Pelton, "Critical Evaluation and Optimization of the Thermodynamic Properties of Liquid Tin Solutions", *Metallurgical and Materials Transactions*, **27B** 810-814 (1996).

[15]S. Tariolle and F. Thévenot, M. Aizenshtein, M.P. Dariel, N. Froumin, and N. Frage, "Boron Carbide-Copper Infiltrated Cermets", *Journal of Solid Chemistry,* in press 2003

[16] N. Frage, N. Froumin, M. Aizenshtein and M. P. Dariel. "Interface reaction in the B_4C/(Cu-Si) system" Acta Met. (Accepted for publication)

[17]L. Levin, N. Frage and M. P. Dariel, "The Effect of Ti and TiO_2 Additions on the Sintering Behaviour of B_4C" *Metallurgical and Materials Transactions*, **30A** 3201-3210 (1999).

[18]B. Drevet, S. Kalogeropoulou and N. Eustathopoulos, "Wettability and Interfacial Bonding in Au-Si/SiC System", *Acta Metallurgica et Materialia,* **41** [11] 3119-3126 (1993).

[19]A.Tsoga, S. Ladas and P.Nikolopoulos, "Correlation Between the Oxidation State of α-SiC its Wettability with Non-reactive (Sn) or Reactive (Ni) Metallic Components and Their Binary Si-alloys", *Acta Materialia*, **45** [9] 3515-3525 (1997).

INTERFACIAL REACTIONS BETWEEN METALS AND CERAMICS AT ELEVATED TEMPERATURES

J.E. Indacochea
University of Illinois at Chicago
College of Engineering
Chicago, IL 60607

S.M. McDeavitt
Purdue University
School of Nuclear Engineering
West Lafayette, IN 47907

G.W. Billings
West Cerac, Inc.
McCarran, NV

ABSTRACT
This study describes the wetting results and the resulting interfaces that develop by exposing beryllia, yttria, and zirconium nitride substrates to pure zirconium metal or stainless steel-15 wt.% Zr at temperatures above 1800°C in a high purity argon atmosphere. After a short stay at the peak temperature, the system was cooled to room temperature. At these elevated temperatures Zr reduces even what are considered very stable oxide ceramics. Be alloys Zr to decrease its melting point. The Zr-Be liquid alloy then dissolves BeO at the interface, and infiltrates the substrate. The reaction is very limited between the Zr and yttria. Zr melts at 1850°C and reduces and partially dissolves Y_2O_3 into the liquid metal. Upon solidification, yttrium later segregates to the grain boundaries. ZrN is found to partially dissolve in liquid Zr, and a new stoichiometry of ZrN precipitates at the interface as well as in the Zr metal.

INTRODUCTION

Ceramic/liquid metal interfaces play an important function in the processing of metal matrix composites, in the development of reusable crucibles for reactive metal melting, and in ceramic-metal joining. The question of interface design and control for these systems is complex. The atomic or electronic structures of the metal are important factors that affect the reactivity of the liquid metal in contact with the ceramic material. As ceramics and liquid metals are exposed to each other at elevated temperatures, chemical reactions and new phases are often formed at the interface.[1]

In a ceramic-metal joint, the interfacial region that connects the metal with the ceramic material may adopt different forms. A simple interface of atomic bonds may form, or a new reaction phase and/or interfaces may form between the metal and the ceramic. Complex oxides, carbides, nitrides, and intermetallics frequently form at the interface between metals and ceramics during processing or service at elevated temperatures.[2,3]

There are applications where wetting and the formation of a strong ceramic-metal interface are not desirable. In metal casting, a refractory crucible is used to handle the liquid metal, and ceramic porous bodies are used as filter materials. For such applications, interface development must be avoided, and the paradigms employed for good brazing and composite processing need to be reversed.

Wetting of a ceramic by liquid metal is considered to be a surface phenomenon, and it has been shown to occur if a reaction region is developed at the ceramic/metal interface.[4] Reactive alloy additions can be added to the metal or applied directly to the interface. Coatings of Ni, Ag, Cu, and Cr have been applied to the ceramic surfaces to enhance wettability, and the Na process is used to prepare Al_2O_3 for wetting by metallic liquids.[5-7] Chemical vapor deposition of titanium coatings onto Al_2O_3 has been shown to enhance wettability.[8] The commercial Moly-Manganese process uses a reduction reaction to obtain a reliable braze.[9,10]

Most wetting investigations are based on the contact angle (Figure 1), first defined empirically by Young [11]:

$$\gamma_{SV} = \gamma_{SL} + \gamma_{LV} \cdot \cos \theta \tag{1}$$

where γ_{SV}, γ_{SL}, and γ_{LV} is the solid-vapor, solid-liquid, and liquid-vapor interfacial energy, respectively. A contact angle $\theta \geq 90°$ is considered a non-wetting condition. If $\theta < 90°$, wetting is said to have occurred. The experimental measurement of the equilibrium contact angle, however, may be complicated by interfacial reactions. The formation of new phases at the interface is not accounted for in Equation 1; under that condition, the results become ambiguous.

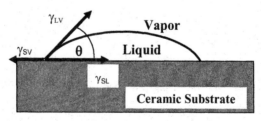

Figure 1. Schematic of a metal drop resting on a flat substrate showing the contact angle for a wetting condition ($\theta < 90°$).

Interfacial reaction may partially dissolve the substrate, resulting in a non-horizontal surface at the triple point.[8] The resulting configuration no longer has the horizontal force balance assumed in the Young's equation, and determination of the contact angle is experimentally difficult. In addition, the nature of the surface is also unknown.

The work presented in this paper describes the high-temperature reactions at ceramic-metal interfaces prior to and during reactive metal melting. The microstructures that develop after solidification are also discussed. Results are presented for the interactions of pure liquid zirconium and a stainless steel-zirconium liquid metal alloy with beryllia, yttria, and zirconium nitride.

EXPERIMENTAL PROCEDURE

The oxide and non-oxide ceramic substrates used were previously fabricated from high purity powders by hot uniaxial pressing (HUP). All materials were cleaned with acetone to remove dirt and other contaminants before the thermal cycle. Two metals were used, pure Zr, and ferritic stainless steel alloyed with 15 wt.% Zr. The composition of the ferritic stainless steel in wt.% is 0.5 Ni, 12.0 Cr, 0.2 Mn, 1.0 Mo, 0.25 Si, 0.5 W, 0.5 V, 0.2 C, and 84.85 Fe.

The metal/ceramic assembly, consisting of a small piece of metal (cubic geometry) placed on top of the ceramic disk, was heated in a tungsten mesh furnace in high purity argon. The sensing thermocouple was located 0.5 cm beneath the samples. The system was preheated to 600°C and held for a short period of time. Then it was heated continuously at ~20°C/min to 1600°C. From there, the heating rate was decreased to 10°C/min, and the whole assembly was heated to a peak temperature of about 2000°C. The system was kept at the peak temperature for 5 minutes, and then cooled to room temperature at ~20°C/min. The wetting and high temperature interactions between the molten metals and ceramic substrates were monitored through an external video camera. In-situ observations were followed by post-test microstructure examinations using scanning electron microscopy (SEM) and energy dispersive spectroscopy (EDS).

RESULTS AND DISCUSSION

Zirconium Metal – Beryllia Interactions

A surface crust developed in the Zr metal near the BeO interface as the temperature approached 1500°C. At about 1550°C, a reaction was noticed at this interface, which intensified with increasing time and temperature. Gas evolution is visible from the solid-liquid boundary as the temperature reaches 1580°C. At ~1605°C a liquid layer is clearly observed at this interface, although the Zr metal is well below its melting point, and the sample maintains its original parallelopiped shape. As the temperature reaches 1735°C, the Zr metal piece distorts, and assumes a trapezoid-like shape. At about 1780°C, the Zr specimen deforms greatly because of the increased amount of liquid within the sample. At 1790°C, the Zr metal collapses, and a hemispherical droplet forms. A wetting angle of ~60° is measured at 1795°C. The contact angle decreases significantly at ~1825°C.

The gas evolution and localized melting at the ceramic-metal interface observed by video at ~1605°C are indications of an interface reaction between beryllia and zirconium metal. Since liquid is observed at a temperature much lower than the melting point of pure Zr (1855°C), the results suggest that Zr must have been alloyed during heating. From the Zr-Be equilibrium phase diagram, it is seen that Be acts as a temperature depressant for Zr, and Be can only come from the reduction of the beryllia. It is likely that Zr reduces the BeO; one possible chemical reaction is:

$$Zr + BeO(s) \rightarrow Zr(O_x) + Be[Zr(O_x)] + yO_2 \qquad (2)$$

where, $x + 2y = 1$. Evidence of such a reaction is confirmed by the changes in the microstructure and composition of the resolidified Zr on top of the BeO after cooling to room temperature. The microstructure, shown in Figure 2, consists of a coarse, proeutectic dark phase, blocky-like particles, and a eutectic structure in a matrix of α-Zr. EDS standardless analysis reveals oxygen levels of about 10 at. % in the α-Zr. No Zr is found in the black proeutectic phase, only oxygen. Beryllium must also be present, but it is too light to be detected by the EDS equipment. The fine acicular particles of the eutectic microstructure contain Zr and oxygen; however, these fine particles are dark like the proeutectic phase and are also expected to contain Be.

Dissolution of the BeO substrate is also evident by the fact that the original Zr/BeO interface recedes into the ceramic substrate. In addition, Zr metal infiltrates the beryllia (Figure 2a). The sudden decrease in the contact angle at ~1825°C may be a consequence of the infiltration of the ceramic substrate by the liquid Zr alloy as it continues to dissolve the beryllia.

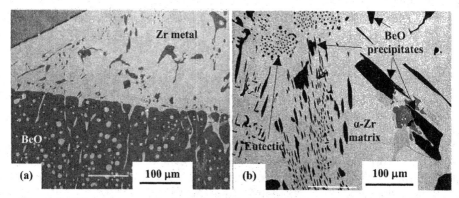

Figure 2. Microstructure features in the Zr/BeO system: a) near the interface where Zr metal infiltrates BeO; and b) re-solidified Zr with BeO precipitates.

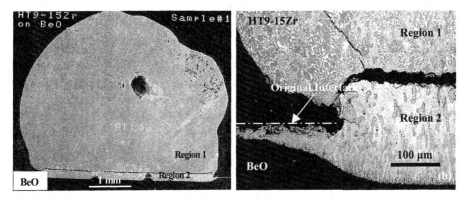

Figure 3. SEM micrographs showing: a) lack of wetting of the BeO substrate by HT9-15Zr alloy; and b) the reaction band (Region 2) that forms and remains attached to the BeO.

Stainless Steel/15 wt. % Zr Alloy – Beryllia Interactions

During the heating cycle, no chemical reactions or partial melting are detected at the interface. Initial melting of the alloy occurs at ~1340°C, but a non-wetting hemispherical cap is formed at ~1350°C. This alloy has a eutectic temperature of 1335°C. The contact angle between the Zr alloy and the BeO is always greater than 90°, even at 2000°C. The interface in this system after cooling and resolidification consists of a ~150-200 μm thick reaction layer (Figure 3). EDS results show the reaction band is rich in Zr (~54.0 at. %) and oxygen (~45.0 at.%); however, beryllium must also be present. The excess Zr present in this region confirms the reduction of BeO by Zr. The Zr alloy infiltrates the BeO, but only to the extent of the reaction band (Figure 3b). This reaction band (Region 2) remains attached to the beryllia, but breads away from the metal alloy droplet (Region 1). This could be expected if the new phase has a different coefficient of thermal expansion than the metal droplet. The microstructure in the solidified metal droplet and in the metal alloy infiltrated into the reaction band is the same, and consists of dendrites and a eutectic structure

Zirconium Metal – Yttria Interactions

The Zr metal melts initially at the yttria interface at ~1850°C. Some gas evolution is also noted, but the reaction does not seem vigorous. The bulk of the metal melts shortly after 1850°C. The ceramic is wetted by the Zr metal almost immediately w upon melting, and a contact angle of ~50° is measured. Similar to the BeO case, this chemical reaction at the interface may be caused by Zr reducing the oxide:

$$3\,Zr(l)+Y_2O_3(s)\rightarrow 3\,Zr(O_x)(l)+2\,Y[Zr(O_x)](l)+y\,O_2(g) \qquad (3)$$

where, $x + 2/3\,y = 1$.

The interface in this system shows trivial changes, and Y_2O_3 is not easily dissolved. Based on lower temperature thermodynamics and energies of formation, Y_2O_3 is more stable than BeO. The maximum solubility of Y and Be in Zr is about 4.0 at.%; however, this level of solubility is reached at 1363°C for Zr-Y, and at 965°C for Zr-Be. Equilibrium phase diagrams also show that Be is more effective than Y as a melting temperature depressant for the Zr. This helps explain why the Zr melts near its true melting point when in contact with Y_2O_3.

The Zr/Y_2O_3 interface is continuous, with no stable transition phases, and the system is strongly bonded. Zr does dissolve some of the Y_2O_3, and produces some discoloration near the interface (~25-μm band), as seen in Figure 4. In addition to the discoloration, there a second phase precipitates at the grain boundaries of the resolidified Zr metal. The oxygen levels in the discolored region are notably higher than those in the α-Zr metal away from the interface. The second phase at the grain boundaries is rich in Y. Most likely, Y produced from the interfacial chemical reaction diffuses into the Zr, and segregates to the grain boundaries on cooling because of its limited solubility in α-Zr. The dark phase at the grain boundaries contains Y (86-90 at.%) and Zr (14-10 at.%), but no oxygen is detected at the grain boundaries or in the α-Zr matrix regions away from the interface. A few cracks are seen in the ceramic substrate near the interface. This is expected due to differences in thermal expansion coefficients between the materials in this system.

Zirconium Metal – Zirconium Nitride Interactions

During heating, Zr does not melt at its melting temperature of 1855°C, but at

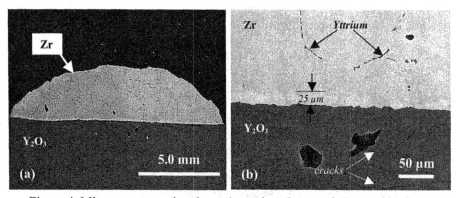

Figure 4. Microstructures showing: a) wetting characteristics; and b) the interface microstructure and Y segregation to the grain boundaries in Zr.

1975°C. Melting is first limited to the interface. The higher melting temperature indicates that pure Zr is likely contaminated with nitrogen from the ZrN as it reacts with the Zr metal before melting. In line with the Zr-N equilibrium phase diagram, β-Zr has a maximum nitrogen solubility of about 0.8 wt.%. This composition should melt at ~1880°C. When the N level in Zr is 3-5 wt.%, the alloy is α-Zr, and the melting temperature ranges between ~1880°C and ~1985°C, respectively. The metal sample melts completely at 2000°C, and the ceramic substrate is wetted with a contact angle of ~37°.

A dense, ~40-50 μm thick reaction layer forms between the Zr metal and the ZrN substrate (Figure 5). The reaction band contains N (~15.0 at.%), and Zr (~15 at.%). It is some type of ZrN_x, with a structure that is different than that of the substrate. The reaction layer shows continuity across the Zr metal interface, and with the original ZrN substrate. This transition band has a large crack parallel to the interface between the metal and ceramic joint, most likely due to the thermal mismatch between Zr and ZrN.

The changes in the ceramic substrate area adjacent to the reaction layer occur in addition to the reaction at the interface (Figure 5b). The thickness of this region is similar to the new ZrN precipitated reaction layer on the metal side of the interface. Its composition includes N (~18.0 at.%) and Zr (~82.0 at.%). N diffuses into the metal droplet at the interface, and into the Zr metal cap to produce lath-type ZrN_x precipitates, as seen in Figure 5.

Stainless Steel/15 wt.% Zr Alloy – Zirconium Nitride Interactions

The alloy HT9-15Zr melts at its eutectic temperature, and the molten metal does not wet the ZrN substrate (contact angle >140°). Bonding develops between the metal droplet and the ZrN (Figure 6), but no transition phase is seen at the

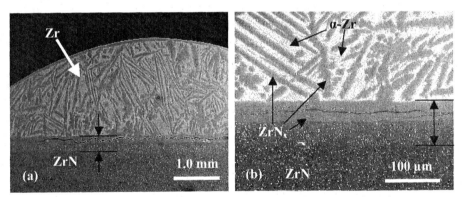

Figure 5. Zr/ZrN system: a) interface characteristics; and b) interface showing the ZrN_x reaction and the ZrN_x lath-like precipitates in the resolidified Zr.

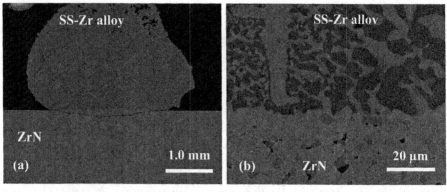

Figure 6. SS-15 % Zr/ZrN system: a) wetting characteristics; and b) interface microstructure.

interface. There is continuity between the bright intermetallic phase of the HT9 alloy, Zr(Fe, Cr, Ni)$_{2+x}$[9], and the ZrN substrate (Figure 6b). EDS spot analysis of the intermetallic near the interface with ZrN reveals a composition in at.%: 42.0 Fe, 6.0 Cr, 22.0 Ni, 25.0 Zr, and 4.7 N. The composition of the spot on the ZrN side of the interface contains just Zr (~53.0 at.%) and N (~46.0 at.%). The dark iron solution matrix near the interface contains in at.%: 71.0 Fe, 23.0 Cr, and 5.0 Ni.

CONCLUSIONS

1. Zr is a reactive element in these metal/ceramic systems. It reduces BeO and both elements dissolved in the Zr metal to form a lower melting point alloy. The liquid Zr alloy formed further dissolves and infiltrates the BeO substrate.
2. Zr also reduces Y$_2$O$_3$ but to a much lesser extent. Both yttrium and oxygen dissolve in the liquid Zr, but they segregate too the grain boundaries on solidification.
3. Zr ultimately wets Y$_2$O$_3$. No transition phases form across the Zr/Y$_2$O$_3$ interface; however, there is continuity across this boundary, and a strong bond develops.
4. Zr wets and partially dissolved the ZrN substrate. A new and denser ZrN band forms at the interface and ZrN laths precipitate in the resolidified Zr metal.
5. The stainless steel-15 wt.% Zr alloy does not wet the BeO or the ZrN. However, a strong bond results at the metal alloy/ZrN interface without any reaction phases.

REFERENCES

[1] L. Espie, B. Drevet, and N. Eustathopoulos, "Experimental Study of the Influence of Interfacial Energies and Reactivity on Wetting in Metal-Oxide Systems", *Metallurgical. Transactions A*, **25A** 599-605 (1994).

[2] R.L. Mehan, and R.B. Bolon, "Interaction Between Silicon Carbide and Nickel-Based Superalloy at Elevated Temperatures," *J. Mater. Sci.,* **14** 2471 (1979).

[3] D. Uphadyaya, M. Wood, C.M. Ward-Close, P. Tsakiropoulos, and F.H. Froes, "Coating and Fiber Effects on SiC-Reinforced Titanium," *J. Metals,* **46** 62 (1994).

[4] Kritsalis, P., L. Coudurier, and N.J. Eustathopoulos, "Contributions to the Study of Reactive Wetting in the CuTi/Al$_2$O$_3$", *J. Mater. Sci.,* **26** 3400-08 (1991).

[5] M. Naka, M. Tsuyoshi, and I. Okamoto, "Ti Precoating Effect on Wetting and Joining of Cu to SiC," *ISIJ International,* **30** 1108-13 (1990).

[6] J.H.Selverian, and S. Kang, "Ceramic-to-Metal Joints Brazed with Palladium Alloys," *Welding Journal,* **71** 25s-33s (1992).

[7] D.A. Weirauch, W.M. Balaba, and A.J. Perrotta, "Kinetics of the Reactive Spreading of Molten Aluminum on Ceramic Surfaces," *J. Mater. Res.,* **10** 640-50 (1995).

[8] D.A.Javernick, P.R Chidambaram, and G.R. Edwards, "Titanium Preconditioning of Al$_2$O$_3$ for Liquid State Processing of Al-Al$_2$O$_3$ Composite Materials," *Metallurgical Transactions A,* **29A** 327-37 (1998).

[9] H. Mizuhara, and K. Mally, "Ceramic-to-Metal Joining with Active Brazing Filler Metal," *Welding Journal,* **64** 27-32 (1985).

[10] A.J Moorhead, and H. Keating, "Direct Brazing of Ceramics for Heavy Duty Diesels," *Welding Journal,* **65** 17-31 (1986).

[11] S.M.McDeavitt, , D.P. Abraham, and Y.P. Park, "Evaluation of Stainless Steel Zirconium Alloys as High-Level Nuclear Waste Forms," *J. Nuclear Materials*, **257** 21-34 (1998).

INTRINSIC WETTABILITY AND WETTING DYNAMICS IN THE Al/α-Al$_2$O$_3$ SYSTEM

Ping Shen, Hidetoshi Fujii, Taihei Matsumoto, and Kiyoshi Nogi
Joining and Welding Research Institute, Osaka University, 11-1 Mihogaoka
Ibaraki, Osaka, 567-0047, Japan

ABSTRACT

The wetting and spreading of molten Al on α-Al$_2$O$_3$ single crystals, R$(01\bar{1}2)$, A$(11\bar{2}0)$ and C$(000\bar{1})$, and on α-Al$_2$O$_3$ polycrystals (PC) were investigated over a wide temperature range using an improved sessile drop method. The effect of the substrate orientation is significant. The intrinsic contact angles are estimated to be 75-85° for the R and A faces, 88-95° for the C face and 75-90° for the polycrystals, depending on the temperature. Spreading at high temperature is reaction-limited, and is characterized not only by a linear relationship of the drop base diameter versus time, but also by an exponential decay law for the dynamic contact angle versus time.

INTRODUCTION

The study on the wettability of Al$_2$O$_3$ by molten Al is of significant importance, not only for the basic knowledge of fabricating Al$_2$O$_3$-reinforced Al composites, but also for a basic understanding of wettability in other Al/ceramic systems. Therefore, the Al/Al$_2$O$_3$ system has been extensively investigated with regard to achieving intrinsic wettability.[1-7] However, because of the oxidizability of Al, an oxide layer usually covers the metal that inhibits the formation of a true Al-ceramic interface. The reported contact angles are so scattered that it is difficult to get a clear picture of the intrinsic wettability in this system.[1] Ranges of the reported contact angles[1] and the possible "intrinsic" contact angles[2] as a function of temperature are illustrated in Figure1. Except for the known effects of the Al surface oxidation at relatively low temperatures (T<1000°C), and the interfacial reaction between Al and Al$_2$O$_3$ at relatively high temperatures (T>1000°C), the influence of substrate surface roughness and crystallographic orientation on the wettability has not been well investigated.

Additionally, although a continuous decrease in the contact angle with time is widely observed at high temperatures in this system,[3-7] the wetting dynamics has never been investigated. The objectives of this study, thus, are to reappraise the intrinsic wettability of α-Al$_2$O$_3$ by molten Al, and to investigate the high temperature wetting dynamics in this system.

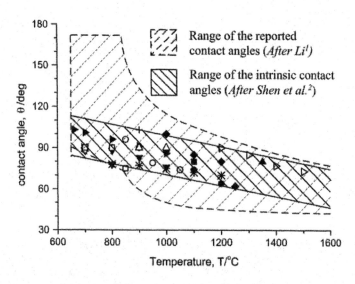

Figure 1. Ranges contact angles of Al on Al_2O_3 reported[1] and the possible "intrinsic" contact angle[2] as a function of temperature.

EXPERIMENTAL MATERIALS AND PROCEDURE

Pure Al and α-Al_2O_3 single crystals (Kyocera Co. Ltd, Japan) as well as polycrystals (Kojundo Chemical Co. Ltd., Japan) were used as the raw materials. The single crystal wafers were cut along R (01̄12), A (112̄0) and C (0001) crystallographic planes, with an orientation error of ±0.3°. One side of these surfaces was polished to an average roughness of 30 Å, as measured by a DEKTAK 3 surface profilometer. The surfaces of the polycrystal (PC) plates were ground and polished using different micron-size diamond pastes to about 1000 Å, except for those specifically designated.

An improved sessile drop method was employed in this study. The sessile drop was formed by extruding Al through an orifice at the bottom of an alumina tube (99.6%purity), and dropping the Al onto a substrate surface.[8] The experiments were conducted in a flowing purified Ar-3%H_2 atmosphere under a pressure of approximately 1.1×10^5 Pa (1.1 atm) to reduce Al evaporation. The droplet photographs captured were analyzed using an axisymmetric-drop-shape-analysis (ADSA) program, in which the contact angle, surface tension, and density, can be simultaneously calculated free of the operator subjectivity.[8]

RESULTS AND DISCUSSION

Effect of Al Surface Oxidation at Different Dropping Temperatures

The variations in the contact angle of the molten Al on the R-face single crystals with temperature (at T≤1100°C) and time (at T= 1100°C) are shown in Figure 2. As can be seen, the dropping temperature has a significant effect on the wetting behavior. When dropping at T<1000°C, a decrease in the contact angle is

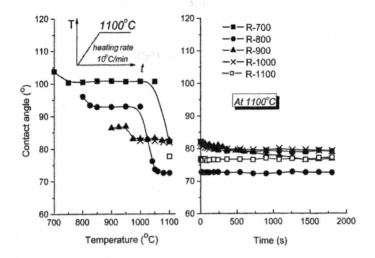

Figure 2. Variation Al contact angle on the single crystal Al_2O_3 with temperature (at T≤1100) and time (at T=1100).

observed during the subsequent heating process. The magnitude of the contact angle decrease, on the whole, decreases with the dropping temperature. Obviously, the decrease is due to the deoxidization of the Al surface oxide film, which is either not completely removed or is newly formed after dropping at relatively low temperatures. The lower the dropping temperature, the more pertinacious the oxide film; thus, the higher the temperature needed to disrupt it. However, when dropping at T≥1000°C, no noticeable decrease in contact angle with temperature is found. The contact angle remains almost constant or exhibits only a slight decrease during subsequent heating and dwell periods, which implies that the Al surface oxide has been removed or has negligible influence on the wettability. Note that a decrease in the contact angle on R-900 (dropped on the R single crystal surface at 900°C) occurs at 950-1000°C, and the decreased contact angle is very close to that of R-1000, which does not change significantly during the subsequent heating process. This implies that a 1000°C dropping temperature may be the threshold to avoid any serious influence of the Al surface oxidation on wettability under our experimental conditions. The contact angles of 82° and 78° (Figure 2) may be the intrinsic contact angles for the Al/R-Al_2O_3 system at 1000°C and 1100, respectively.

Effect of Substrate Crystallographic Orientation
    The variations in the contact angle of the molten Al on the R, A, C and PC α-Al_2O_3 substrates with time at temperatures between 1000°C and 1500°C are plotted in Figure 3. Clearly, the wettability of the α-Al_2O_3 by the molten Al is sensitive to the substrate orientation. Al_2O_3 substrate wettability decreases in the following order: R>A>PC>C. The mechanism for this dependence has been

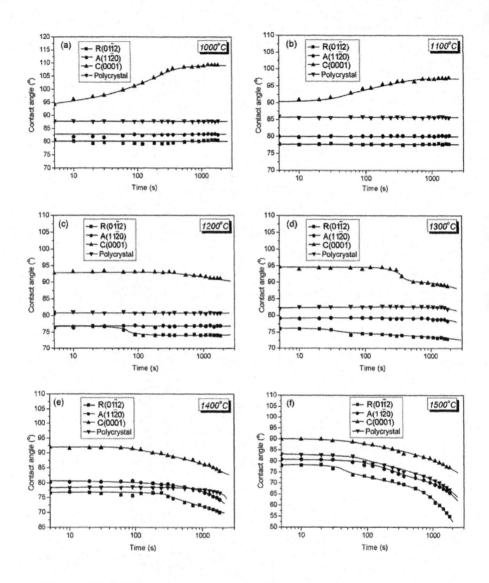

Figure 3. Variations in the contact angle of the molten Al on R, A, C, and PC α-Al$_2$O$_3$ substrates with time at 1000-1500°C.

reported elsewhere.[9] It is interesting to note that, at temperatures lower than 1200°C, the contact angle on the C face first increases with time, and then reaches a constant value. The time needed to reach the constant value decreases with temperature. The increase in contact angle is attributed to a surface structural reconstruction from a C-(1×1) to a C-($\sqrt{31} \times \sqrt{31}$)$R \pm 9°$ structure, which is extremely stable at temperatures higher than 1200°C.[8]

The effect of temperature on wettability does not seem to be very significant. The slight decrease in the contact angle with time at temperatures higher than 1200°C is a result of an interfacial reaction, which will be described in detail later.

Effect of Substrate Surface Roughness
The effect of substrate surface roughness on the wettability of the Al/(R,PC) α-Al$_2$O$_3$ system at 1100°C is shown in Figure 4. As indicated, the effect is not very pronounced for a relatively smooth substrate surface under a clean Al surface condition (i.e., Al free of the surface oxidation). The contact angle on the rougher PC (Ra = 3880Å) and R surface (Ra = 4610Å) is 95° and 82°, respectively. On the flat PC (Ra = 980Å) and R surface (Ra = 30Å), the contact angle is 78° and 77°, respectively. In this context, the substrate surface roughness may account less for the large scatter in the reported contact angles, since the substrates are usually prepared quite carefully before the experiment.

High Temperature Wetting Dynamics
Figure 5 shows plots of the relative droplet base diameter (D$_t$/D$_0$) vs. time (t). Figure 6 shows the plots of ln(θ$_t$/θ$_0$) vs. time (t) at high temperatures, where D$_0$ and θ$_0$ represent the droplet base diameter and the contact angle at 30s, respectively. As can be seen, contact angle continuously decreases during the isothermal dwell. The rate of decrease significantly increases with temperature. The wetting or spreading is essentially reaction-limited, as evidenced by the linear relationship of the relative base diameter (D$_t$/D$_0$) vs. time (t)

$$D_t / D_0 = 1 + k_1 t \tag{1}$$

and by the exponential decay law of the dynamic contact angle (θ$_t$) vs. time (t)

$$\theta_t = \theta_0 \exp(k_2 t) \tag{2}$$

where k_1 and k_2 are coefficients corresponding to the wetting/spreading rates.

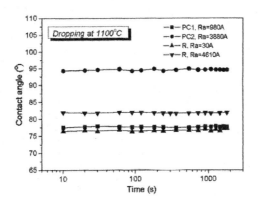

Figure 4. Effect of substrate surface roughness on R and PC α-Al$_2$O$_3$ wettability by Al at 1100°C.

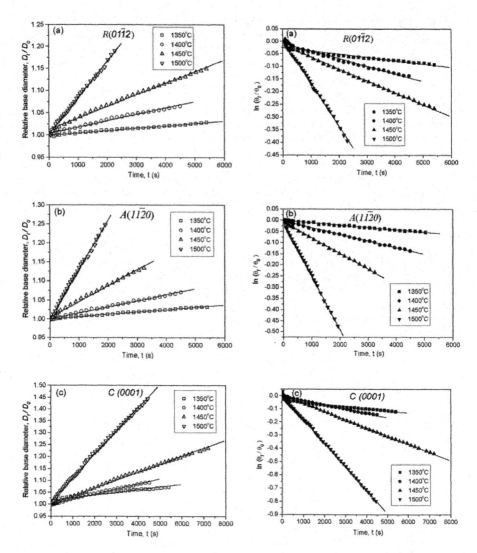

Figure 5. Plots of relative base diameter D_t/D_0 *vs.* time (t) for molten Al on α-Al_2O_3 single crystals at high temperatures: a) R; b) A; and c) C.

Figure 6. Plots of $\ln(\theta_t/\theta_0)$ *vs.* time (t) for molten Al on α-Al_2O_3 single crystals at high temperatures: a) R; b) A; and c) C.

CONCLUSIONS

(1) Al surface oxidation has a pronounced effect on wettability in the Al/α-Al$_2$O$_3$ system at relatively low temperatures. Dropping should be performed over a critical temperature ($\sim$1000°C) to obtain the intrinsic contact angle.

(2) The crystallographic orientation of the α-Al$_2$O$_3$ substrate has a noticeable affect on wettability; however, substrate surface roughness and the temperature are less significant. The intrinsic contact angle for the Al/α-Al$_2$O$_3$ system at high temperatures (T$\geq$1000°C) is estimated to be 75-85° for the R and A faces, 88-95° for the C face, and 75-90° for the polycrystals, depending on temperature.

(3) The wetting/spreading of the molten Al on α-Al$_2$O$_3$ single crystals at high temperatures is reaction-limited. It is characterized by a linear relationship of the drop base diameter versus time, as well as an exponential decay law for the dynamic contact angle versus time.

REFERENCES

[1]J.G. Li, "Wetting of Ceramics Materials by Liquid Silicon, Aluminium and Metallic Metals Containing Titanium and Other Reactive Elements: A Review," *Ceram. Int.*, **20**, 391-412 (1994).

[2]P. Shen, H. Fujii, T. Matsumoto, and K. Nogi, "The Critical Factors Affecting the Wettability of α-Alumina by Molten Aluminum," submitted to *J. Am. Ceram. Soc.*

[3]R.D. Carnahan, T.L. Johnston, and C.H. Li, "Some Observations on the Wetting of Al$_2$O$_3$ by Aluminum," *J. Am. Ceram. Soc.*, **41** [9] 343-47 (1958).

[4]J.J. Brennan, J.A. Pask, "Effect of Nature of Surfaces on Wetting of Sapphire by Liquid Aluminum," *J. Am. Ceram. Soc.*, **51** [10] 569-73 (1968).

[5]J.A. Champion, B.J. Keen, and J.M. Sillwood, "Wetting of Aluminium Oxide by Molten Aluminium and Other Metals," *J. Mater. Sci.*, 4, 39-49 (1969).

[6]P.D. Ownby, K. Wen, K. Li, and D.A. Weirauch Jr., "High-Temperature Wetting of Sapphire by Aluminum," *J. Am. Ceram. Soc.*, **74** [6] 1275-81 (1991).

[7]G. Levi and W. D. Kaplan, "Oxygen Induced Interfacial Phenomena During Wetting of Alumina by Liquid Aluminium," *Acta Mater.*, **50**, 75-88 (2002).

[8]P. Shen, H. Fujii, T. Matsumoto, and K. Nogi, "Wetting of (0001) α-Al$_2$O$_3$ Single Crystals by Molten Al," *Scripta Mater.*, **48**, 779-84 (2003).

[9]P. Shen, H. Fujii, T. Matsumoto, and K. Nogi, "The Influence of Surface Structure on Wetting of α-Alumina by Al in a Reduced Atmosphere," *Acta Mater.*, **51** [16] 4897-906 (2003).

WETTING IN THE TIN-SILVER-TITANIUM/SAPPHIRE SYSTEM

Laurent Gremillard, Eduardo Saiz, and Antoni P. Tomsia
Materials Sciences Division
Lawrence Berkeley National Laboratory
Berkeley, CA 94720, USA

ABSTRACT

The wetting of tin-silver-based alloys on Al_2O_3 has been studied using the sessile-drop configuration. Small additions of Ti decrease the contact angle of Sn-3 wt% Ag alloys on alumina from 150° to 25°; however, a wide variability in contact angle and spreading rate is observed. The variability is related to the kinetics of Ti dissolution in the alloy, and the formation of triple-line ridges. Enhanced spreading is not accompanied by the formation of a continuous reaction layer at the metal/ceramic interface. Furthermore, no reaction product is detected after tests performed at temperatures below 800°C.

INTRODUCTION

New low-temperature brazing alloys are required in many applications to integrate components that decompose or degrade above typical brazing temperatures.[1-2] Traditionally, a key component in the design of brazing alloys for ceramic joining is a reactive element such as Ti, Cr, Zr, etc., that enhances spreading.[1] The improved wetting resulting from the addition of a reactive elements is usually associated with the formation of new compounds at the solid/liquid interface.[1] However, it is unproven whether compound formation is actually necessary for enhanced wetting, or mechanistically, how the potential for compound formation translates into the capillary forces that specifically drive spreading. Recently, an alternative hypothesis has been proposed that focuses on the adsorption of the reactive element at the solid-liquid interface before the nucleation of the reaction phase. This is proposed as the critical step that reduces the interfacial energy and drives spreading.[1]

Tin-silver-based alloys have emerged as a lead-free substitute to the traditional solders used in the microelectronics industry, and as an alternative for low-temperature brazing.[2-3] In this work the effect of Ti additions on the wetting and bonding of Sn3Ag-based alloys to alumina is analyzed. Sn-Ag-Ti alloys are also of theoretical interest. It has been observed that the wetting of Sn-based alloys on ceramics can be greatly enhanced by the addition of titanium; however, in our work, no reaction products are detected at the metal-ceramic interface, as is commonly observed in other systems.[1] These observations indicate that spreading to produce low contact angles can be driven by adsorption at the interface, either without a reaction phase, or before one is formed. *In situ* neutron reflectometry studies of Sn-Ti/alumina interfaces have shown segregation of titanium at the ceramic-metal interface without the formation of reaction products at the experimental wetting temperature.[4] Derby et al.[4] have shown that interfacial oxides of Ti appear at the Ag-Cu-Ti/sapphire interface only after solidification.

The present work focuses on the study of the wetting behavior of Sn-Ag-Ti alloys on sapphire. The spreading of the molten metals at different temperatures is analyzed, and the corresponding interfacial microstructures are examined.

EXPERIMENTAL

The wetting of Sn3Ag and Sn3Ag1Ti alloys (1 wt% Ti) on Al_2O_3 was studied using sessile-drop experiments. The alloys have a typical tin-silver eutectic microstructure, with long Ag_3Sn platelets immersed in a tin matrix. In the titanium-containing alloys, Sn_5Ti_6 platelets (~100 µm in size) are homogeneously dispersed throughout the metal.

Wetting experiments were conducted under vacuum (pressure ranging from 0.1 to 1 mPa) at temperatures ranging from 600 and 1000°C. A metal piece was placed on a flat sapphire substrate (random orientation, chemically polished), and the assembly was heated to the test temperature at 50°C/min. Either metal cubes (1 mm^3) or flat disks were used in order to explore hysteresis of the contact angle. For the cubes the initial contact angle is higher than the equilibrium one, so the liquid front advances (the contact angle decreases) during the experiments. Conversely, the front recedes, and the contact angle increases when using the flat disks. Images of the drop were recorded *in situ* through a window in the furnace,

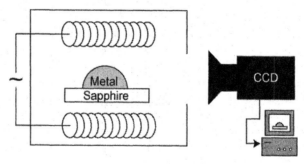

Figure 1. Experimental apparatus used to measure of contact angle.

with a CCD camera (Figure 1). The contact angles were then measured using a program developed by our group. At least three experiments were performed at each temperature.

The interfacial microstructures were analyzed using cross sections and "drop-free" specimens. The cross sections were polished to a 1 µm diamond finish and, subsequently etched using a suspension of 15 wt%, 1 µm size alumina powder, and 15 wt% $FeCl_3$ in H_2O/HCl (85/15 vol%). After some experiments, the metallic drop was dissolved from the sapphire substrate using aqua regia, and the area under the drop was analyzed by optical microscopy, scanning electron microscopy with associated energy-dispersive X-ray spectroscopy (SEM-EDS), atomic force microscopy (AFM), and X-ray Photoelectron Spectroscopy (XPS).

RESULTS
Wetting Experiments.

Although the melting temperature of the Sn3Ag-based alloys is around 220°C, at temperatures below 550°C the metal drop is encapsulated in a solid oxide surface layer that impedes contact between the liquid and the ceramic. At temperatures above 550°C in vacuum, this layer disappears, the drop exhibits a shiny metallic surface, and the liquid spreads on the ceramic. Thus, sessile-drop experiments were performed at temperatures between 600 and 1000°C. Figure 2a shows the equilibrium contact angle measured for Sn3Ag1Ti after 60 minutes at different temperatures. It is clear that the temperature has only a secondary effect on the contact angle. A clear dependence of contact angle with temperature does not emerge. The lowest measured contact angles are independent of temperature.

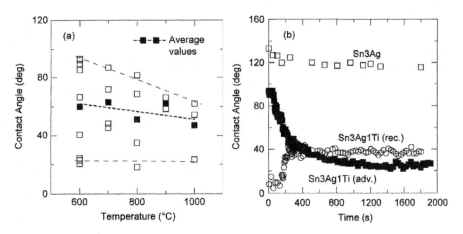

Figure 2. a) Contact angle as a function of temperature for Sn3Ag1Ti on sapphire after 60 minutes, b) Wetting kinetics for experiments performed at 600°C (adv = advancing drops of Sn3Ag1Ti on sapphire; rec = receding drops in the same system).

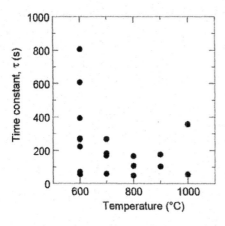

Figure 3. Evolution of the time constant τ (a measure of the spreading time required to reach a stationary contact angle) versus wetting temperature.

The wetting kinetics at 600°C is summarized in Figure 2b. The final contact angle is strongly influenced by the addition of titanium. Indeed, an addition of 1 wt% titanium to the Sn3Ag alloy leads to a decrease in the final contact angle from 150° to 25°. However, the variability of the final contact angle for the Sn3Ag1Ti alloy is clear. It should also be pointed out that, in this system, the final contact angle measured for a receding drop can be larger than the lower value measured for an advancing front.

The time needed to reach an equilibrium contact angle is also very variable. We can assume that the change in contact angle versus time follows an exponential relationship:

$$\theta = a + b \exp\left[-\left(\frac{t}{\tau}\right)^n\right] \qquad (1)$$

where τ is a characteristic spreading time. The values of τ for advancing liquid fronts are plotted in Figure 3. Typically τ varies between 50 and 800 seconds. However, during recent drop-transfer experiments performed in our laboratory, τ decreases to a few milliseconds, which is close to the times measured for low-temperature liquids such as water or other organics.[5]

Microstructural Characterization.

Optical microscopy or SEM-EDS show no new phases present at the metal/ceramic interface of the samples heated at temperatures lower than 800°C. At higher temperatures, isolated islands of reaction product are observed at the interface (Figure 4a). The reaction product forms primarily on scratches on the sapphire surface that act as favorable nucleation sites, and that are close to the triple line. EDS analyses indicate that the reaction product is a Ti-rich phase.

A continuous reaction product covering the interface is never observed. In some cases, triple-line ridges are observed at the triple line after cooling

(Figure 4b). The chemical composition of the ridges is not constant. In some cases, it is pure alumina, but in others, the analyses also show the presence of titanium.

Two typical microstructures of metallic drops after the wetting experiments at 600°C are shown in Figure 5. Sn_5Ti_6 platelets are observed. The platelets are present in the starting alloy, and the picture suggests that they are not completely dissolved during the experiments. The greatest difference between the metal drops is in the distribution of the Sn_5Ti_6 platelets; the smaller contact angle corresponds to a lower number of platelets in the alloy, which seems to be the general tendency for all of the experiments.

DISCUSSION

The equilibrium oxygen partial pressure $p(O_2)$ for the reaction $Sn+O_2 \rightarrow SnO_2$ ranges from $\sim 10^{-45}$ Pa at 250°C to $\sim 10^{-8}$ Pa at 1000°C. Therefore, the surface oxide layer is always stable at temperatures below 450°C. Titanium also has a strong affinity for oxygen, and can contribute to the formation of a surface oxide layer that encapsulates the metal. The formation of resilient oxide layers on the surface of low-temperature melting metals in vacuum is well documented.[6-8] Typically, there is a critical temperature at which the oxide layer disappears. Three reasons have been proposed for this: 1) erosion through the formation of volatile species;[6-8] 2) cracking due to volume changes and thermal-expansion mismatches;[6,7,9] and 3) dissolution of the oxide in the metal.[10] Above a critical temperature, it is expected that the metal vapor reacts with the oxygen in atmosphere such that the oxygen activity around the drop is equal to or lower than the equilibrium $p(O_2)$ for the oxidation of the metal.[8]

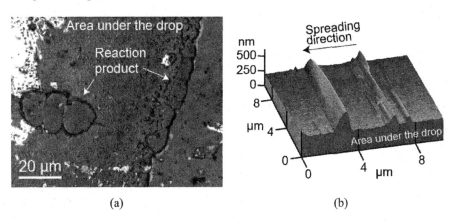

(a)   (b)

Figure 4. a) Optical micrograph showing a discontinuous reaction product in a "drop-free" sample after a wetting experiment at 1000°C, b) AFM image showing a consecutive series of ridges at the triple line, suggesting a series of attachment-detachment events, in a "drop-free" sample after a wetting experiment at 600°C.

The observed slow spreading velocities during the sessile-drop experiments are not consistent with fluid-flow or adsorption-controlled spreading (i.e., where spreading times on the order of few milliseconds).[1] Two main factors contribute to the observed slow spreading rate, and to the large variability in contact angle: 1) the kinetics of Ti dissolution into the liquid; and 2) the formation of triple-line ridges. The starting alloy has a homogeneous distribution of Sn_5Ti_6 platelets that dissolve slowly in the molten metal, enriching the liquid in Ti. The variability observed in the final contact angle is related to the different amounts of Ti dissolved in the liquid during the tests. As can be observed in Figure 5, fewer platelets in the drop (more titanium in the liquid) can result in a lower contact angle. Consequently, spreading kinetics can be controlled either by the speed of Ti dissolution from the platelets, or by the diffusion of titanium through the liquid to the triple junction. Additionally, triple-line ridges can form on the triple line by local diffusion or solution-precipitation. The ridge forms to achieve full two-dimensional equilibrium of the interfacial forces at the triple junction.[1] Our results indicate that, in some cases, ridges can nucleate at the triple line (for example, a scratch can act as a nucleation site for a ridge), effectively stopping spreading, and contributing to the wide variability in recorded contact angle. This interpretation is also consistent with the fast spreading observed in drop-transfer experiments on other metal-metal and metal-ceramic systems,[3, 13-14] and with the measured difference in the advancing and receding contact angle. In the drop-transfer setup, the liquid and the sapphire are not in contact during heating, which minimizes the chances of ridge formation. Additionally, the liquid alloy is allowed to homogenize for a long time on an inert substrate before it comes in contact with the sapphire. For a receding drop, the front recedes on the sapphire substrate until there is enough Ti dissolved in the liquid to promote wetting. After that, the formation of triple-line ridges can effectively stop the liquid front. This explains why, in some cases, the contact angle measured for a receding front is higher than that measured for an advancing front.

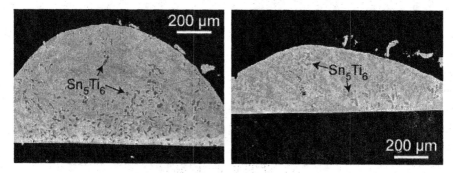

Figure 5. Microstructure showing the different distribution of the Sn_5Ti_6 platelets (in black) in two drops after wetting at 600°C.

The experiments clearly show that the addition of Ti enhances the wetting of Sn3Ag alloys. It has often been proposed that the effect of reactive elements (such as Ti, Cr, and others) on the wetting behavior of liquid alloys on ceramics is related to the formation of interfacial reaction layers.[11] Our results imply that the formation of an interfacial reaction product is not a necessary condition to improve spreading. Low contact angles can be reached at temperatures below 800°C where no reaction is detected. Even at higher temperatures, the observed reaction product does not form a continuous layer. This suggests that there is a finite barrier to nucleation and the formation of the reaction phase, and that the reaction product and the liquid do not extend together. If this is the case, the large decrease in contact angle observed with the addition of titanium should be a consequence of interfacial adsorption of Ti-species prior to the formation of the reaction product.[1] The discontinuous reaction product is rich in Ti and Sn. A comparison of the EDS analysis with other results reported in the literature regarding reactions between Al_2O_3 and Ti-containing brazing alloys suggests that the reaction phase is a Ti oxide or a mixed Ti-Sn oxide.[1]

CONCLUSIONS

Additions of 1 wt% Ti decrease the contact angle of Sn3Ag alloys on alumina from 150° to 25°. Spreading times of the order of 1 to 20 minutes and a wide variability of measured contact angle are observed during sessile-drop experiments. Two phenomena contribute to the observed behavior: 1) triple-line ridging; and 2) the kinetics of Ti dissolution and diffusion in the molten alloy. There has been a perception that good wetting is only attained with concentrations of reactive element so high that a reaction product inevitably forms. This is demonstrably not true for the experiments with the Sn3Ag1Ti system presented in this paper. At temperatures below 800°C, the decrease in contact angle is not accompanied by the formation of an interfacial reaction product. Additionally, only discontinuous islands of a new phase are observed after the tests performed at higher temperatures. These results show that chemical reaction and the formation of a new phase is not a necessary step to enhance the wetting of a reactive alloy. It is proposed that, in many reactive cases, the large decrease in contact angle observed may be a consequence of adsorption effects.

ACKNOWLEDGEMENTS

This work was supported by the Director, Office of Science, Office of Basic Energy Sciences, Division of Materials Sciences and Engineering, of the U.S. Department of Energy under Contract No. DE-AC03-76SF00098. This work was in part, supported by a NEDO International Joint Research Grant supervised by the Ministry of Economy, Trade and Industry of Japan. L. Gremillard wishes to acknowledge the Regional Council of Rhone-Alpes and the Physical Metallurgy and Materials Science Group (GEMPPM-INSA Lyon, France) for their financial support.

REFERENCES

[1]E. Saiz, R.M. Cannon and A.P. Tomsia, "Reactive Spreading: Adsorption, Ridging and Compound Formation," *Acta Materialia*, **48** [18-19] 4449-62 (2000).

[2]M. Abtew and G. Selvaduray, "Lead-free Solders in Microelectronics," *Materials Science & Engineering. R*, **27** [5-6] 95-141 (2000).

[3]E. Saiz, C.-H. Hwang, K. Suganuma and A.P. Tomsia, "Spreading of Sn-Ag Solders on FeNi Alloys," *Acta Materialia*, **51** [11] 3185-97 (2003).

[4]B. Derby and J.R.P. Webster, "Neutron Reflection Studies of the Composition of Interfaces between Titanium Containing Active Braze Alloys and Sapphire," Transactions of the Japanese Welding Research Insititue, **30** 233-38 (2001).

[5]L. Gremillard, E. Saiz and A. P. Tomsia, unpublished work.

[6]V. Laurent, D. Chatain, C. Chatillon and N. Eustathopoulos, "Wettability of Monocrystalline Alumina by Aluminium between its Melting Point and 1273 K," *Acta Metallurgica*, **36** [7] 1797-803 (1988).

[7]J.J. Brennan and J.A. Pask, "Effect of Nature of Surfaces on Wetting of Sapphire by Liquid Aluminum," *Journal of the American Ceramic Society*, **51** [10] 569-73 (1968).

[8]E. Ricci and A. Passerone, "Review: Surface Tension and its Relations with Adsorption, Vaporization and Surface Reactivity of Liquid Metals," *Materials Science and Engineering*, **A161** [1] 31-40 (1993).

[9]W. Jung, H. Song, S.W. Park and D. Kim, "Variation of Contact Angles with Temperature and Time in the Al-Al_2O_3 System," *Metallurgical and Materials Transactions B*, **27B** [1], 51-55 (1996).

[10]I. Rivolet, D. Chatain and N. Eustathopoulos, "Wettability of Alumina Single Crystals with Gold and Tin Between Their Melting Point and 1673 K," *Acta Metallurgica*, **35** [4] 835-44 (1987).

[11]N. Eustathopoulos, "Dynamics of Wetting in Reactive Metal Ceramic Systems," *Acta Materialia*, **46** [7] 2319-27 (1998).

[12]Metals Handbook Desk Edition, ed. J.R. Davis, ASM International, Materials Park, OH (1998), p.598-601.

[13]N. Rauch, E. Saiz and A.P. Tomsia, "Spreading of Liquid Ag and Ag-Mo Alloys on Mo Substrates," *Zeitschrift für Metallkunde*, **94** 233-37 (2003).

[14]N. Grigorenko, V. Poluyanskaya, N. Eustathopoulos, N. and Y. Naidich, "Kinetics of Spreading of Some Metals Melts Over Covalent Ceramic Surfaces," pp 57-67 in *Interfacial Science of Ceramic Joining*. Edited by A. Bellosi, T. Kosmac, and A.P. Tomsia. Kluwer Academic Publishers, Dordrecht, The Netherlands, 1998.

TENSILE PROPERTIES OF A FRICTION STIR WELDED THIN-SHEET OF 1050-H24 ALUMINUM ALLOY

H. J. Liu
National Key Laboratory of Welding
Harbin Institute of Technology
96 West Dazhi, Nangang, Harbin
150001, P. R. China

H. Fujii and K. Nogi
Joining and Welding Research
Institute, Osaka University
11-1 Mihogaoka, Ibaraki, Osaka
567-0047, Japan

ABSTRACT

Friction stir welding (FSW) is a promising welding process that can produce low cost and high quality joints with aluminum alloys. To examine the friction stir weldability of a 3 mm thick 1050-H24 aluminum alloy sheet, joint tensile properties and fracture locations in the joints have been studied. The experimental results show that FSW clearly softens the aluminum alloy sheet; thus, the tensile properties of FSW joints are lower than those of the base material, and the ultimate strength of the joints is only 70% that of the base material. The FSW parameters do not significantly affect the tensile properties of the joints, thus the range of the welding parameters that can be used to form a joint is comparatively wide. However, the welding parameters influence the fracture locations in the joints. When the revolutionary pitch (RP) is smaller than a critical value, the joints fracture on the retreating side (RS) of the weld. When the RP is greater than the critical value, a lack-of-penetration root defect is formed in the joints, and the joints fracture in the weld center.

INTRODUCTION

Aluminum alloys are commonly welded using conventional fusion welding processes; however, the mechanical properties of such joints are comparatively low due to excess heat input into the joints.[1-3] A friction stir welding (FSW) process has been developed to produce high-quality and low-cost joints of aluminum alloys.[4,5] Recent studies of the mechanical properties of friction stir welded joints indicate that different types of aluminum alloys have different FSW characteristics. Precipitate-hardened aluminum alloys such as the 2xxx,[6-11] 6xxx[1-3,10-15], and 7xxx series [8-11,16] undergo a process of dissolution or growth of the strengthening precipitates during the FSW thermal cycle, which results in a degradation of the mechanical properties. The strain-hardened aluminum alloys such as the 1xxx[17-20] and 5xxx series[3,21] are also softened by FSW due to a decrease in the dislocation density in the weld and the heat-affected zone.

The mechanical properties of a welded joint are different when different welding parameters are used to weld the same type of aluminum alloy.[6,9,17] In addition, some experimental results indicate that the mechanical properties of the joints of precipitate-hardened aluminum alloys such as 2024-T3,[6] 2195-T8[7] and 6082-T6[15] change with the thickness of the base materials. However, it is still unknown whether or not the friction stir weldability of strain-hardened aluminum alloys are influenced by the thickness of the base materials; thus, it is necessary to further study this topic.

In this study, FSW of a 3mm thick, strain-hardened 1050-H24 aluminum alloy sheet is examined. The emphasis of this study is on the joint tensile properties, and on the fracture locations in FSW joints. The experimental results are compared to results obtained previously for a 5 mm thick 1050-H24 plate[17].

EXPERIMENTAL PROCEDURE

The base material used in this study was a 3 mm thick 1050-H24 aluminum alloy sheet with the chemical composition and mechanical properties listed in Tables I and II. The sheet was cut and machined into rectangular welding samples, 300 mm long by 80 mm wide, and they were longitudinally butt-welded using an FSW machine (Hitachi, SHK207-899). The welding tool size and welding parameters used in the experiments are listed in Table III.

After welding, an electrical-discharge cutting machine was used to cross-section the joints perpendicular to the welding direction to prepare samples for metallographic analysis and tensile testing. The cross-sections of the metallographic specimens were polished with an alumina suspension, etched with Keller's reagent (150 ml water, 3 ml nitric acid, 6 ml hydrochloric acid and 6 ml hydrofluoric acid) at zero °C for about thirty seconds, and characterized using optical microscopy.

Table I. Chemical composition of 1050-H24 aluminum alloy (wt.%)

Al	Si	Fe	Cu	Mg	Ti	V
Bal.	0.04	0.32	0.02	0.01	0.02	0.01

Table II. Mechanical properties of 1050-H24 aluminum alloy

Ultimate strength (MPa)	0.2% proof strength (MPa)	Elongation (%)
123	86	11.6

Table III. Tool size and welding parameters used in the FSW experiments

Tool size (mm)			Welding parameters		
Shoulder diameter	Pin diameter	Pin length	Tool tilt (°)	Rotation speed (rpm)	Welding speed (mm/min)
12	4	2.8	3	1500	100-1000

The configuration and size of the transverse tensile specimens were prepared according to Figure 1, where RS and AS denote the retreating side and advancing side of a weld, respectively. Prior to tensile testing, the Vickers hardness profiles across the weld, heat affected zone, and partial base material were measured along the centerlines of the cross-sections of the tensile specimens under a load of 0.98 N for 10 s using an automatic micro-hardness tester (Akashi, AAV-502). Vickers indentations, with a spacing of 1 mm, were also used to determine the fracture locations and plastic strain distributions in the joints. The tensile tests were carried out on a screw-driven testing machine (Baldwin, SS-207D-UAD) at room temperature using a crosshead speed of 1 mm/min. The tensile properties of each joint were evaluated using three tensile specimens cut from the same joint. After tensile testing, the distance, L (mm), between two adjacent Vickers indents was measured and compared to the (1 mm) distance before tensile testing to determine the corresponding plastic strain: $(L-1)/1$, i.e. $L-1$.

RESULTS AND DISCUSSION

Tensile Properties of Joints

Figure 2 shows the tensile properties of FSW joints welded at different revolutionary pitches (RP), which is defined as the welding speed divided by the rotation speed. The tensile strength of the FSW joint is lower than that of the base material (Table II). This result indicates that a softening effect has occurred in the 3mm thick 1050-H24 aluminum alloy sheet during FSW, just as it did in the 5mm thick sheet.[17]

When the RP is smaller than 0.4 mm/r, the ultimate strength and 0.2% proof strength increase slightly with RP. The elongation is relatively high, but decreases with increasing RP. When the RP is greater than 0.4 mm/r, the ultimate strength and proof strength decrease slowly with increasing RP, and the elongation dramatically decreases. These results indicate that the welding parameters affect the tensile properties of the joints, but the effect is not as significant as that for the 5 mm thick 1050-H24 plate.[17] This implies that the greater the thickness of the aluminum alloy plate, the more significant the effect of the welding parameters. With respect to the 3 mm thick 1050-H24 sheet, the ultimate strength of the joint is 85 MPa, which is 70% that of the base material. When the RP is gradually changed from 0.07 mm/r to 0.67 mm/r, the variation in the ultimate strength is very small, only ± 2 MPa. Therefore, the range of welding parameters that can be used to produce a joint is relatively wide.

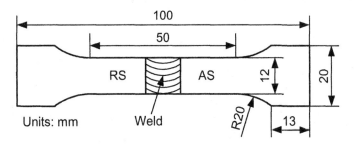

Figure 1. Configuration and size of the tensile test specimens used to characterize FSW joints in 1050-H24 aluminum alloy.

Fracture Locations of Joints

The fracture location in any joint is a direct reflection of the weakest part of the joint. Figure 3 shows the fracture locations in joints welded at different RP values. The fracture location is expressed by the distance between the fracture surface and the weld center. The distance is marked as a minus if the fracture occurs on the RS of the weld. Figure 3 shows that the fracture location in the joint is near the weld center, and that it changes with RP. When the RP is smaller than 0.4 mm/r, the joints fracture on the RS, and at the locations that are 2.6-2.9 mm from the weld center. When the RP is greater than 0.4 mm/r, fracture occurs within 0.8 mm of the weld center; that is, the joints fracture in the weld center.

The aforementioned results indicate that the location of fracture in FSW joints is influenced by the welding parameters. When the RP is smaller than 0.4 mm/r, all of the joints fracture on the RS; thus, the tensile properties on the RS are inferior to those on the AS. This result is different from that observed for the 5 mm thick 1050-H24 plate.[17] This implies that, when the RP is smaller than a critical value, joints in a thinner 1050-H24 sheet tend to fracture on the RS. Joints in thicker 1050-H24 plate are prone to fracture on the AS.

Decisive Factors for Tensile Properties

In nature, the tensile properties and the location of fracture in a joint are dependent on the internal structure of the joint, which is dependent on the welding parameters. Figure 4 shows typical cross-sections of FSW joints produced using different RP values. When the RP is smaller than 0.4 mm/r, FSW produces a defect-free joint (Figure 4a and b). When the RP is greater than 0.4 mm/r, a root defect (i.e. a lack-of-penetration) is formed in the joint due to insufficient heat input into the joint (Figure 4c and d).

When a joint is free of defects, the tensile properties and the location of fracture in the joint are dependent only on the microhardness profile accross the joints. Figures 5 and 6 show the microhardness profile and strain distribution in a typical joint, respectively. A hardness degradation region (i.e., softened region) is observed in the FSW joint (Figure 5), and the plastic strains in the joint are mainly concentrated in this softened region (Figure 6). Thus, the tensile properties of a FSW joint are all lower than those of the base material. There is a minimum

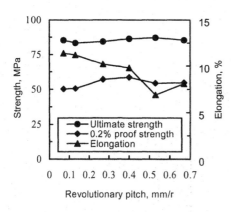

Figure 2. Tensile properties of FSW joints as a function of RP.

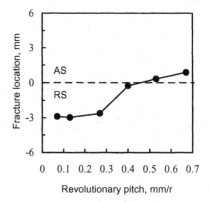

Figure 3. Fracture locations in FSW joints as a function of RP.

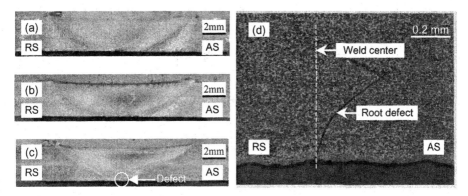

Figure 4. Cross sections of FSW joints produced with different revolutionary pitches: a) 0.07 mm/r; b) 0.27 mm/r; and c) 0.67 mm/r, and d) a magnified root defect in the joint welded at 0.67 mm/r.

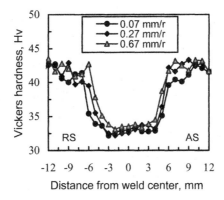

Figure 5. Microhardness profiles in FSW joints as a function of RP.

hardness value on the RS of each joint, and there is a maximum strain value at the corresponding minimum-hardness location. Therefore, the joint fractures on the RS instead of the AS. The minimum hardness value of the joint does not change significantly with RP. Accordingly, there is not much variation in the tensile strength, including the ultimate strength and the proof strength, with RP.

When a root defect exists in the joint, the joint tensile properties and the location of fracture in the joint are more or less affected by the defect. The root defect generally occurs in the weld center (strictly speaking, close to the weld center). It acts as a cracking source during tensile testing, which results in fracture in the weld center. However, the root defect does not significantly affect the tensile strength of the joints. The reason for this is that the defect size is much smaller than the thickness of the base material.

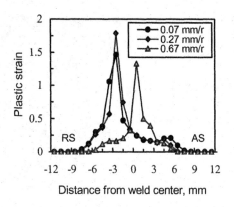

Figure 6. Strain distributions in FSW joints as a function of RP.

CONCLUSIONS
(1) FSW clearly results in softening of 3 mm thick 1050-H24 aluminum alloy sheet; thus, the tensile properties of the FSW joint are lower than those of the base material, and the ultimate strength of the joint is only 85 MPa.
(2) The welding parameters do not significantly affect the tensile properties of the FSW joint, especially the ultimate strength. Because of this, the welding parameters for the 3 mm thick 1050-H24 aluminum alloy sheet can be varied over a relatively wide range.
(3) The welding parameters influence the location of fracture in the joint. When the RP is smaller than a critical value, the joint fractures on the RS of the weld. When the RP is greater than the critical value, a lack-of-penetration root defect is formed in the joint, and the joint fractures in the weld center.

ACKNOWLEDGEMENTS
This work was performed in the Joining and Welding Research Institute (JWRI), Osaka University, Japan. Dr. H. J. Liu would like to express his gratitude to JWRI for the financial support.

REFERENCES
[1] Y. Nagano, S. Jogan, and T. Hashimoto, "Mechanical Properties of Aluminum Die Casting Joined by FSW," *Proc. 3rd Int. FSW Symp.*, Kobe, Japan, TWI Ltd, Paper No. Post-12, Sept. 2001.
[2] J. Hagstrom and R. Sandstrom, "Mechanical Properties OF Welded Joints of in Thin Walled Aluminum Extrusions," *Sci. Technol. Weld. Join.*, **2** [5] 199-208 (1997).
[3] M. Kumagai and S. Tanaka, "Properties of Aluminum Wide Panels by Friction Stir Welding," *Proc. 1st Int. FSW Symp.*, California, USA, TWI, Paper No. S3-P2, June 1999.
[4] W.M. Thomas, E.D. Nicholas, J. C. Needham, M. G. Murch, P. Temple-Smith, and C. J. Dawes, *International Patent Application PCT/GB92/02203 and GB Patent Application 9125978.8*, UK Patent Office, London, December 6, 1991.

[5] C.J. Dawes and W.M. Thomas, "Friction Stir Process Welds Aluminum Alloys," *Weld. J.,* **75** [1] 41-45 (1996).

[6] G. Biallas, R. Braun, C.D. Donne, G. Staniek, and W.A. Kaysser, "Mechanical Properties and Corrosion Behavior of Friction Stir Welded 2024-T3," *Proc. 1st Int. FSW Symp.,* California, USA, TWI, Paper No. S3-P3, June 1999.

[7] D.G. Kinchen, Z.X. Li, and G.P. Adams, "Mechanical Properties of Friction Stir Welds in Al-Li 2195-T8," *Proc. 1st Int. FSW Symp.,* California, USA, TWI, Paper No. S9-P2, June 1999.

[8] M.G. Dawes, S.A. Karger, T.L. Dickerson, and J. Przyoatek, "Strength and Fracture Toughness of Friction Stir Welds in Aluminum Alloys," *Proc. 2nd Int. FSW Symp.,* Gothenburg, Sweden, TWI Ltd and IVF, Paper No. S2-P1, June 2000.

[9] T. Hashimoto, S. Jyogan, K. Nakada, Y.G. Kim, and M. Ushio, "FSW Joints of High Strength Aluminum Alloy," *Proc. 1st Int. FSW Symp.,* California, USA, TWI, Paper No. S9-P3, June 1999.

[10] L. Magnusson and L. Kallman, "Mechanical Properties of Friction Stir Welds in Thin Sheet of Aluminum 2024," *Proc. 2nd Int. FSW Symp.,* Gothenburg, Sweden, TWI Ltd and IVF, Paper No. S2-P3, June 2000.

[11] A. V. Strombeck, J. F. D. Santos, F. Torster, P. Laureano, and M. Kocak, "Fracture Toughness Behavior of FSW Joints on Aluminum Alloys," *Proc. 1st Int. FSW Symp.,* California, USA, TWI, Paper No. S9-P1, June 1999.

[12] H. Okamura, K. Aota, M. Sakamoto, M. Ezumi, and K. Ikeuchi, "Behavior of Oxide during Friction Stir Welding of Aluminum Alloy and Its Influence on Mechanical Properties," *Q. J. Jap. Weld. Soc.,* **19** [3] 446-456 (2001).

[13] Y. S. Sato and H. Kokawa, "Distribution of Tensile Property and Microstructure in Friction Stir Weld of 6063 Aluminum," *Metall. Mater. Trans. A,* **A32** [12] 3023-3031 (2001).

[14] B. Heinz and B. Skrotzki, "Characterization of a Friction Stir Welded Aluminum Alloy 6013," *Metall. Mater. Trans. B,* **B33** [6] 489-498 (2002).

[15] L. E. Svensson, L. Karlsson, H. Larsson, B. Karlsson, M. Fazzini, and J. Karlsson, "Microstructure and Mechanical Properties of Friction Stir Welded Aluminum Alloys with Special Reference to AA5083 and AA6082," *Sci. Technol. Weld. Join.,* **5** [5] 285-296 (2000).

[16] M. W. Mahoney, C. G. Rhodes, J. G. Flintoff, R. A. Spurling, and W. H. Bingel, "Properties of Friction Stir Welded 7075-T651 Aluminum," *Metall. Mater. Trans. A,* **A29** [7] 1955-1964 (1998).

[17] H. J. Liu, M. Maeda, H. Fujii, and K. Nogi, "Tensile Properties and Fracture Locations of Friction Stir Welded Joints of 1050-H24 Aluminum Alloy," *J. Mater. Sci. Lett.,* **22** [1] 41-43 (2003).

[18] O. V. Flores, C. Kennedy, L. E. Murr, D. Brown, S. Pappu, B. M. Nowak, and J. C. Mcclure, "Microstructural Issues of in a Friction Stir Welded Aluminum Alloy," *Scripta Mater.,* **38** [5] 703-708 (1998).

[19] Y. S. Sato, M. Urata, H. Kokawa, and K. Ikeda, "Effect of Friction Stir Processing on Microstructure and Hardness Profile of Equal Channel Pressed Aluminum Alloy 1050," *Proc. 7th Int. Weld. Symp.,* Kobe, Japan, JWS, 633-638, Nov. 2001.

[20] Y. S. Sato, M. Urata, H. Kokawa, K. Ikeda, and M. Enomoto, "Retention of the Grained Microstructure of Equal Channel Pressed Aluminum Alloy by Friction Stir Welding," *Scripta Mater.,* **45** [1] 109-113 (2001).

[21] H. Jin, S. Saimoto, M. Ball, and P. L. Threadgill, "Characterization of Microstructure and Texture in Friction Stir Welded Joints of 5754 and 5182 Aluminum Alloy Sheets," *Mater. Sci. Technol.,* **17** [12] 1605-1614 (2001).

SURFACE TENSION OF 304 STAINLESS STEEL UNDER PLASMA CONDITIONS

Hidetoshi Fujii, Teruhiko Misono, Taihei Matsumoto, and Kiyoshi Nogi
Joining and Welding Research Institute, Osaka University
11-1 Mihogaoka, Ibaraki, Osaka, 567-0047, Japan

ABSTRACT
Using the sessile drop method, the surface tension of 304 stainless steel was measured in Ar plasma between 1773 and 2073 K. Type 304 stainless steel with different sulfur contents of 10 ppm, 110 ppm and 250 ppm was examined. The substrate used was $(La_{0.9}Ca_{0.1})CrO_3$ (>99.5%). Surface tension was calculated from photographs of sessile drops using the Laplace equation. At low-temperature, the plasma does not significantly affect surface tension. The temperature coefficient of the surface tension of 304 stainless steel with a sulfur content of 10 ppm is negative. It is positive for the sulfur contents of 110 ppm and 250 ppm.

INTRODUCTION
Arc welding is one of the most widely used processes to join materials[1]; hence, it is very important to understand the molten pool phenomena during arc welding. During arc welding, convection is generated in the molten pool, and this affects the properties of the products as well as the efficiency of the welding. Because surface tension is the most significant driving force for the convection[2,3], a precise value of the surface tension is required to control convection.

Several investigations have been conducted on the surface tension of the molten materials.[4,5] However, almost all of the experiments have been completed in vacuum or in an inert gas. These data have been used to understand convection during arc welding. Because the molten pool is covered with plasma during the actual arc welding process, it is really necessary to measure surface tension under the plasma.

In this study, the surface tension of the 304 stainless steels was measured both with and without plasma, and the effect of the plasma on surface tension was investigated. Type 304 stainless steel with a sulfur content of 10 ppm, 110 ppm and 250 ppm was characterized to determine the effect of a surface-active element[6-12] on surface tension. Density was also measured, simultaneously.

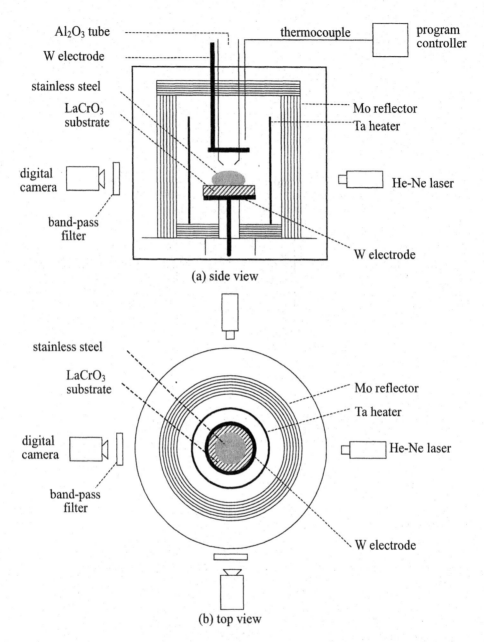

Figure1. Schematic of the experimental setup used to measure surface tension using the sessile drop method.

EXPERIMENTAL PROCEDURE

Samples of 304 stainless steel (ss) with sulfur contents of 10 ppm, 110 ppm, and 250 ppm were examined. The substrate was $(La_{0.9}Ca_{0.1})CrO_3$ (>99.5%), which has a high electric conductivity.[13,14] This property enabled the substrate to be an electrode to generate the plasma. The ss sample and the substrate were cleaned in acetone using an ultrasonic cleaner before each experiment.

A schematic of the equipment used to measure surface tension using the sessile drop method is shown in Figure1. It consists of a vacuum chamber, a tantalum heater with seven concentric molybdenum reflectors, a sample dropping device[15,16], an image recording system, and a plasma generating device. The vacuum chamber is connected to a turbo-molecular pump and a rotary pump.

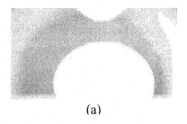

(a)

(b)

Figure 2. Photo of a sessile drop taken: a) without; and b) with a laser and a band-pass filter

The image recording system consists of two 7 mW He-Ne lasers, two band-pass filters, and two digital cameras with 2000×1312 pixels. The band-pass filter eliminates the light from all wavelengths except that of the laser beam (632 nm); therefore, the radiation from the sample and the heater filtered out, and a high definition drop profile is obtained in our experiments (Figure 2). The droplet shape was observed in two right-angled directions to confirm the symmetry of the droplet.

Before heating the furnace, the substrate, the He-Ne lasers, the band-pass filters, and the digital cameras were horizontally adjusted, and a stainless steel rod was placed in the dropping device outside the chamber. The furnace was evacuated using the rotary pump and the turbo-molecular pump. When the pressure reached approximately 10^{-5} Pa, heating was started. The temperature was measured with a tungsten thermocouple (W-5%Re / W-26%Re) and controlled using a PID digital programmable controller. To remove any adsorbed impurities from the furnace and heater, the temperature was first raised 30°C higher than the experimental temperature, and then lowered to 1200°C. The temperature was then raised again to the experimental temperature under an Ar gas atmosphere. When the requisite temperature was reach, the pressure was reduced to approximately 100 Pa, and the stainless steel rod was inserted into the bottom of the Al_2O_3 tube. After the sample was held at temperature for 90 s to melt, it was dropped onto the substrate using a small pressure difference between the chamber and the Al_2O_3 tube. The plasma was then generated in a low pressure Ar gas by applying a DC voltage between the substrate and a tungsten electrode placed above the droplet. A current density of 64 mA/cm² was used. As soon as the ss sample came into contact with the substrate and produced a sessile drop shape, a photo was taken. After taking the photos, the sample was then cooled to a temperature below the melting point. Using these photographs, the coordinates of ninety- nine points from the droplet profile were determined. The surface tension was calculated by fitting these points to the Laplace equation using a computer program. The volume of droplet was also determined from the sessile drop photographs, so the density was simultaneously calculated.

The drop symmetry significantly affects the surface tension determined by the sessile drop method; therefore, each droplet was observed in two directions so that the symmetry of the droplet could be verified during the experiment. A measured value was adopted only when measurements from the two different directions were within 3% of one another.

After the experiments, the sulfur and oxygen contents of the sample were measured using the combustion infrared absorption method and the non-dispersible infrared absorption method, respectively.

RESULTS AND DISCUSSION

Figure 3 shows the change in the surface tension with time for type 304 stainless steel with a sulfur content of 110 ppm at 1873 K under Ar plasma. The plasma was generated 60 seconds after the sample was dropped onto the substrate. There is no significant difference before and after the plasma is generated.

Figure 4 shows the surface tension of the three different 304 stainless steels under an Ar plasma in the temperature range of 1773 K to 2073 K. The sulfur content of the 304 stainless steels with 10 ppm and 110 ppm does not change during the experiment. The sulfur content of the ss with 280 ppm sulfur changed to approximately 250 ppm after the experiments. The sulfur content of this sample is defined as 250 ppm because the surface tension was calculated using the data for the last 60 seconds of the experiments.

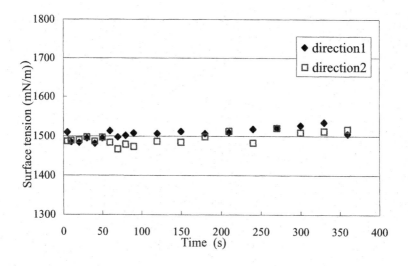

Figure 3. Surface tension of 304 stainless steel on $(La_{0.9}Ca_{0.1})CrO_3$ at 1873 K under Ar plasma as a function of time.

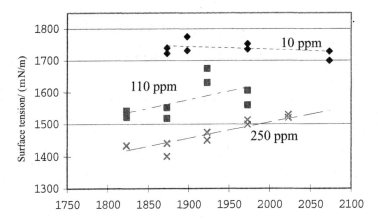

Figure 4. Surface tension as a function of temperature for 304 stainless steel with 10, 110, and 250 ppm sulfur.

The surface tension of the 304 stainless steel with a sulfur content of 10 ppm slightly decreases with the increasing temperature. On the other hand, the surface tensions of the 304 stainless steels with a sulfur content of 110 ppm and 250 ppm increases with the increasing temperature. The measured values of the surface tensions are expressed as follows:

$$\gamma = -0.12 \times T + 2030, \text{ mN/m (with 10 ppm sulfur)}$$

$$\gamma = 0.54 \times T + 550, \text{ mN/m (with 110 ppm sulfur)}$$

$$\gamma = 0.49 \times T + 519, \text{ mN/m (with 250 ppm sulfur)}$$

The calculated temperature coefficients, $d\gamma/dT$, of the 304 stainless steel are in good agreement with the results of Mills et al.[8] They measured the surface tensions of 304 stainless steels using the levitated drop method and found that $d\gamma/dT$ changes from negative to positive when the sulfur content exceeds approximately 50 ppm.

The oxygen content of the 304 stainless steels after the experiments varies from 15 ppm to 40 ppm.

Figure 5 shows the density calculated for the 304 stainless steels in this study. Because the sulfur contents are very low, from 10 ppm to 250 ppm, the density is not affected by the change in the sulfur content. Therefore, in Figure 4, the data for all three 304 stainless steels are plotted. The temperature dependent density is calculated from all of the data, and can be expressed as follows:

$$\rho = 8.11 \times 10^3 - 0.616 \times T, \text{ kg/m}^3$$

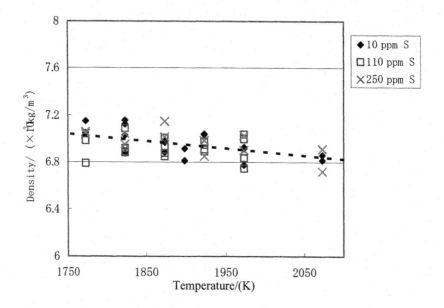

Figure 5. Density as a function of temperature of 304 stainless steel with 10, 110, and 250 ppm sulfur.

CONCLUSIONS

The surface tension of 304 stainless steel with a sulfur content of 10 ppm, 110 ppm or 250 ppm was determined in the presence of and in the absence of a low pressure Ar plasma, and the following was determined.

1) Ar plasma does not significantly affect the surface tension of the 304 stainless steel.

2) The temperature dependent surface tension of 304 stainless steel is expressed as follows:

$$\gamma = -0.12 \times T + 2030, \text{ mN/m (with 10 ppm sulfur)}$$
$$\gamma = 0.54 \times T + 550, \text{ mN/m (with 110 ppm sulfur)}$$
$$\gamma = 0.49 \times T + 519, \text{ mN/m (with 250ppm sulfur)}$$

3) The temperature dependent density of 304 stainless steel with 10-250 ppm is expressed as follows:

$$\rho = 8.11 \times 10^3 - 0.616 \times T, \text{ kg/m}^3$$

ACKNOWLEDGEMENTS

This work is the result of the "Development of Highly Efficient and Reliable Welding Technology", which is supported by the New Energy and Industrial Technology Development Organization (NEDO) through the Japan Space Utilization Promotion Center (JSUP) in the program from the Ministry of Economy, Trade and Industry (METI), and of the Grant-in-Aid for Scientific Research (c) from the Ministry or Education, Sports, Culture, Science and Technology of Japan.

REFERENCES

[1] G.J. Dunn, C.D. Allem, and, and T.W. Eagar, "Metal Vapors in Gas Tungsten Arcs," *Metall Trans A*, **17A**, 1851-1863(1986)

[2] S. Kou and D.K. Sun, "Fluid Flow and Weld Penetration in Stationary Arc Welds," *Metall. Trans A*, **16**, 203-213 (1985).

[3] C.R. Heiple and J.R. Roper, "Mechanism for Minor Element Effect on GTA Fusion Zone Geometry," *Welding Journal*, **61**, 97s-102s(1982)

[4] B.F. Dyson, "The Surface Tension of Iron and Some Iron Alloys," *Trans. Met. Soc. AIME*, **227**, 1098-1102(1963)

[5] F.A. Halden and W.D. Kingery, "Surface Tension at Elevated Temperatures," *J. Phys. Chem*, **59**, 557-559(1955)

[6] K. Ogino, K. Nogi, and C. Hosoi, "Surface Tension of Molten Fe-O-S Alloy," *Tetsu to Hagane*, **69**, 1989-1994 (1983).

[7] K. Nogi and K. Ogino, "Effect of Sulfur on Surface Tension of Liquid Iron and Wettability of Solid Oxides by Liquid Iron," *J. High Temp. Soc*, **16**[1], 20-26 (1989)

[8] P.R. Scheller, R.F. Brooks, and K.C. Mills, "Influence of Sulfur and Welding Conditions on Penetration in Thin Strip Stainless," *Weld. Res. Sppl*, 69-s-75-s (1995).

[9] M.J. McNallan and T. Debroy, "Effect of Temperature and Composition on Surface Tension in Fe-Ni-Cr Alloys Containing Sulfur," *Metall. Trans B*, **22**, 557-560 (1991)

[10] S. Ban-ya and J. Chipman, "Sulfur in Liquid Iron Alloy: II -Effects of Alloying Elements," **245**, 133-143 (1969)

[11] M. Noguchi and T. Narita, "Sulfidation Behavior of Fe-Cr Alloy in Low Sulfur Pressures," *J. Jap. Inst. Met*, **60**[2], 198-204 (1996)

[12] K. Ogino, K. Nogi, and O. Yamase, "The Effect Selenium and Tellurium on the Surface Tension of Molten Iron and Wettability of Solid Oxide," *Tetsu to Hagane*, **66**[2], 179-185 (1980)

[13] T. Brylenski, K. Przybylski, and J. Morgiel, "Microstructure of Fe-Cr/(La,Ca)CrO$_3$ Composite Interconnector in Solid Oxide Fuel Cell Operating Conditions," *Mater. Chem. Phys*, **81**, 434-437 (2003).

[14] D.-H. Peck, M. Miller, and K. Hilpert, "Vaporization and Thermodynamics of La$_{1-x}$Ca$_x$CrO$_{3-\delta}$ Investigated by Knudsen Effusion Mass Spectrometry," *Solid State Ionics*, **143**, 391-400 (2001).

[15] K. Nogi, K. Ogino, "Role of Interfacial Phenomena in Deoxidation Process of Molten Iron,"*Can. Inst. Mine. Metall*, **22**, 19-28 (1983).

[16] H. Fujii, H. Nakae, and K. Okada, "Interfacial Reaction Wetting in the Boron Nitride/Molten Aluminum System," *Acta Metall*, **41**, 2963-2971 (1993).

THE EFFECT OF MINOR ELEMENTS IN ARGON SHIELDING GAS ON CONVECTION IN A MOLTEN WELD POOL

Hidetoshi Fujii, Yasuyuki Morikawa, Masayoshi Kamai, and Kiyoshi Nogi
Joining and Welding Research Institute, Osaka University, 11-1 Mihogaoka, Ibaraki, Osaka, 567-0047, Japan

ABSTRACT

To investigate the effect of minor elements in the Ar shielding gas on the convection in a molten weld pool, tungsten inert gas (TIG) welding was performed under conventional and microgravity conditions. The effect of small amounts of H_2, N_2, and He in the Ar shielding gas was investigated. Butt-welding of pure aluminum (5N) and an aluminum-6%copper alloy (A2219) was performed. Specimen cross-sections were analyzed using electron probe microanalysis (EPMA) after TIG welding to evaluate the copper distribution. To differentiate between convection due to the plasma stream and convection due to the electromagnetic force, the arc pressure and the current density were also measured on water–cooled copper under the same welding conditions. There is very little convection in an aluminum molten weld pool under microgravity, showing that there is a mutual balance between the convection due to the plasma stream and the convection due to the electromagnetic force. A mixture of 1% H_2, 10% N_2, or 10% He in the Ar shielding gas significantly changes the magnitude of the convection.

INTRODUCTION

Various parameters such as welding current, arc length, and shielding gas affect convection in a molten weld pool.[1-3] It is well known that convection in a molten weld pool plays an important role in many welding properties, such as the molten weld pool shape and the penetration depth.[4,5] Accordingly, various methods have been developed to investigate the convection phenomena in a molten weld pool.[6,7] In a molten weld pool during TIG welding, there are four driving forces for convection[8,9]: the plasma stream, electromagnetic force, surface tension, and buoyancy. However, it is difficult to distinguish the effect of each driving force in an actual welding experiment. To redress this problem, TIG welding can be performed under microgravity. Under microgravity, convection due to buoyancy is negligible.[10] In addition, when an aluminum alloy is welded, convection due to the surface tension is negligible.[10] Under these conditions, the convection in the molten weld pool is affected only by the plasma stream and the electromagnetic force. Additionally, the directions of the convection due to the electromagnetic force and due to the plasma stream are opposite. Taking these characteristics into account, the effect of the minor elements in the Ar shielding gas on convection is investigated in this study.

EXPERIMENTAL PROCEDURE
Welding Conditions
Tungsten inert gas (TIG) welding was performed under microgravity and conventional conditions. The experiment under microgravity was performed at the Japan Microgravity Center. Figure 1 illustrates the microgravity system. The system attained a microgravity level of 10^{-5}G for 10 seconds. The welding position was horizontal under the conventional conditions. Butt-welding was performed using pure aluminum (5N) and an aluminum-6%copper alloy (A2219). The sample dimensions were $100^l \times 25^w \times 3^t$mm.

Table I shows the welding conditions used. Ar, Ar-1%H$_2$, Ar-10%N$_2$, and Ar-10%He were used as the shielding gas. The welding speed was 5 mm/s for the Ar shielding gas, and it was controlled for the other shielding gases so that the bead width would be the same as that for the Ar shielding gas. The copper distribution in cross-sections of each specimen was analyzed

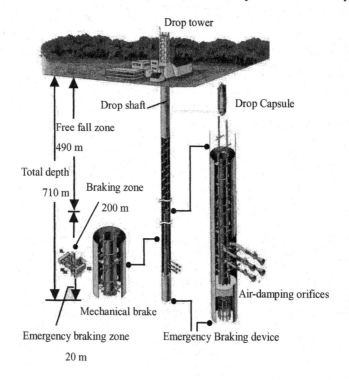

Figure 1. Schematic of the microgravity system used to complete the TIG welding experiments.

using electron probe microanalysis (EPMA) after TIG welding to evaluate the convection.

Arc Pressure Measurement

The arc pressure was measured under conventional welding conditions. Figure 2 illustrates the system used to measure the arc pressure. The orifice

Table I.   Welding conditions

Sample	Pure aluminum, Al-6%Cu			
Shielding gas	Ar	Ar-1%H$_2$	Ar-10%N$_2$	Ar-10%He
Welding current	DCEN 100A			
Arc length	3 mm			
Gas flow rate	10 L/min			

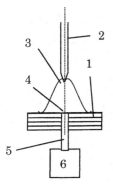

Figure 2. Schematic of the system used to measure the arc pressure:
(1) water-cooled copper anode
(2) cathode (3) arc discharge
(4) gating orifice
(5) communication channel
(6) pressure sensor

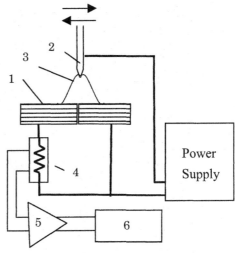

Figure 3. Schematic of the system used to measure the current density:
(1) water-cooled copper anode
(2) cathode (3) arc discharge
(4) current shunt (5) amplifier
(6) computer

diameter in the water-cooled copper anode was 1.0 mm. The communication channel connected the output of the gating orifice with the input of a pressure sensor. The output signal from the pressure sensor was recorded using a computer. The resolution for this measurement was 1 Pa.

Current Density Measurement
The current density was also measured under conventional welding conditions. Nestor's[11] experimental method was used. Figure 3 illustrates the system used to measure the current density. The system consisted of two water-cooled copper anodes with a 0.1 mm gap. The currents for both anodes were measured as a function of the arc position relative to the gap. At 0.5 mm intervals. The current was measured 200 times with a sampling interval of 0.1 seconds, and averaged. The resolution of the measurement was 0.03A. The radial current density distribution was calculated using the Abel inversion.[12]

RESULTS AND DISCUSSION
Balance of Plasma Stream and Electromagnetic Force
Figure 4 shows the EPMA results for pure aluminum and the A2219 samples butt-welded using a 100 A welding current and a 3 mm arc length in Ar shielding gas. Figures 4a and b show the results under conventional

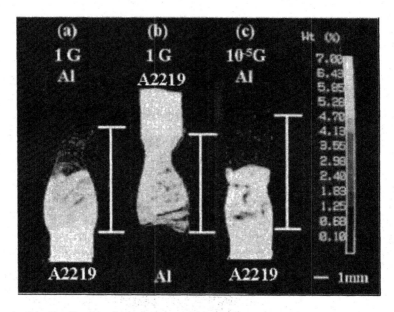

Figure 4. Cu distribution in the TIG weld of pure aluminum and A2219 alloy under microgravity and conventional conditions. (for a 100 A welding current and a 3 mm arc length in Ar shielding gas)

welding conditions. Pure Al is positioned on the top of the joint in the Figure 4a, and the A2219 is positioned on the top of the joint in Figure 4b. Figure 4c shows the result under microgravity. The bar next to the results shows the size of the molten weld pool. The molten weld pool is barely stirred under microgravity. This result indicates that the convection due to the plasma stream and the convection due to the electromagnetic force are mutually balanced under microgravity conditions. By comparison, the molten weld pools were better stirred under conventional conditions. Kou et al.[8] showed, in a computer simulation, that convection due to buoyancy is weaker relative to convection due to the other driving forces. However, under the present balanced conditions, even the convection due to buoyancy is significant.

Effect of Shielding Gas

Figure 5 shows the EPMA analysis results of butt-welding A2219 and pure aluminum under microgravity with the shielding gases: a)Ar-1%H$_2$; b)Ar-10%N$_2$; and c)Ar-10%He. In all cases, the molten weld pools are uniformly mixed compared to the molten weld pool produced using Ar shielding gas, (Figure 4c).

The effect of the plasma stream can be roughly estimated from the arc pressure. Table II shows the arc pressure at the arc center on the water-cooled copper with each shielding gas. The arc pressures for the Ar-1%H$_2$ and the Ar-10%N$_2$ shielding gases are approximately 1.26 times higher than that for Ar alone. The arc pressure for Ar-10%He is approximately 0.85 times higher than that for Ar alone. The magnitude of

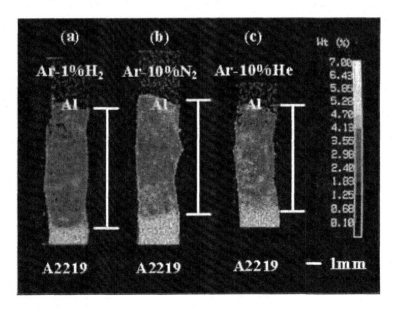

Figure 5. Effect of shielding gas on Cu distribution in a TIG weld of pure aluminum and A2219 alloy under microgravity. (100 A welding current, 3 mm arc length)

Table II. Arc pressure for the different kinds of shielding gas.

	Ar	Ar-1%H$_2$	Ar-10%N$_2$	Ar-10%He
Arc Pressure (Pa)	272	344	343	233

the convection due to the plasma stream is related to the velocity of the plasma stream. If it is assumed that the pressure near the electrode is much larger than the pressure near the anode, and that the pressure is completely converted to the plasma stream, then the plasma stream velocity can be estimated using Bernoulli's equation.[13]

$$p = (1/2) \rho v^2 \qquad (1)$$

where, p = arc pressure, ρ = density of the plasma, and v = velocity of the

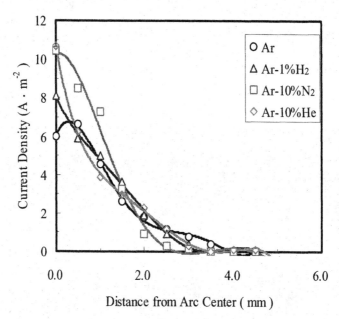

Figure 6. Anode current density distribution as a function of distance from the arc center for different kinds of shielding gas.

Table III. Root radius for the different kinds of shielding gas.

	Ar	Ar-1%H$_2$	Ar-10%N$_2$	Ar-10%He
Root Radius (mm)	3.0	2.6	2.0	2.6

plasma stream. Thus, the velocity of the plasma stream is proportional to the square root of the arc pressure. Consequently, the magnitude of the convection due to the plasma stream for Ar-1%H_2 and Ar-10%N_2 is estimated to be approximately 1.1 times greater than that for pure Ar. Similarly, convection for the Ar-10%He plasma stream is approximately 0.9 times higher than that for Ar alone.

Figure 6 shows the current density distribution on the water-cooled copper anode for the different kinds of shielding gas. The current density distribution was calculated using the Abel inversion, assuming that the arc discharge is axially symmetric. The current density decreases as the distance from the arc center increases. Near the center of the arc, the current density for the Ar-10%N_2 and the Ar-10%He shielding gas is high. In particular, the current density with Ar-10%N_2 is the highest below 1.5 mm.

Table III shows the root radius for each shielding gas. The root area is defined as the area where 90% of the total arc current is supplied. The root radius for the Ar-1%H_2, Ar-10%N_2, and Ar-10%He shielding gas are approximately 0.86, 0.66, and 0.86 times greater than that for Ar shielding gas, respectively. These results indicate that the arc for the Ar-1%H_2, Ar-10%N_2, and Ar-10%He shielding gas is contracted toward the arc center as compared to pure Ar.

The electromagnetic force is determined by the vector product of the current density and the magnetic field. Accordingly, the magnitude of the convection due to the electromagnetic force is proportional to the product of the current density and the current; consequently, it is inversely proportion to the root area when the welding current is constant. The magnitude of the convection due to the electromagnetic force for Ar-1%H_2, Ar-10%N_2 and Ar-10%He is estimated to be approximately 1.4, 2.3, and 1.4 times than that for Ar, respectively. Thus, the balance of convection due to the plasma stream and due to the electromagnetic force changes significantly by adding minor elements such as 1% H_2, 10% N_2, or 10% He to the Ar shielding gas. In other words, convection in the molten weld pool is sensitive to the shielding gas.

CONCLUSION

To investigate the effect of shielding gas composition on convection in a molten weld pool, aluminum alloys were TIG welded under micrograviy conditions. The following conclusions were drawn:

1) For a 100 A welding current with a 3 mm arc length in Ar shielding gas, convection in the molten weld pool is very weak. This result indicates that the convection forces due to the plasma stream and due to the electromagnetic force have similar magnitudes, and are balanced.

2) The balance of the convection forces due to the plasma stream and the electromagnetic force can be changed when minor elements such as 1%H_2, 10%N_2, and 10%He are added to the Ar shielding gas.

ACKNOWLEDGEMENT

This work is supported by the "Development of Highly Efficient and Reliable Welding Technology", from New Energy and Industrial Technology Development Organization (NEDO) through the Japan Space Utilization Promotion Center (JSUP) in the program of Ministry from the Economy, Trade and Industry (METI) and "Priority Assistance of

Formation of Worldwide Renowned Centers of Research —The 21st Century COE Program (Project: Center of Excellence for Advanced Structural and Functional Materials Design)" from the Ministry of Education, Sports, Culture, Science and Technology in Japan.

REFERENCE
[1]C.R. Heiple and P. Burgardt, "Effects of SO_2 Shielding Gas Additions on GTA weld Shape," *Welding J.*, **64**, 159-162 (1985).
[2]M.L. Lin and T.W. Eagar, "Influence of Arc Pressure on Weld Pool Geometry", *Welding J.*, **64**, 163-169 (1985).
[3]M. Onsoien, R. Peters, D.L. Olson, and S.Liu, "Effect of Hydrogen in an Argon GTAW Shielding gas:Arc Characteristics and Bead Morphology," *Welding J.*, **74**, 10-15 (1995).
[4]C.R. Heiple and J.R. Roper, "Effects of Selenium on GTAW Fusion Zone Geometry," *Welding J.*, **60**, 143-145 (1981).
[5]C.R. Heiple and J.R. Roper, "Mechanism for Minor element Effect on GTA Fusion Zone Geometry," *Welding J.*, **61**, 97-101 (1982).
[6]R.A. Woods and D.R. Milner, "Motion in the Weld Pool in Arc Welding," *Welding J.*, **50**, 163-173 (1971).
[7]C. Limmaneevichitr and S. Kou, "Visualization of Marangoni Convection in simulated weld pools," *Welding J.*, **79**, 126-135 (2000).
[8]S. Kou and D.K. Sun, "Fluid Flow and Weld Penetration in Stationary Arc Welds," *Metall. Trans. A*, **16A**, 203-213 (1985).
[9]S.Yokoya and A.Matsunawa : *Trans. JWRI, Osaka Univ.*,**16**[2], 1, (1987).
[10]F. Fujii, N. Sogabe, M. Kamai, and K. Nogi, "Convection in Molten Pool under Electron Beam Welding," *Preprints of National Meeting of Japan Welding Soc.*, **69**, 306-307 (2001).
[11]O.H. Nestor, "Heat Intensity and Current Density Distributions at the Anode of High Current, Inert Gas Arcs," *J. Appl. Phys.*, **33**[5], 1638-1648 (1962).
[12]O.H. Nestor and H.N. Olsen, "Numerical Methods for Reducing Line and Surface Probe Data," *SIAM Rev.*, **2**[3], 200-207 (1960).
[13]T.Ohji, "Arc Plasma and Arc Phenomena for Welding", 123-125, "Fundamentals of Welding and Joining Processes," 1, Sanpou, Tokyo, (1996)

Particulate Materials

CHARACTERIZATION OF A PHOTOCATALYST PREPARED BY A NEW METHOD TO INTRODUCE Ti SITES ON THE SURFACE OF SILICA

Masayoshi Fuji and Minoru Takahashi
Ceramics Research Laboratory
Nagoya Institute of Technology
Asahigaoka 10-6-29, Tajimi
Gifu 507-0071, Japan

Nanami Maruzuka, Takashi Takei, and
Masatoshi Chikazawa
Department of Applied Chemistry
Graduate School of Engineering
Tokyo Metropolitan University
Minami-ohsawa 1-1, Hachioji
Tokyo 192-0397, Japan

ABSTRACT

A photocatalyst was prepared using a new method to introduce Ti sites at specific spacing intervals on the surface of SiO_2. The treated sample shows photocatalytic activity. There are some Si-O-Ti structures on the surface, but no Ti-O-Ti structures. It is postulated that the Si-O-Ti structure may contribute to the production of hole generated oxidants (e.g., hydroxyl radicals) to increase photocatalytic activity. The Si-O-Ti structure, and surface adsorption on SiO_2 are considered to be important factors in the photocatalytic activity of the treated SiO_2 sample.

INTRODUCTION

Many photocatalytic materials have been used in various industrial fields since the work of Honda and Fujishima.[1] For example, due to its photocatalytic activity and the strong oxidizing potential of photogenerated holes and hole generated oxidants (e.g., hydroxyl radicals)[2,3], TiO_2 is used as a catalyst for environmental purification.[4] Many studies have been carried out to determine the mechanism responsible for this photocatalytic behavior.[5-7]

Anderson and Bard[10] reported that the photocatalytic activity of TiO_2 supported on the surface of SiO_2 is greater than that of TiO_2 alone. The increase in the photocatalytic activity has been attributed to the presence of the Si-O-Ti structure at the TiO_2/SiO_2 interface. Van Dyk and Heyns[12] reported that TiO_2 treated with SiO_2 also has a high photocalytic activity. These studies indicate that SiO_2 improves the photocatalytic activity of TiO_2.

TiO_2 forms a film on materials, or forms mixed oxide phases with materials.[10,11,13,14] SiO_2 is often chosen as the material to support the TiO_2 photocatalyst.

We would like to know the size of a TiO_2 cluster that results in photocatalytic activity on the surface of SiO_2, and what the role of the Si-O-Ti structure is on the TiO_2/SiO_2 photocatalyst. Anpo et al.[13] previously reported that the photocatalytic activity increases with decreasing TiO_2 particle. Thus, if the two issues described above can be resolved, it may be possible to develop a new photocatalytic material that is more photoactive and more useful.

In this study, a new method to make a photocatalyst is proposed. This new method introduces Ti atoms on the surface of SiO_2 at controlled interval spacing. The new process is used to fabricate samples to determine how TiO_2 cluster size and the Si-O-Ti structure affect photocatalytic activity.

EXPERIMENTAL SECTION

Samples were made using SiO_2 powder (Aerosil OX50, Nippon Aerosil, Ltd.) with a reported surface area of $\sim 50 m^2/g$. The surface modification reagent used was $(CH_3)_2CHOTi(OCOC_{17}H_{35})_3$ (KEN-REACT KR-TTS, Kenrich Petrochemicals, Inc.). The surface treatment process used to introduce the Ti sites onto the SiO_2 surface consisted of three steps as shown Figure 1. Surface modification was carried out through a topo-chemical reaction, using a reflux method for 1 hour in normal hexane solvent (Kanto Chemical, Inc.). Then, the solution containing the sample was filtered, the powder was washed with normal hexane, and the powder was dried under reduced pressure. The surface was oxidized at 400°C in an electric furnace (Rigaku, Ltd.) under flowing oxygen gas to remove the organics introduced with the Ti. Finally, surface hydroxylation was completed by placing the sample in a desiccator at 85%RH for about 10 hours. In this way, Si-O-Ti-(OH)n sites were introduced onto the sample surface.

By repeating the process, the number of Ti sites on the surface could be controlled. When the process was repeated, it was predicted that there would be both Si-O-Ti-(OH)n and Si-(O-Ti)m-(OH)n on the sample surface. We had intended to prepare TiO_2 clusters from the Si-(O-Ti)m-(OH)n.

The specific surface area and the weight loss during oxidation were used to determine the density of the Ti sites introduced onto the SiO_2 surface. The specific surface area was measured by N_2 adsorption using the BET method. The weight loss was calculated from TGA-DTA curves (Thermo plus, Rigaku, Ltd.). The

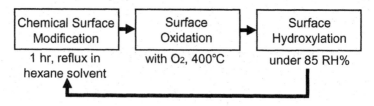

Figure 1. The procedure used to modify the surface of SiO_2 with Ti.

treated sample was pressed into a pellet, and then was characterized using photoelectron spectroscopy (ESCA3400, Shimadzu, Ltd.) and UV-vis diffuse reflectance spectroscopy (UV-3100PC, Shimadzu, Ltd.).

The photocatalytic activity of the treated sample was determined by characterizing the degradation of methylene blue, which can be decomposed by photogenerated oxidants. Treated samples (0.04 g) were added to 25 ml of methylene blue solution (2×10^{-5} M) in a pyrex flask and stirred with a magnetic stirrer. The absorbance at 2×10^{-5} M is about 1.35. The experiment was carried out in an apparatus capable of exposing the solution to UV irradiation. The inside of the apparatus is completely shielded from outside lights. The UV source consisted of two blacklight-blue fluorescent bulbs (6 W) with a peak wavelength at 352 nm. Measurements were made after waiting 12 hours to reach the adsorption equilibrium of methylene blue molecule. Control tests were completed with irradiated methylene blue solution in the presence of an untreated SiO_2 sample. Additional control tests also were completed on treated samples that were not irradiated (a sample in which the surface modification process was repeated four times). The treated sample was filtered out to separate it from the solution. The UV-vis absorption spectrum of the solution was then measured at 660 nm. Titanium Oxide (P 25, Nippon Aerosil, Ltd.) was used as a reference for both measurements.

RESULTS

Figure 2 shows the photodegradation of methylene blue in the presence of treated photocatalyst samples, and relative to control tests. The open square at zero time in Figure 2 indicates the light absorbance for the starting concentration of methlene blue in solution without powder. The difference between this value and the absorbance for each powder sample at zero time indicates the decrease in the concentration of the methylene blue in solution due to its physical adsorption onto the powder sample surface. The control test with the irradiated methylene blue solution in the presence of an untreated SiO_2 sample did not show a significant change in light absorbance. The other control test on the treated sample that was not UV irradiated shows a slight but insignificant decrease in light absorbance with time from 0.5 – 6 h (i.e., over the period of time the other test samples were irradiated). By comparison, the decrease in absorbance for the treated samples with irradiation time is much greater. In the presence of the UV irradiation, the concentration of methylene blue significantly decreases in all of the treated samples. The treated samples degraded the methylene blue in conjunction with the UV irradiation, which indicates the treated samples exhibit photocatalytic activity. In the treated samples that underwent repeated surface treatments to introduce more Ti sites, the degradation of the methylene blue is accelerated. The number of Ti atoms on the sample surface increases with every surface treatment, and the number of surface Ti atoms influence the photo degradation.

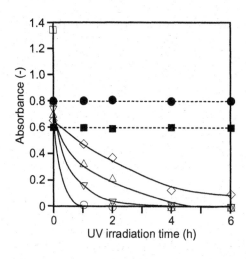

Figure 2. Photo degradation of methylene blue with UV irradiation. ($\Diamond$=1, $\triangle$=2, ∇=3, and $\bigcirc$=4. ■=untreated silica, ●=4 without irradiation.) "Untreated sample" is a result of the control test with the untreated sample. Sample 4 without irradiation" is a another control test with a surface treated sample on which the process was repeated four times in the absence of the irradiation. The symbol "$\square$" represents the absorbance of starting solution of methylene blue without powder.

The weight loss, measured by TGA-DTA (data not shown here), and the specific surface area (which did not change with surface hydroxylation) of the treated samples were used to determine the concentration of Ti atoms on the SiO_2 surface. For a single surface treatment, about 0.3 Ti atoms/nm^2 are introduced onto the sample surface. About 0.3 Ti atoms/nm^2 is the maximum concentration of Ti that can be introduced onto the sample surface. The increase in the concentration of the Ti atoms on the surface of the SiO_2 after the surface modification process is also observed in the X-ray photoelectron spectra (Figure3a), and in the UV-vis diffuse reflectance spectra for the treated samples (Figure 4).

After a single surface treatment, the Ti atoms on the treated sample surface are spaced at specific intervals controlled by the steric hindrance of the surface modification reagent. Thus a sample reacted only once must have absolutely no Ti-O-Ti structures. The surface binding state was investigated using X-ray photoelectron spectroscopy to identify the Si-O-Ti structures and the Ti-O-Ti structures. Figure 3 shows the X-ray photoelectron spectra for surface modified samples and an untreated SiO_2 sample. In the Ti2p spectra, two peaks appear for the treated samples. The two peaks shift to a higher binding energy (about 3eV) compared with the binding energy of 459eV for the TiO_2 reference sample. This binding energy is almost the same as that reported in previous work.[16,17] The intensity of the peaks gradually increases with the increasing number of surface treatment steps. The result proves that the number of Ti atoms on the sample surface increases. In the O1s spectra, the main peak is assigned to oxygen in the Si-O-Si structures on the sample. Compared to the O1s peak for the untreated SiO_2 sample, a shoulder is observed on the lower binding energy side of the O1s peak of the treated samples. The shoulder gets larger with the number of surface treatment steps. The shoulder is caused by the existence of oxygen in the Si-O-Ti

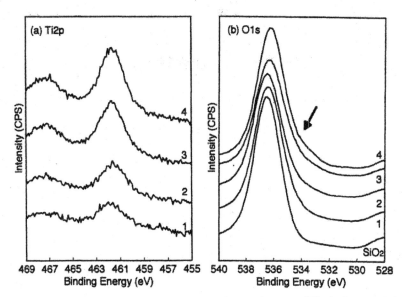

Figure 3. X-ray photoelectron spectra for surface modified SiO_2 samples and for untreated SiO_2: a) Ti2p spectra; and b) O1s spectra. The numbers in the figures indicate the number of times the hydroxylation process was repeated. An arrow marks the shoulder of the treated samples in the O1s spectra.

structure; the binding energy of oxygen in the Si-O-Ti structure is larger than in the Si-O-Si. The slight change in skewness shown by the arrow in   Figure 3b is consistent with a chemical shift between the Si-O-Ti structure and the Si-O-Si structure.[16] The Si-O-Ti structure is also consistent with the result of the XPS spectra studied in TiO_2 thin films prepared by the evaporation of Ti onto a SiO_2 substrate.[19] The intensity of the Si-O-Ti peak increases with an increase in the number of Ti atoms, but peaks for the Ti-O-Ti structures are not observed in the O1s spectra. The Ti-O-Ti structure in the O1s spectra is present at about 530 eV.[16,17] Thus, the treated samples have no Ti-O-Ti structures. However, although the treated samples have only Si-O-Ti structures, they still exhibit photocatalytic activity.

Figure 4 shows the UV-vis diffuse reflectance spectra for the surface modified SiO_2 samples, and for a reference sample of TiO_2. Figure 4b shows the band gap region of each spectrum. In Figure 4a, the light absorbance of the treated SiO_2 sample gradually increases with the number of surface treatment steps, indicating that the number of Ti atoms increases on the sample surface.[20] Compared to TiO_2, the treated samples have only a very few Ti atoms; however, they still show photocatalytic activity. In the band gap region of Figure 4b, band edges in the spectra shift to a longer wavelength with an increase in the number of Ti atoms.

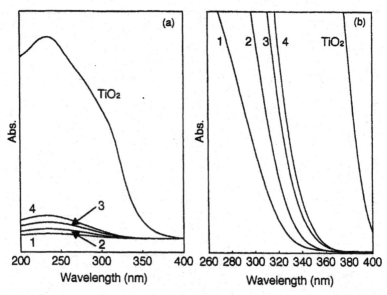

Figure 4. UV-vis diffuse reflectance spectra for surface modified SiO_2 samples, and for a TiO_2 reference standard: a) the peaks in the spectra; and b) the band gap region in the spectra. The numbers in the Figures indicate the number of times that the hydroxylation process was repeated.

The photocatalytic activity increases as the band edge of the treated samples gets closer to that of TiO_2.

DISCUSSION

The treated SiO_2 samples have a band gap as shown in the Figure 4b. Only the pairing of O and Ti in the Si-O-Ti structure could make a band structure capable of producing a hole and an electron with the observed excitation energy. When the Ti-O structure is excited by the UV irradiation, a hole and an electron will appear on the treated sample surface. The hole and the electron can attack the methylene blue causing it to decompose.[5,6] Increasing the number of Ti atoms increases the photocatalytic activity of the treated samples. Additionally, the band edge shifts to a longer wavelength, and the band gap becomes smaller. This is attributed to the increase in the concentration of the Si-O-Ti structures on the SiO_2 surface. Ti-O-Ti structures are not observed. It was found that even samples without the Ti-O-Ti structure exhibit photocatalytic activity. In the treated samples in this study, the Si-O-Ti structure plays an important role in determining where the band gap is located, and consequently, in determining the photocatalytic activity.

The properties of the SiO_2 surface may also contribute to the photocatalytic activity. In SiO_2/TiO_2 photocatalysts, photogene rated intermediates are sufficiently mobile to react with species adsorbed onto the SiO_2.[10] In our study, methylene blue appears to adsorb onto SiO_2, which may also be a factor in determining the photocatalytic activity of the treated samples.

CONCLUSIONS

A new method was developed to make a photocatalyst that involves putting Ti onto the surface of SiO_2. Photocatalytic activity is observed after a single surface treatment that produces only Si-O-Ti structures. As a result, we now have new knowledge that photocatalytic activity is observed without TiO_2 clusters. Treated samples have photocatalytic activity due to both the Si-O-Ti structure and due the ability of the SiO_2 surface to adsorb species.

REFERENCES

[1]A. Fujishima and K. Honda, "Electrochemical photolysis of water at a semiconductor electrode," *Nature*, **238** [5358] 37-8 (1972).

[2]K. Kosuge and P.S. Singh, "Synthesis of Ti-containing Porous Silica with High Photocatalytic Activity," *Chemistry Letters* [1] 9-10 (1999).

[3]H. Kominami, S.Y. Murakami, Y. Kera and B. Ohtani, "Titanium(IV) Oxide Photocatalyst of Ultra-High Activity: A New Preparation Process Allowing Compatibility of High Adsorptivity and Low Electron-Hole Recombination Probability," *Catalysis. Letters* **56** [2,3] 125-29 (1998).

[4]R. Wang, K. Hashimoto, A. Fujishima, M. Chikuni, E. Kojima, A. Kitamura, M. Shimohigoshi and T. Watanabe, " Light-induced amphiphilic surfaces," *Nature*, **388** [6641] 431-32 (1997).

[5]C.S. Turchi and D.F. Ollis, "Photocatalytic degradation of organic water contaminants: mechanisms involving hydroxyl radical attack," *J. Catalysis*, **122** [1] 178-92 (1990).

[6]E. Pelizzetti and C. Minero, "Mechanism of the Photooxidative Degradation of Organic Pollutants Over Titanium Dioxide Particles," *Electronchimica Acta* **38** [1] 47-55 (1993).

[7]G. Lu, A. Linsebigler, J.T. Yates, Jr., "Photooxidation of CH3Cl on TiO2(110): A Mechanism Not Involving H2O," *J. Phys. Chem.* **99** [19] 7626-31 (1995).

[8]C.C. Chuang, C.C. Chen and J.L. Lin, "Photochemistry of Methanol and Methoxy Groups Adsorbed on Powdered TiO2," *J. Phys. Chem. B* **103** [13] 2439-44 (1999).

[9]P.A. Mandelbaum, A.E. Regazzoni, M.A. Blesa and S.A. Bilmes, "Photo-Electro-Oxidation of Alcohols on Titanium Dioxide Thin Film Electrodes," *J. Phys. Chem. B* **103** [26] 5505-11 (1999).

[10]C. Anderson and A.J. Bard, "An Improved Photocatalyst of TiO2/SiO2 Prepared by a Sol-Gel Synthesis," *Journal of Physical Chemistry* **99** [24] 9882-5 (1995).

[11]C. Anderson and A.J. Bard, "Improved Photocatalytic Activity and Characterization of Mixed TiO2/SiO2 and TiO2/Al2O3 Materials," *Journal of*

Physical Chemistry B **101**[14] 2611-16 (1997).

[12]A.C. Van Dyk and A.M. Heyns, "Dispersion Stability and Photoactivity of Rutile (TiO_2) Powders," *J. Colloid and Interface Sci.* **206** [2] 381-91 (1998).

[13]M. Anpo, N. Aikawa, Y. Kubota, M. Che, C. Louis, and E. Giamello, "Photoluminescence and Photocatalytic Activity of Highly Dispersed Titanium Oxide Anchored Onto Porous Vycor Glass," *J. Phys. Chem.* **89** [23] 5017-21 (1985).

[14]J. Rabini and J. Stark, "Photocatalytic Dechlorination of Aqueous Carbon Tetrachloride Solutions in TiO2 Layer Systems: A Chain Reaction Mechanism," *J. Phys. Chem. B* **103** [40], 8524-31 (1999).

[15]M. Anpo, T. Shima, S. Kodama, Y. Kubokawa, "Photocatalytic Hydrogenation of Propyne with Water on Small-Particle Titania: Size Quantization Effects and Reaction Intermediates," *J. Phys. Chem.* **91** [16] 4305-10 (1987).

[16]V.M. Jimenez, G. Lassaletta, A. Fernandez, J.P. Espinos, and A.R. Gonzalez-Elipe, "Substrate Effects and Chemical State Plots for the XPS Analysis of Supported TiO_2 Catalysts," *Surface and Interface Analysis* **25** [4] 292-94 (1997).

[17]R. Wang, N. Sakai, A. Fujishima, T. Watanabe, and K. Hashimoto, "Studies of Surface Wettability Conversion on TiO2 Single-Crystal Surfaces," *J. Phys. Chem. B* **103** [12] 2188-94 (1999).

[18]H. Nasu, J. Heo, and D. Mackenzie, "XPS study of nonbridging oxygens in sodium oxide-silica (Na2O-SiO2) gels," *J. Non-Crystalline Solids* **99**[1] 140-50 (1988).

[19]G. Lassaletta, A. Fernandez, J.P. Espinos, and A.R. Gonzalez-Elipe, "Spectroscopic Characterization of Quantum-Sized TiO2 Supported on Silica: Influence of Size and TiO2-SiO2 Interface Composition," *J. Phys. Chem.* **99** [5] 1484-90 (1995).

[20]K. Kosuge and P.S. Singh, "Titanium-Containing Porous Silica Prepared by a Modified Sol-Gel Method," *J. Phys. Chem. B* **103**[18] 3563-69 (1999).

PARTICLE ORIENTED BISMUTH TITANATE CERAMICS PREPARED IN A MAGNETIC FIELD

A. Makiya, D. Kusano, S. Tanaka,
N. Uchida, and K. Uematsu
Nagaoka University of Technology
Kamitomioka, 1603-1
Nagaoka, 940-2188, Japan

T. Kimura
Tokyo Metropolitan University
1-1 Minami Ohsawa, Hachioji
Tokyo, 192-0397, Japan

K. Kitazawa
Japan Science and Technology
Corporation
4-1-8 Moto-machi, Kawaguchi
Saitama 332-0012, Japan

Y. Doshida
Taiyo Yuden Co., Ltd
6-16-20 Ueno, Taito-ku
Tokyo 110-0005, Japan

ABSTRACT

High density, highly oriented bismuth titanate ceramics were prepared by drying slurries in a high magnetic field, and sintering. X-ray diffraction shows that the b-axes of the bismuth titanate particles are aligned normal to the applied magnetic field in the green and sintered bodies. The degree of orientation in the sintered ceramic is 0.9 as determined using the Lotgering method. SEM micrographs also show the anisotropy in sintered bismuth titanate prepared in a magnetic field.

INTRODUCTION

Increasing concern for the environment makes the lead-free, bismuth titanate family very promising as next generation ferroelectric materials.[1-3] High performance comparable to PZT is expected[4], but bismuth titanate has highly anisotropic properties. The spontaneous polarization is about 50 $\mu C/cm^2$ at 25°C along the a-axis of the monoclinic crystal structure.[5] This high anisotropy means that special processing is required to attain useful properties in the bulk material. In application as a bulk electric material, the direction of high spontaneous polarization in the microstructure must be oriented parallel to the electric field.[3] This is very difficult to accomplish with conventional methods such as hot forging[3], uniaxial pressing[5] and doctor blade forming.[6-7] Recently, a novel method has been developed in which the interaction between a material and a magnetic field is used to orient particles in the green body.[8] Another very attractive merit of this technique is that regular powders having an equiaxial shape can be oriented under a magnetic field. Conventional methods require needle- or plate-shape particles to get orientation in the green body, which often makes it difficult to achieving high density on sintering.

When the magnetic field is very strong, i.e., super-conducting magnet, the interaction of a material with a magnetic field is appreciable, even for "non-magnetic" materials such as para- and dia-magnetic materials. The induced magnetization differs along the principle axes in a non-cubic material in a magnetic field. In para- and dia-magnetic materials, the axes of the largest magnetization and the smallest anti-magnetization tend to align along the magnetic field, respectively. Application of this principle to orient particles in superconductors[8] and alumina[9] has been well documented.

This paper applies the use of a magnetic field to develop high sintered density particle oriented bismuth titanate ceramics. More detailed research is in progress to apply this technique to develop grain-oriented microstructures for other materials in the bismuth titanate family as well, and to evaluate their ferro-electric properties.

EXPERIMENT

Powders of bismuth oxide (Soekawa-Rikagaku, Ltd) and titanium oxide (Ishihara Sangyo, Ltd) were used to synthesize the bismuth titanate powder. The appropriate amounts of the powders, distilled and deionized water, and a dispersant were mixed in a nylon ball mill with alumina balls for 24 h. After rapid drying, the mixture was pulverized with a mortar and pestle. This powder was placed in an alumina crucible and heated at $5^{o}C$/min to $900^{o}C$ and held for 1 h. The calcined powder was examined using x-ray diffraction analysis (MO3XHF22, Mac Science) with CuKα radiation at an acceleration voltage 40 kV, and with a current of 40 mA. A scan speed of 2 deg/min, and scan step 0.02° was used. The powder was also characterized using a scanning electron microscope (JEOL 5300LV).

The synthesized powder was placed in a ball mill with distilled and deionized water, dispersant, and alumina balls, and mixed for 24 h to prepare a slurry with a solids loading of 30 vol%. The flow characteristics of the slurry were characterized with a viscometer (VT550, HAAKE, Switzerland). For the test, the shear rate was increased from 0 to 250 s^{-1} in 3 minutes, and then returned to 0 in 3 minutes. After resting for 3 minutes, the flow characteristics of the slurry were measured by increasing and then decreasing the shear rate in 10 minutes over the same range of shear rates. The temperature of the slurry was kept constant at $20^{o}C$ during the measurements.

After passing through a mesh opening of 60 μm, the slurry (4-6 $x10^{-6}$ m^{3}) was placed in a shallow container made of TEFLON. The container was placed in a 10 T magnetic field in a super conducting magnet (TM-10VH10, Toshiba, Japan), and left for 2 days at the room temperature to dry. The magnetic field was applied in a direction horizontal to the container. For comparison, slurry was also dried without the applied magnetic field.

The resultant green compact was heated at $5^{o}C$/min to $1200^{o}C$ and held for 1 h, followed by cooling to the room temperature. The orientation of the particles in the compact was determined using x-ray diffraction analysis with the Lotgering method. Sintered density was measured using the Archimedes method, with distilled water as the immersion liquid. To examine the microstructure with SEM, the specimen was polished and thermally etched at $1150^{o}C$ for 1 h.

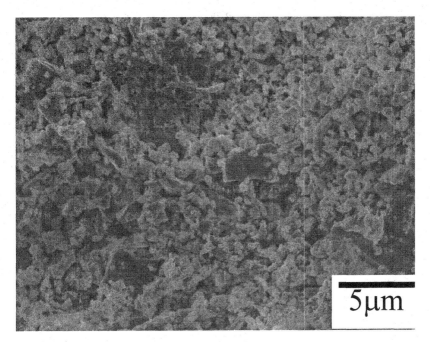

Figure 1. SEM micrograph of the bismuth titanate powder.

RESULTS AND DISCUSSION

Figure 1 shows an SEM micrograph of the bismuth titanate powder synthesized in this study. The particles are elongated, and the particle size varies widely. The smaller particles are nearly equiaxial in shape, but the larger particles tend to have a platelet shape. The major face of the large plate-like particles should be the *b* face of bismuth titanate. In single crystal bismuth titanate, the *b* axis of the lies perpendicular to the major face of the plate-like particles.[6]

Figure 2 shows the x-ray diffraction pattern for the powder synthesized in this study. The material is single phase $Bi_4Ti_3O_{12}$.

Figure 3 shows the rheological behavior of the bismuth titanate slurry prepared in this study. The condition required to produce the best dispersion was determined by changing the concentration of the dispersant. Figure 3 shows the characteristics of the lowest viscosity slurry prepared in this study. The scatter in the data is largely due to the low viscosity. The slurry shows Newtonian behavior. The yield stress is almost zero, and the shear stress increases linearly with increasing shear rate. No appreciable interaction is expected between the particles in this slurry. Each particle should be able to move and rotate freely in the applied magnetic field.

Figure 4 shows the x-ray diffraction pattern taken from the top surface of the sintered specimen; that is, the plane of analysis is in the direction parallel to the applied magnetic field for the specimen prepared in the magnet. In the specimen

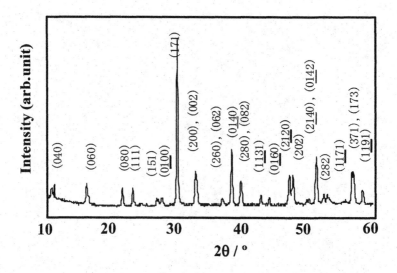

Figure 2. X-ray diffraction pattern for the synthesized bismuth titanate powder.

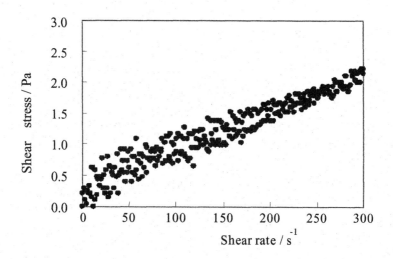

Figure 3. The shear stress-shear rate curves of the slurry

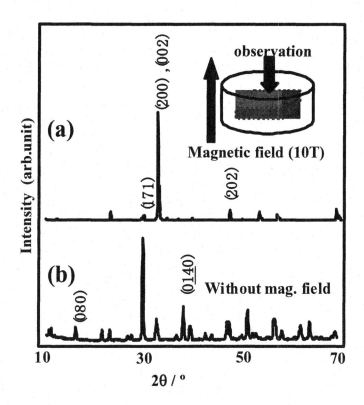

Figure 4 X-ray diffraction patterns of: a) the grain oriented ceramic surface normal to the applied magnetic field; and b) a randomly oriented sample prepared without magnetic field.

prepared in the high magnetic field, the strong diffraction peaks are those associated with the *c* planes of the crystal, such as (200) and (002) in the specimen prepared in the magnetic field (Figure 4a). The strongest peak in the raw powder (171) is almost absent from specimen prepared in the magnetic field. In the specimen prepared without the magnetic field, strong diffraction peaks also are observed for (0140) and (080).

Figure 5 shows microstructures of bismuth titanate compacts before firing. Figure 6 shows the sintered microstructure of the bismuth titanate parallel to and perpendicular to the magnetic field. The relative density after sintering is 98% of theoretical density. This high density was expected for a compact made from the fine powder. Clearly, different microstructures are evident in the two different directions, suggesting that the ceramic has an anisotropic structure. The majority of particles appear to have an elongated shape. Additionally, the particles are randomly oriented in the microstructure parallel to the applied magnetic field (Figure 6a). In contrast, elongated- and plate-shape particles are observed in the direction perpendicular to the applied magnetic field (Figure 6b). The elongated grains tend to orient with their longest axis parallel to the direction of applied magnetic field.

X-ray results and the anisotropic microstructure suggest that the *b* axes of particles are aligned normal to the applied magnetic field. The degree of orientation in the *a* and/or *c* planes is calculated to be 0.9 by the Lotgering

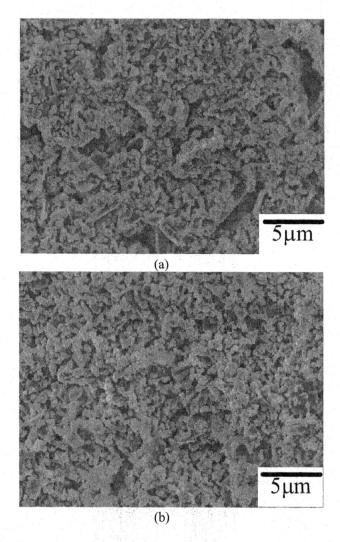

Figure 5. Microstructure and particle orientation in parallel to the magnetic field for a specimen prepared at: a) 10 T; and b) 0 T.

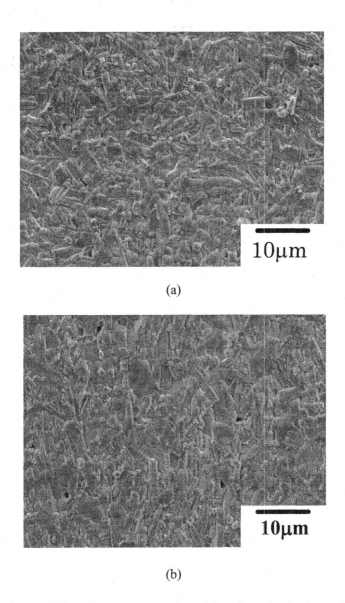

(a)

(b)

Figure 6. The microstructure and particle orientation in sintered bismuth titanate: a) parallel to the magnetic field; and b) perpendicular to the field.

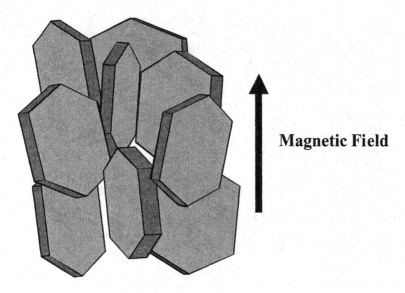

Magnetic Field

Figure 7. Schematic of the oriented particle structure in the bismuth titanate.

method.   Bismuth titanate has a tendency to form a plate shape, with the *b* plane being the largest face of the grain.   The grains appear elongated when viewed from the direction of the *a* and *c* axes.   They appear plate-shaped when viewed from the *b* axis.   Elongated and plate shape particles are observed in the direction perpendicular to the applied magnetic field, in which the direction of the *b* plane is randomly oriented.

Figure 7 shows a schematic of the particle orientation in sintered bismuth titatate ceramic prepared in a magnetic field.   The *a* and *c* axes of the crystals are aligned along the applied magnetic field.   Within the plane normal to the applied magnetic field, however, the *b* axes are randomly oriented.

The detailed mechanism responsible for particle orientation in bismuth titanate still needs to be determined.   There are two possibilities: 1) the material is diamagnetic, and the magnetic susceptibility is the smallest in the *b* axis; or 2) the material is paramagnetic, and the susceptibility is the largest in the *b* axis.

CONCLUSION

Dense, particle oriented bismuth titanate ceramics have been produced using a novel processing technique involving the use of a high magnetic field. The *b* axes of the bismuth titanate powder particles are oriented perpendicular to the applied magnetic field in the green compact.   After sintering to 98% density, the microstructure is highly anisotropic, with a high degree of orientation (0.9).

REFERENCES
1.   T. Takenaka, "Grain Orientation Effects on Electrical Properties of Bismuth Layer-structured Ferroelectric Ceramics", *J. Ceram. Soc. Japan*, **110**

[4] 215-224 (2002).

2. S. Ikegami and I. Ueda, "Piezoelectricity in Ceramics of Ferroelectric Bismuth Compound with Layer Structure," *Jpn. J. Appl. Phys.*, **13** [10] 1572-77 (1974).

3. T. Takenaka and K. Sakata, "Grain Orientation and Electrical Properties of Hot-forged $Bi_4Ti_3O_{12}$ ceramics," *Jpn. J. Appl. Phys.*, **19** [1] 31-39 (1980).

4. I.-S. Yi and M. Miyayama,"Electrical Properties of Layer-Strkuctured $Pb_2Bi_4Ti_5O_{18}$ Single Crystals," *J. Ceram. Soc., Japan,* **106** [3] 285-89 (1998).

5. S.E. Cummins and L.E. Cross, "Electrical and Optical Properties of Ferroelectric $Bi_4Ti_3O_{12}$ Single Crystals," *J. Appl. Phys.* **39** [5] 2268-74 (1968),

6. Y. Inoue, T. Kimura and T. Yamaguchi,"Sintering of Plate-Like $Bi_4Ti_3O_{12}$ Powders," *Am. Ceram. Soc. Bull.*, **62**[6]704-707,711(1983).

7. H. Watanaba, T. Kimura, and T. Yamaguchi,"Sintering of Platelike Bismuth Titanate Powder Compacts with Preferred Orientation", *J. Am. Ceram. Soc.*, **74** [1] 139-47 (1991).

8. M.M. Seabaugh, I.H. Kerscht and G.L. Messing, "Texture Development by Templated Grain Growth in Liquid-Phase-Sintered α-Alumina", *J. Am. Ceram. Soc.*, **80** [5] 1181-88 (1997).

9. S.C. Peterson and M.J. Cima, "Magnetic Inducement of Texture in $Ba2YCu3O6.9$ Particle Assemblies under Cryogenic Conditions, "*J. Am. Ceram. Soc.*, 71 [11] C458-C459 (1988).

10. T.S. Suzuki, Y. Sakka, and K. Kitazawa, "Preferred Orientation of the Texture in the SiC Whisker-dispersed Al2O3 Ceramics by Slip Casting in a High Magnetic Field," *J. Ceram. Soc. of Japan*, **109** [10] 886-90 (2001) [in Japanese].

FORMATION OF NANOSTRUCTURE COMPOSITES USING ADVANCED MECHANICAL PROCESING

D. Tahara, Y. Itoh, and T. Ohmura
Hamamatsu Research Center,
NICHIAS Corporation, Hamamatsu
Shizuoka 431-2103, Japan

H. Abe and M. Naito
Joining and Welding Research
Institute
Osaka University, Ibaraki
Osaka 573-1132, Japan

ABSTRACT

Nanostructure composites were successfully fabricated using an advanced mechanical processing method in a dry ambient temperature environment. Glass fibers were mechanically processed with fumed silica powder to produce a nanoporous fiber composite consisting of fumed silica on glass fibers. The nanoporous fiber composite is easily formed into larger, bulk ceramic bodies by dry pressing. The advanced mechanical processing method can easily be used to produce nano-scale composite particles for advanced materials.

INTRODUCTION

Recently, there has been increasing interest in nanostructured materials. The hope is that their properties can be superior to conventional materials that have a coarser phase or grain structure. The availability of a novel class of materials with properties mainly determined by their tremendously high surface to volume ratio gives nanomaterials the potential to impact applications ranging from electrochemical reactors[1] to ceramic nano-composites with significantly improved properties.[2]

Nanoporous structures have many different applications.[3] For instance, silica aerogels have been intensively investigated due to the unique properties resulting from their nanoporosity. Silica aerogels are often fabricated by the sol-gel method. This synthesis route, however, has practical disadvantages such as the inability to produce thick or large bodies. Although alternatives to sol-gel processing have been pursued, most are based on wet processing, which involves drying and the associated high energy costs.[4]

Here, we present a novel approach to produce nanoporous materials in a dry ambient temperature environment (Figure 1). Nanostructured porous composites consisting of fumed silica and glass fiber composite particles were successfully formed using an advanced mechanical processing method. Then, large bulk ceramic bodies were fabricated by dry-pressing the nanoporous composites. In this paper, we discuss this novel method to make bulk nanoporous materials. Additionally, the properties of the nanoporous composites and the resultant bulk ceramic body are evaluated.

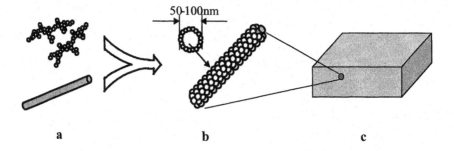

50-100nm

a　　　　　　　　b　　　　　　　　c

Figure 1. Schematic of a dry processing method to fabricate nanoporous materials: a) mix raw materials; b) mechanically process to coat glass fiber with fumed silica; and c) dry press nanoporus composites from (b) to produce bulk ceramic bodies

EXPERIMENTS

The starting materials used in this study include fumed silica powder and glass fibers. Figure 2 a shows a TEM image of the fumed silica powder. It has a three-dimensional nanoscale chain-like agglomerated structure with 50 to 100 nm pore spaces. Its BET specific surface area is 300 m^2/g, which corresponds to an average particle diameter of about 10 nm. Figure 2b shows an SEM image of the glass fibers. Their diameter and length are 10 µm and 6 mm, respectively. The as- received glass fibers were coated with a polymeric binder, so dewaxing was performed at 673 K for 36 h prior to mechanical processing.

The powder mixture consisted of 80 wt% fumed silica powder and 20 wt% glass fibers. This mixture was mechanically processed using an advanced mechanical method. An attrition type device was used to mix and combine the raw materials (Mechanofusion System[5], Hosokawa Micron Corp., Japan), using a rotation speed of at 1000 rpm for 5 min in a dry ambient temperature environment. The processed powder mixture was removed from the mixing

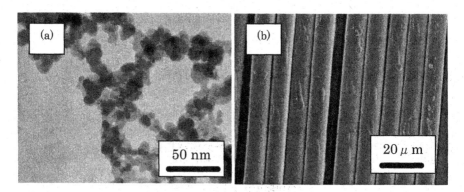

(a)

(b)

50 nm

20 µ m

Figure 2. The raw materials used to form the nanoporous composites: a) TEM image of fumed silica powder; b) SEM image of glass fibers.

device for analysis. and subsequently dry pressed. BET specific surface area was used to characterize the resultant nanoporous composites. Their morphology was characterized using SEM. The processed powder mixture was uniaxially pressed at 0.7 MPa in a square die (100×150 mm) to make bulk ceramic bodies. The apparent density of the bulk bodies was calculated from their volume and weight. Three-point fracture strength was measured on 15 × 30 × 100 mm specimens with an inner span length of 80 mm. Testing was completed using a crosshead speed of 10mm/min.

RESULTS AND DISCUSSION

Figure 3 shows SEM micrographs of the resultant powder mixture. The processed nanoporous composite consists of a glass fiber coated with fumed silica. The thickness of the fumed silica layer on the surface of glass fiber is about 2 μm. A significant difference in BET specific surface area is measured between the powder mixture before and after the mechanical processing (Table I). The lower surface area after mechanical processing is believed to be due to the fine particles of fumed silica adhering onto the core glass fiber surface.

Figure 4 shows a photograph of a uniaxially pressed green compact (150×100×15 mm) formed using the nanoporous composites fibers. The apparent density of the compact is 347 kg · m^{-3}, and the three-point bending strength is 0.22 MPa.

The present study demonstrates that the mechanical processing method can produce nanoporous composite fibers from fumed silica powder and glass fibers without any binders. It is believed that the thickness of fumed silica particles can be adjusted by controlling the mass ratio of silica powder in the formulation.

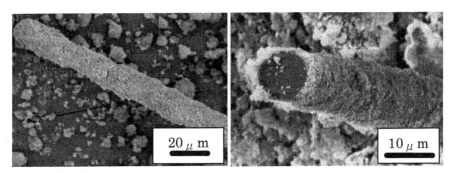

Figure 3. SEM micrographs of the composite particles consisting of fumed silica coating glass fiber.

Table I. Effect of mechanical processing on the specific surface area of the fumed silica/glass fiber mixture.

Sample	BET specific surface area [m^2/g]
Powder mixture (before processed)	247
Powder mixture (after processed)	183

Figure 4. Photomicrograph of a nanoporous plate fabricated by dry pressing nanoporous composite fibers (scale is in cm)

Mechanical processing has advantages over more conventional processes in the production of nanocomposites and engineered particles. With this new processing method, using mechanical energy without any binders, and in a dry environment can control the nanostructure of materials.

One potential issue is the fracture of the glass fibers during mechanical processing. The average length of the resultant nanoporous fibers is about 1 mm in this study, which is considerably shorter than the original fiber length of 6 mm. We are currently developing softer bonding technology to produce longer nanoporous fiber composites.

CONCLUSION

Nanoporous composites were produced from fumed silica powder and glass fibers using an ambient temperature advanced mechanical processing method without any binders. During mechanical processing, the fumed silica powder coats the glass fibers to form nanoporous composites. Large bulk ceramic bodies can be formed from the resultant nanoporous composite fibers. Die pressing easily produced a thick plate with a nanoporous structure of ≤100 nm. It has been demonstrated that advanced mechanical processing is a promising approach to produce bulk ceramic bodies with a nano-scale structure.

ACKNOWLEDGEMENTS

This work was partially supported by The Thermal & Electric Technology Foundation.

REFERENCES

[1] T. Fukui, S. Ohara, and K. Mukai, "Long-Term Stability of Ni-YSZ Anode with a New Microstructure Prepared from Composite Powder," *Electrochemical and Solid-State Letters*, **1** [3] 120-22 (1998).

[2] B. Kim, T. Sekino, T. Nakayama, T. Kusunose, J. Lee, and K. Niihara,

"Mechanical and Magnetic Properties of Alumina/Nickel Nanocomposites Prepared by Pulse Electric Current Sintering", *J. Ceram. Soc. Japan,* **111** [7], 457-460 (2003).

[3] J. Fricke, T. Tillotson, "Aerogels: Production, Characterization, and Applications," *Thin Solid Films* **297** 212-223 (1997).

[4] Y. Kwon, S. Choi, "Ambient-Dried Silica Aerogel Doped with TiO2 Powder for Thermal Insulation," *J. Mater Sci.* **35** 6075-6079 (2000).

[5] M. Naito, A. Kondo and T. Yokoyama, "Applications of Comminution Techniques for the Surface Modification of Powder Materials," *ISIJ International*, Vol. 33, No. 9, 915-924 (1993).

SYNTHESIS OF AlN BY GAS-REDUCTION-NITRIDATION OF TRANSITION ALUMINA POWDER

Tomohiro Yamakawa, Junichi Tatami, Takeshi Meguro, and Katsutoshi Komeya
Yokohama National University
79-7 Tokiwadai Hodogayaku
Yokohama City, 240-8501 Japan

ABSTRACT

Aluminum nitride (AlN) powder was synthesized by gas-reduction-nitridation of γ-Al(OH)$_3$ and γ-Al$_2$O$_3$ powder using a mixture of NH$_3$ and C$_3$H$_8$ gases. Nothing but AlN was identified in powders processed at 1100-1400°C for 120 min in the NH$_3$-C$_3$H$_8$ gas. Although the specific surface area of the product powders decrease with increasing firing temperature and soak time, they were much higher than the surface areas of commercial AlN powders. The reason why such fine grain AlN powders are obtained from γ-Al$_2$O$_3$ and Al(OH)$_3$ is believed to be due to the higher surface area of transition alumina precursor powders.

INTRODUCTION

Aluminum nitride (AlN) has attracted much interest in recent years, particularly in the electronics industry as a substrate material. AlN has high intrinsic thermal conductivity ($\sim$320Wm^{-1}K^{-1}), high electrical insulation resistance, and has a thermal expansion coefficient close to that of silicon.[1,2] Some researchers have reported on various investigations to synthesize AlN powders from aluminum hydroxide (Al(OH)$_3$) and transition alumina. These studies concluded that the nitridation reaction rate is fast [3-5], and the resultant AlN has low oxygen content [6] because of the high reactivity of transition alumina or Al(OH)$_3$. More recently, it was determined that the nitridation reaction occurs at a lower temperature in the Al$_2$O$_3$-NH$_3$-C$_3$H$_8$ system in comparison to the conventional Al$_2$O$_3$-N$_2$-C system.[7-9] The purpose of this study is to synthesize AlN powder by firing transition alumina powder in a mixture of NH$_3$-C$_3$H$_8$ gas, to determine the phases present after nitridation, and to determine the size and morphology of the powder after nitridation.

EXPERIMENTAL

In this study, commercial high-purity γ-Al(OH)$_3$ (Powder A) and γ-Al$_2$O$_3$ (Powder B) were used. Table I shows the characteristics of the

Table I.   Characteristics of the Raw Al_2O_3 Powder

Characteristic	Powder (A)	Powder (B)
Purity (%)	>99.6	>99.99
Crystalline phase	γ-Al(OH)$_3$	γ-Al$_2$O$_3$
Specific Surface area (m^2/g)	6	145
XRD crystallite size (nm) †	51	5
Transition to α-Al$_2$O$_3$ ($^\circ$C) ‡	1293	1281

† Calculated from the γ-Al(OH)$_3$(002) peak and γ-Al$_2$O$_3$(440) peak.
‡ Determined by differential thermal analysis.

powders. The powders were spread onto the surface of a high-purity alumina boat (99.999%) and placed in alumina tube electric furnace (inner diameter of 42 mm). After filing with argon gas, the reactor was heated at 5°C/min from room temperature to 700°C. At 700°C, a gas mixture of NH$_3$ (99.999% purity) and C$_3$H$_8$(99.99% purity) with a molar ratio of C$_3$H$_8$/ NH$_3$ = 5×10^{-3} was flowed through the alumina tube at 4 L/min. The furnace was subsequently heated to 1100-1400°C at 8°C/min. After the gas-reduction nitridation process, the samples cooled naturally in NH$_3$. Composition phases were identified by X-ray diffraction (XRD; RINT 2500 Rigaku, Tokyo, Japan), microstructures were observed using a scanning electronic microscope (SEM; JSM-5200, JEOL, Tokyo Japan), and specific surface area was measured using the BET single-point method (Quantasorb, Quantachrome Boynton Beach, FL). The nitridation rate was calculated from the change in weight with time measured using a thermogravimetric analysis system (TGA; TG-8120, Rigaku).

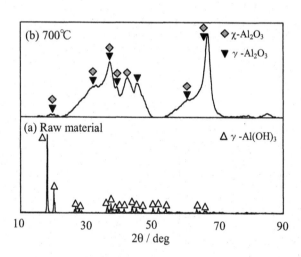

Figure 1. XRD profiles of the: a) raw γ-Al(OH)$_3$ powder; and   b) the powder synthesized at 700°C in Ar after 0 min.

RESULTS AND DISCUSSION

Figure 1 shows the XRD profiles for the raw Powder (A), and for the powder produced after firing at 700°C for 0 min in Ar. This is the preliminary stage of the reduction-nitridation reaction, and the product generated is a mixture of γ and χ -Al_2O_3 (Figure1b). The specific surface area of the product powder is at $330m^2g^{-1}$. The γ-Al(OH)$_3$ is converted to a transition alumina by the combined effect of the heat treatment accompanied by dehydration.[10]   As such, the high specific surface area may be due to the blowholes formed by dehydration.

Figure 2 shows XRD profiles for raw Powder (B) and for the powder produced after firing at 700°C for 0 min in Ar. Only single-phase γ -Al_2O_3 can be seen after the heat treatment, and the specific surface area is also nearly the same as that of the starting raw material.

Figure 3 shows the XRD profiles for the products obtained after firing Powder (A) and Powder (B) at 1100 to 1400°C for 120 min. After firing at temperatures at or above 1200°C the product powder from Powder (A) is single phase AlN; however some of the original γ-Al_2O_3 remains after firing at 1100°C. γ-Al_2O_3 is also detected in the XRD profile of the product powder obtained from Powder (B) after firing at 1100°C. This result demonstrates the similar nature of the nitridation reaction with theγ-Al(OH)$_3$ and the γ-Al_2O_3 starting raw material.

Figure 4 shows the percent of Powder (A) and Power (B) that is converted to AlN after firing at 1100 to 1400°C. Consistent with the XRD results, the nitridation rate increases significantly as the reaction temperature increases. For both raw powders, 80% of the starting powder is converted to AlN after firing at 1300°C. Complete (100%) conversion to AlN was not achieved with either of the raw powders after 120 min., even after firing at high temperature. This may be due to the formation of Al_2O_3 during a reaction between the high surface area AlN and the oxygen present in the air. [11]

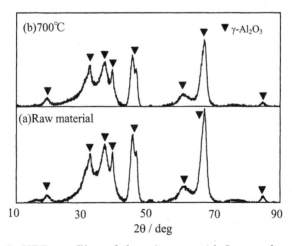

Figure 2. XRD profiles of the; a) raw γ-Al_2O_3 powder; and b) the powder synthesized at 700°C in Ar after for 0 min.

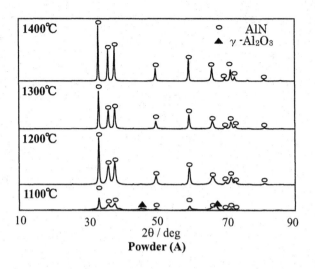

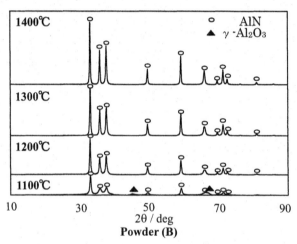

Figure 3. XRD profiles of the product powders obtained from Powder (A) (γ-Al(OH)$_3$) and from Powder (B) (γ-Al(OH)$_3$) after firing at 1100-1400°C for 120 min.

Figure 5 shows SEM photographs of Powder (A) after firing at 1100 to1400°C for 120 min. The product powder is comprised of plate-like grains that are approximately 500 nm in size. With increasing firing temperature, finer grains can be observed on the surface of the larger grains. Grain cohesion and grain growth are observed in the final product obtained after firing at 1400°C for 120 min.

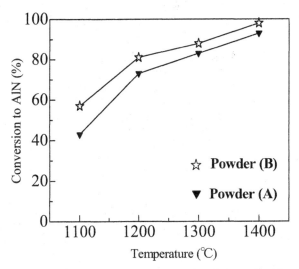

Figure 4. The conversion of Powder (A) and of Power (B) to AlN during firing at 1100 -1400°C for 120min.

500nm

Figure 5. SEM photographs of the product powders synthesized from Powder (A) after firing for 120 min. at: a) 1100°C; b) 1200°C; c) 1300°C; and d) 1400°C.

Figure 6 shows SEM photographs of Powder (B) after firing at 1100 to 1400°C for 120 min. Again, grain growth is observed with increasing temperature.

Figure 7 shows the specific surface area of the product powder obtained by firing Powder (A) and Powder (B) at 1100 to 1400°C for 120 min. The surface area of the products synthesized from Powder (A) fired at 1400°C is 12.8 m^2g^{-1}. Comparatively, the product of powder (B) fired at 1300°C has a very large surface area of 28.0 m^2g^{-1}. The powder specific surface area decreases as the firing temperature increase. This is most likely a result of sintering between primary grains and grain growth during nitridation.

The surface area of the AlN powder produced in this study is very high compared to commercial Al_2O_3 and AlN powder. High specific surface areas was achieved by nitridation at a low temperature.

The above results demonstrate that the gas-reduction-nitridation of transition Al_2O_3 using γ-Al(OH)$_3$ and γ-Al$_2O_3$ facilitates nitridation at a very low temperature to synthesize fine AlN powder with a high specific surface area.

500nm

Figure 6. SEM photographs of the product powders synthesized from Powder (B) after firing for 120 min. at: a) 1100°C; b) 1200°C; c) 1300°C; and d) 1400°C.

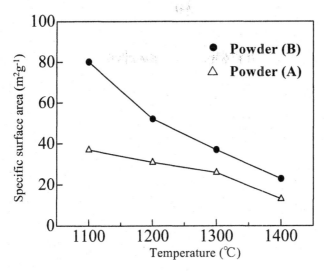

Figure 7. Specific surface area of the product powder synthesized from Powder (A) and Powder (B) after nitridation at 1100-1400°C.

CONCLUSION

Gas-reduction-nitridation of γ-Al$_2$O$_3$ and γ-Al(OH)$_3$ powder in a mixture NH$_3$-C$_3$H$_8$ gas enabled us to synthesize nearly single-phase AlN powder at 1200°C. SEM and TEM analysis confirm that nitridation starts at 1100°C. BET measurements of the specific surface area of the product powder reveal that gas-reduction-nitridation of γ-Al$_2$O$_3$ and γ-Al(OH)$_3$ powder produces is a very fine grain size AlN powder with a high specific surface area. Thus, we determined that gas-reduction-nitridation of transition Al$_2$O$_3$ is an effective means of synthesizing AlN.

REFERENCES

[1]L.M. Sheppard, "Aluminum Nitride: A Versatile but Challenging Material," *Bull. Am. Ceram. Soc.*, **69**[11] 1801-12 (1990).

[2]Y. Baik, R.A.L. Droew, "Aluminum Nitride: Processing and Applications," *Key Engineering Materials*, **553** 122-24 (1996).

[3]Y.W. Cho and J.A. Charles, "Synthesis of Nitrogen Ceramic Powders by Carbothermal Reduction and Nitridation, Part 3 Aluminum Nitride," *Materials Science Technology*, **7** 495-504 (1991).

[4]W.H. Tseng and C.I. Lin, "Carbothermal Reduction and Nitridation of Aluminum Hydroxide," *J. Mater. Sci.*, **31** 3559-65 (1996).

[5]M. Mitomo and Y. Yoshioka, "Preparation of Si$_3$N$_4$ and AlN Powders from Alkoxide-Derived Oxides by Carbothermal Reduction and Nitridation," *Advanced Ceramic Materials*, **2**[3A] 253-56 (1987).

[6]A. Tuge, H. Inoue, M. Kasori and K. Shinozaki, "Raw Material Effect on AlN Powder Synthesis from Al$_2$O$_3$ Carbothermal Reduction," *J. Mater. Sci.*, **25** 2359-61 (1990).

[7]T. Suehiro, J. Tatami, T. Ide, T. Meguro, S. Matsuo and K. Komeya, "Synthesis of Spherical AlN Particles by Gas-Reduction-Nitridation Method," *J. European Ceram. Soc.*, **22** 521-26 (2002).

[8]T. Suehiro, J. Tatami, T. Meguro, K. Komeya and S. Matsuo, "Aluminum Nitride Fibers Synthesized from Alumina Fibers Using Gas-Reduction-Nitridation Method" *J. Amer. Ceram. Soc.*, **85**[3] 715-17 (2002).

[9]T. Suehiro, N. Hirosaki, R. Terao, J. Tatami, T. Meguro and K. Komeya, "Synthesis of Aluminum Nitride Nanopowder by Gas-Reduction-Nitridation Method" *J. Amer. Cer. Soc.*," **86**[6] 1046-48 (2003).

[10]K. Wefers "Nomenclature, Preparation, and Properties of Aluminum Oxides, Oxide Hydroxides, and Trihydroxides"; pp13-22 in *Alumina chemicals*, L.D. Hart. Editor, The American Ceramic Society Inc, Westerville, Ohio, 1990.

[11]E. Ponthieu, P. Grange, B. Delmon, L. Lonnoy, L. Leclercq, R. Bechara and J. Grimblot, "Proposal of a Composition Model for Commercial AlN Powders," *J. European Ceram. Soc.*, **8** 233-41 (1991).

EFFECTS OF α PHASE CONTENT ON THE SINTERING BEHAVIOR OF Si_3N_4 POWDER

Ryoichi Nishimizu, Junichi Tatami,
Katsutoshi Komeya, and Takeshi Meguro
Yokohama National University
Hodogayaku
Yokohama 240-8501, Japan

Masahiro Ibukiyama
Denki Kagaku Kogyo Corporation
Machida 194-8560, Japan

ABSTRACT

The sintering behavior of Si_3N_4 powders with different concentrations of α-phase, Y_2O_3, and Al_2O_3 sintering aids was studied. Shrinkage during sintering was measured directly using a dilatometer within an electric furnace. The shrinkage on sintering to 1800°C, where the α to β phase transformation is complete, was greater in the Si_3N_4 powders with the highest α content. This indicates that the α to β phase transformation in Si_3N_4 influences shrinkage during sintering.

INTRODUCTION

Silicon nitride has excellent mechanical characteristics such as high strength, high toughness, high hardness, and abrasion resistance.[1-3] However, it is difficult to densify by solid-state sintering. Therefore, liquid-phase sintering, using sintering aids such as Y_2O_3 and Al_2O_3 is commonly employed.[1-4] Excellent mechanical characteristics of silicon nitride are attributed to its sintered body texture, a composite texture where prismatic grains are entangled together.[1-3] The texture varies with the Si_3N_4 powder, sintering aids, and sintering conditions.[5-8] Controlling texture is important for improving the mechanical characteristics of silicon nitride, and the influence of processing factors on mechanical properties has been examined in past studies.[1-8]

Recently, studies have been completed to improve thermal conductivity in Si_3N_4 with excellent mechanical characteristics.[9] A major result of these studies is that high thermal conductivity can be obtained by firing for a long time, through the alignment of the elongated grains, and through the use of appropriate additives. To determine the mechanisms to improve each characteristic, it is necessary to understand the sintering mechanism for the liquid-phase sintering of silicon nitride.

Liquid-phase sintering of silicon nitride involves grain rearrangement, solution-precipitation, and grain growth in which viscous flow, the α to β phase transformation, and Ostwald growth are occurring simultaneously. The effects of the α to β phase transformation on sintering behavior, has yet to be determined, and is the focus of this study.

There are two methods to evaluate sintering shrinkage. One is to measure the dimensions of the sintered body after firing. The other is to measure the dimensions of the specimen in real time during firing. The former enables one to evaluate the characteristics of the sintered body at each temperature, but a large number of specimens and a long time are required to complete a study. The latter can determine the shrinkage ratio and the shrinkage rate at any time and temperature during sintering. Although special equipment is needed, one can get a continuous shrinkage curve from a single sample. Therefore, in-situ shrinkage measurements are useful to characterize sintering behavior in detail.

The objective of this study was to determine how the α to β phase transformation affects silicon nitride sintering by evaluating the in-situ sintering shrinkage of silicon nitride powers with different concentrations of the α phase.

EXPERIMENTAL

Five Si_3N_4 powders with different α contents, and with Y_2O_3 and Al_2O_3 sintering aids were studied. The powders are designated A-E. The characteristics of the five Si_3N_4 powders are shown in Table I. Si_3N_4, Y_2O_3, and Al_2O_3 were mixed such that their final ratios were 93:5:2 by weight. These powder mixtures were ball-milled with 2 mass% of dispersant in ethanol for 96 h. After adding the binder and drying, the powder mixture was uniaxially molded at 50 MPa for 15 s. The pressed parts were then cold isostatic press (CIP) molded at 200 MPa for 1 min. The organic processing aids were removed by heating in air at 500°C for 3 h. The bodies were fired at 1800°C for 2 h in 0.9 MPa nitrogen. Sintering shrinkage was measured in-situ using a dilatometer. The density of the sintered body was measured using the Archimedes' method. The crystalline phases were identified using x-ray diffraction (XRD), and the microstructure was evaluated using scanning electron microscopy (SEM).

RESULTS AND DISCUSSION

Figures 1-5 show the sintering shrinkage of Si_3N_4 powders A-E. A comparison of all of the sintering data confirms that the different Si_3N_4 powders exhibit different sintering behavior. In Figure 1, the shrinkage of powder A starts at around 1400°C, and rapid shrinkage is observed at temperatures up to 1800°C. Then, after a 30 min pause during the hold at 1800°C, some additional shrinkage occurs.

In Figure 2, rapid shrinkage is observed for powder B at temperatures up to 1800°C, and there was no pause in shrinkage at 1800°C. The observed shrinkage in powder B does not show the pause reported by Koga during the α to β phase transformation[10]. The XRD spectra for powder B after sintering at 1800°C

Table I. Properties of the raw Si_3N_4 powders examined in this study

	A	B	C	D	E
α content (%)	98.0	91.2	40.8	28.4	3.0
Average particle size (μm)	0.52	0.80	0.52	0.53	0.62
Specific surface area (m^2/g)	11.5	10.0	14.8	15.0	13.2
Aluminum impurity (ppm)	>0.001	500	180	90	420
Oxygen impurity (ppm)	1.19	0.80	0.81	0.71	0.90

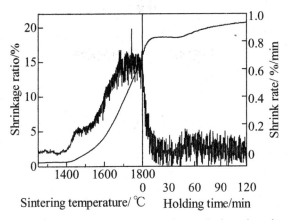

Figure 1. Shrinkage of Si_3N_4 powder A during sintering

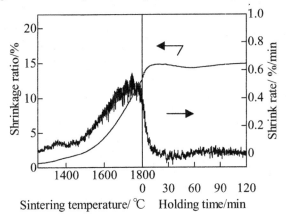

Figure 2. Shrinkage of Si_3N_4 powder B during sintering

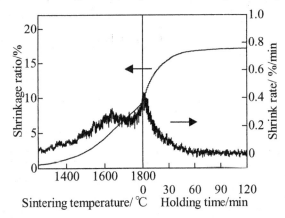

Figure 3. Shrinkage of Si_3N_4 powder C during sintering

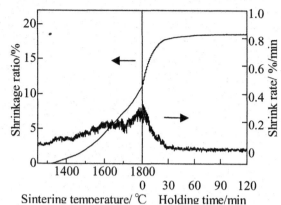

Figure 4. Shrinkage of Si₃N₄ powder D during sintering

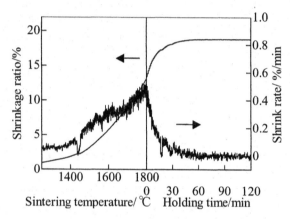

Figure 5. Shrinkage of Si₃N₄ powder E during sintering

for 0-2 h (Figure 6) confirms that the α to β phase transformation is complete after 2 h at 1800°C, and is almost complete after 0 h at 1800°C. Therefore, the rapid shrinkage up to 1800°C seems to be influenced by the α to β phase transformation. Shrinkage at temperatures over 1800°C, including pauses, is not influenced by the α to β phase transformation.

For specimens with a relatively low α content (Figures 3-5), the rapid shrinkage noted for powders A and B is not observed at temperatures up to 1800°C, and at temperatures over 1800°C, shrinkage proceeds without pause.

To investigate relationships between shrinkage and microstructure, we examined a plasma-etched face of the sintered bodies using SEM (Figure 7). The most grain growth is observed for powder B, but the other powders have similar structures after sintering. Therefore, we are unable to clearly show any relationship between microstructure and shrinkage.

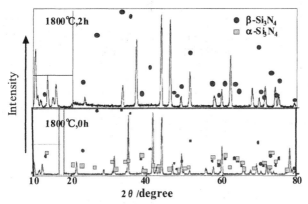

Figure 6. XRD patterns for powder A sintered at 1800°C for 0 and 2 h.

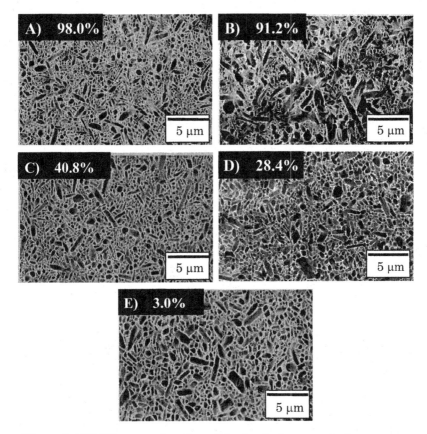

Figure 7. SEM Photographs of a polished and plasma-etched surface of Si_3N_4 A-E after sintering at 1800°C for 2 h.

The results show that rapid shrinkage occurs around 1700°C, and shrinkage pauses at temperatures over 1800°C for Si_3N_4 with a high α content. The shrinkage rate around 1700°C is slower for Si_3N_4 with a relatively low α content, and shrinkage occurs without pause at temperatures over 1800°C.

We cannot determine the specific reason for the differences in sintering behavior from the results of this study. The changes in sintering behavior may be related to changes in texture and microstructure during the sintering process. Additionally shrinkage behavior may differ significantly depending on the properties of the liquid phase, such as solubility and viscosity. We intend to continue our studies to consider such points.

SUMMARY

Rapid shrinkage was identified at temperatures up to 1800°C in Si_3N_4 powder with a high α content. We believe that this is due to the influence of the α to β phase transition because this phase transition is virtually complete upon reaching 1800°C. In powders with a high α content, pauses in shrinkage were observed at temperatures over 1800°C. The reason for this pause is not clear, but we speculate that it may be caused by changes in the microstructure or in the liquid phase during the sintering process.

REFERENCES

1.   A. Tsuge, K. Nishida, and M. Komatsu, "Effect of Crystallizing the Grain Boundary Glass Phase on the High-Temperature Strength of Hot-Pressed Si_3N_4, Containing Y_2O_3," *J. Am. Ceram. Soc.*, **58** [7– 8] 323–26 (1975).
2.   A. Tsuge and K. Nishida, "High-Strength Hot-Pressed Si3N4 with Concurrent Y_2O_3 and Al_2O_3 Additions," *Am. Ceram. Soc. Bull.*, **57** [4] 424–31 (1978).
3.   F.F. Lange, "Fabrication and Properties of Dense Phase Relations in the System, Si_3N_4–SiO_2–MgO and Their Interrelations with Strength and Oxidation," *J. Am. Ceram. Soc.*, 61 [1–2] 53–56 (1978).
4.   M. Mitomo, "Pressure Sintering of Silicon Nitride," *J. Mater. Sci.*, **11**, 1103–107 (1976).
5.   J. Szepvolgyi, F.L. Riley, I. Mohai-Toth, I. Bertoti, and E. Gilbart, "Composition and Microstructure of Nanosized, Amorphous, and Crystalline Silicon Nitride Powders Before, During, and After Densification," *J. Mater. Chem.*, **6** [7] 1175–86 (1996).
6.   M. Mitomo, H. Hirotsuru, T. Nishimura, and Y. Bando, "Fine-Grained Silicon Nitride Ceramics Prepared from b-Powder," *J. Am. Ceram. Soc.*, 80 [1] 211–14 (1997).
7.   I.C. Huseby and G. Petzow, "Influence of Various Densifying Additives on Hot-Pressed Si_3N_4," *Powder Metall. Int.*, **6**, 17–19 (1974).
8.   V. Vandeneede, A. Leriche, F. Cambier, H. Pickup, and R. J. Brook, "Sinterability of Silicon Nitride Powders and Characterization of Sintered Materials," 53–68 in *Non-Oxide Technical and Engineering Ceramics*. Ed. S. Hampshire. Elsevier Applied Science, London, U.K., 1986.
9.   H. Yokota and M. Ibukiyama, "Microstructure Tailoring for High Thermal Conductivity of β-Si_3N_4 Ceramics," *J. Am. Ceram. Soc.*, **86**, [1] (2003).
10.   K. Koga, "Effect of Composition and Microstructure on the Characteristics of Silicon Nitride," Ceramics, 25, No.2, 107-111 (1990).

FABRICATION, MICROSTRUCTURE, AND CORROSION RESISTANCE OF β-SIALON NANO-CERAMICS

Qiang Li, Katsutoshi Komeya,
Junichi Tatami, and Takeshi Meguro
Yokohama National University
79-5 Tokiwadai, Hodogayaku
Yokohama 240-8501, Japan

Mamoru Omori
Institute for Materials Research,
Tohoku University
Aobaku Katahira 2-1-1
Sendai, Japan

Lian Gao
The State Key Lab
Shanghai Institute of Ceramics
Dingxi Rd. 1295
Shanghai 200050, China

ABSTRACT

This paper describes the fabrication and characterization of β-sialon nano-cemramics. Using as-prepared ultrafine β-sialon powder, we fabricated β-sialon nano-ceramics by spark plasma sintering (SPS). The density, toughness, and hardness of the specimens densified by SPS at 1650-1750^{0}C were determined and compared. Nearly fully dense β-sialon ceramics (relative density, 99.0%) were obtained after sintered at 1700^{0}C for 5 minutes without any sintering aid. XRD analysis confirmed that the specimens are pure β-sialon after sintering. AFM and HREM were employed to analyze sintered microstructure. Finally, corrosion testing revealed that β-sialon has better corrosion resistance than commercial Si_3N_4.

INTRODUCTION

Recently, a great deal of effort had been made to prepare nano-ceramics for engineering applications, and to study nano-scale materials for industrial applications.[1,2] β-sialon ($Si_{6-z}Al_zO_zN_{8-z}$) has been recognized as a potential candidate for high temperature applications because of its attractive properties. This paper describes the fabrication of β-sialon nano-ceramics and their properties.

Our goal is to fabricate ultrafine grain size β-sialon ceramics. Typically, to densify sialon or Si_3N_4, sintering aids are necessary to form the liquid phase that accelerates the sintering process. Consequently it is difficult to avoid grain growth. SPS is a new technology for superfast sintering that can densify difficult-to-sinter materials in a very short time at high temperature.[4,5] This is why we chose this technology to fabricate β-sialon ceramics.

EXPERIMENTAL

The preparation of the β-sialon nano powders used in this study is described in another paper.[3] Using as-prepared ultrafine β-sialon powder, we fabricated β-sialon nano-ceramics by spark plasma sintering (SPS). Figure1 shows the SPS device, and a typical sintering profile. Specimens of β-sialon ($Si_{6-z}Al_zO_zN_{8-z}$, z=3) were sintered at 1650-1750 C. The sintered specimens were then examined by x-ray diffraction (XRD). Scanning electron microscopy (SEM) was used to examine the fracture surface of the as-sintered β-sialon ceramics. Atomic force microscopy (AFM) and high-resolution electron microscopy (HREM) were employed to study the microstructure. Finally, the mechanical properties of as-fabricated β-sialon ceramics were analyzed. Hardness and toughness were determined by indentation using a pyramidal indenter with a load of 10 kg for 10 s.

Corrosion tests were conducted in an isothermal bath, where specimen bars were dipped into aqueous solutions of HNO_3 or H_2SO_4 at 80^0C.

RESULTS AND DISCUSSION

The density, toughness, and hardness of sintered β-sialon were determined and are compared in Table I. 99.4% dense β-sialon ceramics are obtained by sintering at 1700^0C for 5 minutes.

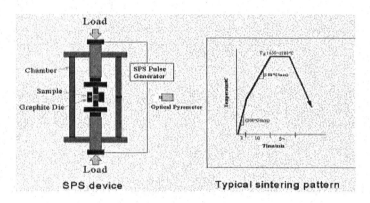

Figure1. A schematic of the SPS equipment, and a typical sintering profile.

Table I. The characteristics of SPS sintered β-sialon ceramics.

Temp.(°C)	Time (min)	Density (g/cm³)	Relative density (%)	Hv	KIC (MPa · m^{1/2})
1650	5	2.77	90.5	768	-
1700	5	3.08	99.4	1548	3.0
1750	5	3.05	98.4	1307	2.3

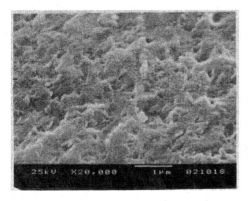

SEM

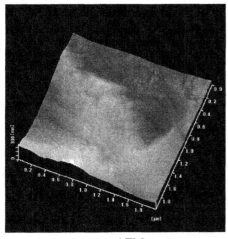

AFM

Figure 2. SEM and AFM photo of a typical β-sialon ceramic.

Figure 2 shows the sintered microstructure. Both SEM and AFM analysis of the sintered β-sialon ceramics determined the grain size to be 0.1-0.2 μm, indicating that the grains remain small even though growth of the nano particles cannot be avoided.

The sintering of Si_3N_4 and β-sialon, which are isostructural, has long been studied. Sintering aids can be used to accelerate the densification process, but they also form a glassy phase at the grain boundary that affects the properties of the ceramic. Our work on rapid sintering of β-sialon ceramic nano powders offers a possible alternative to the use of sintering aids.

Figure 3 shows the shrinkage of β-sialon as a function of temperature during SPS. Rapid shrinkage occurs from 1600 to 1700°C. This phenomenon is generally attributed to the presence of a liquid phase at high temperature. It can also be due to superplasticity, or due to the high curvature of nano particles at high temperature under high pressure.[6-9] Examination by HREM reveals a clean grain boundary without any glass phase (Figure 4), which is expected for the sintering of nano β-sialon ceramic without any sintering aids. In the SPS process, the superplasticity of starting nano β-sialon powders may be the main reason for the rapid densification from 1600-1700°C. It appears that the superplasticity of nano β-sialon powders makes it possible to fabricate β-sialon ceramics quickly at relative low temperature (1700°C) without sintering aids, and without the formation of a glass grain boundary interface. The sintering results confirm the good sinterability of the nano particles.

The chemical stability of Si_3N_4 and sialon in acid is strongly influenced by the grain boundary phase.[10,11] So a clean grain boundary means higher corrosion resistance. Figure 5 shows the results of corrosion tests on Si_3N_4 and β-sialon in different acid solutions. It confirms that β-sialon has better corrosion resistance than commercial Si_3N_4 because there is no grain boundary

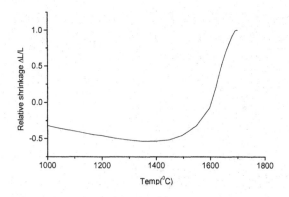

Figure 3.   Shrinkage of β-sialon as a function of temperature during SPS.

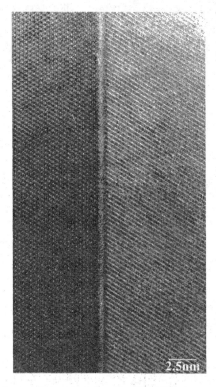

Figure 4. HREM photo of a grain boundary in SPS sintered β-sialon ceramic.

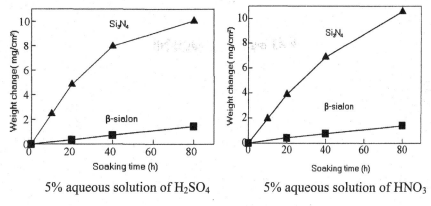

5% aqueous solution of H_2SO_4 · · · · · · · · · · 5% aqueous solution of HNO_3

Figure 5. Corrosion of Si_3N_4 and β-sialon in 5% H_2SO_4 and 5% HNO_3.

glass phase in the β-sialon processed without sintering aids.

SUMMARY

As-synthesized β-sialon powder was successfully densified by spark plasma sintering without sintering aids. HREM examination of the microstructure confirms that the grain boundary is free of any glass phase. As-fabricate β-sialon ceramics have good corrosion resistance relative to commercial Si_3N_4.

REFERENCES

[1]B. Cina, and I. Eldror, "Bonding of Stabilised Zirconia (Y-TZP) by Means of Nano Y-TZP Particles," *Mater. Sci. Eng . A* **301** [2] 187-195 (2001).

[2]P. Ajgalík, A.M. Hnatkoa, F. Lofajb, P. Hvizdob, J. Duszab, P. Warbichlerc, F. Hoferc, R. Riedeld, E. Lecomted, and M.J. Hoffmann, "SiC/Si_3N_4 Nano/Micro-Composite - Processing, RT and HT Mechanical Properties," *J. Eur. Ceram. Soc.*, **20** [4] 453-462 (2000).

[3] Q. Li, C. Zhang, K. Komeya, J. Tatami, T. Meguro, L. Gao, "Nano Powders of β-Sialon Carbothermally Produced Via a Sol-Gel Process," *J. Mater. Sci. Lett.* **22**, 885-887 (2003).

[4] Mamoru Omori, "Sintering, Consolidation, Reaction and Crystal Growth by the Spark Plasma System (SPS)," *Mater. Sci. Eng. A* **287** [2] 183-188 (2000).

[5]L. Gao, J.S. Hong, H. Miyamoto and S.D.D.L. Torre, "Bending Strength and Microstructure of Al_2O_3 Ceramics Densified by Spark Plasma Sintering," *J. Eur. Ceram. Soc .* **20** [12] 2149-2152 (2000)

[6]S. Kwon, "Consolidation of Nano-Crystalline Alumina," *Metal Powder Report*, **52** [4] 42-42 (1997).

[7]D.S. Yan, H.B. Qiu, Y.S. Zheng, and L. Gao, "Bulk Nanostructured Oxide Materials and the Superplastic Behavior under Tensile Fatigue at Ambient Environment," *Nanostructured Materials*, **9** [1-8] 441-450 (1997).

[8]C-L. Huang, C-L. Pan, and S-J Shium, "Liquid Phase Sintering of MgTiO$_3$–CaTiO$_3$ Microwave Dielectric Ceramics," *Materials Chemistry and Physics,* **78** [1] 111-115 (2003).

[9]P.E. McHugh and H. Riedel, "A Liquid Phase Sintering Model: Application To Si$_3$N$_4$ And WC-Co," *Acta Materialia,* **45**[7] 2995-3003 (1997).

[10]K.R. Mikeska, .J. Bennison, and S.L. Grise, "Corrosion of Ceramics in Aqueous Hydrofluoric Acid," *J. Am. Ceram. Soc.* 83 [5] 1160-1164 (2000)

[11]C-H Lin, K. Komeya, T. Meguro, J. Tatami, Y. Abe, and M. Komatsu, "Corrosion Resistance of Wear Resistant Silicon Nitride Ceramics in Various Aqueous Solutions," *J. Ceramic Society of Japan,* **111** [1295] 452-456 (2003)

CHARACTERISTICS OF DEHUMIDIFIER SHEETS FOR AN ADSORPTIVE DESICCANT COOLING SYSTEM

Y. Tashiro, M. Kubo and Y. Katsumi
Matsushita Ecology Systems Co., Ltd.
4017, Aza-Shimonakata,
Takakicho, Kasugaishi
Aichi, 486-8522 Japan.

T. Meguro, K. Komeya, and J. Tatami
Graduate School of Artificial
Yokohama National University
79-5 Tokiwadai, Hodogaya-ku,
Yokohama 240-8501, Japan

ABSTRACT

This study assessed the performance of dehumidifier sheets made of silica gel, pulp, and fiber in terms of water vapor adsorption and the BET surface area. The objective was to find a suitable material for an economical adsorptive desiccant cooling system. Dehumidifier sheets were prepared by spraying raw materials dispersed in water onto a stainless-steel wire netting in a papermaking machine. The water adsorption of sheets composed of silica gel and pulp tends to be lower than values calculated assuming that adsorption is proportional to the silica gel content. This may be because the pulp covers the silica gel surface. No deterioration in the performance of the dehumidifier sheet containing ceramic fiber was observed, and the practical dehumidification performance of the sheet was proved to be almost the same as that of a commercial dehumidifier element.

INTRODUCTION

Adsorptive desiccant cooling systems and dehumidifiers that use the heat exchange process are being considered as alternatives to air conditioners that work with refrigerants. The parameters to assess the performance of desiccant elements at lower temperatures and/or lower heat quantities have been investigated.[4-6] However, only a few studies have been performed to evaluate commercial dehumidifier elements.[4-6] We lack information on how performance is affected by the pulp, fiber, and binder materials that are used to fabricate dehumidifier elements. This paper addresses the preparation of sheet-like dehumidifier elements that contain silica gel, and evaluates their performance. In particular the effects of the pulp and ceramic fiber on the adsorption characteristics of silica gel were investigated.

EXPERIMENT

Preparation of Dehumidifier Element

A previous paper[3] determined that silica gel produced by Fuji Silisia Chemical Co., Ltd. (Japan) provides a favorable water vapor adsorption isotherm. Consequently, silica gel (S(a) type A) was used as the adsorbent in this work. The BET surface area of this material is 783 m^2/g. This silica gel is characterized by an adsorption isotherm in which the difference in water adsorption between 0.65 and 0.07 relative humidity is relatively large. Coniferous pulp, polyester fiber, and ceramic fiber composed of SiO_2 and Al_2O_3 were used as the support materials for the silica gel in the dehumidifier sheet.

Silica gel with sizes of 1 to 10 μm, the pulp, and the fiber were mixed in water. Slurries were prepared with 3 wt% solids (the silica gel, the pulp, and the fiber) in water. Dehumidifier sheets from 0.07 to 0.25 mm thick were prepared by spraying the slurry onto a stainless-steel wire netting in a papermaking machine. After drying at about 100°C, the products were rolled up. The mixing weight ratios of silica gel to pulp were adjusted to 30:70, 50:50, 70:30, and 80:20. When polyester fiber or ceramic fiber was added to the pulp, the ratios of silica gel: pulp: polyester fiber and silica gel: pulp: ceramic fiber were 70:20:10. The raw material mixtures investigated in this study are listed in Table I.

Characterization of Dehumidifier Sheets
To evaluate adsorption characteristics of the dehumidifier sheets, water vapor adsorption isotherms were determined at 25°C using BELLSORP18 (Japan Bell Co., Ltd., Japan). Before the adsorption measurement, the sample was heat-treated at 150°C for 5 h under vacuum. Nitrogen isotherms were determined at -196°C, and the surface area was calculated from the BET plot.[7]

Measurement of Water Vapor Adsorption Rate
Elements 92 mm × 92 mm and 20 mm thick were fabricated using corrugated samples No. 3 and No. 6 (3.4 mm pitch and 1.9

Table I. Dehumidifier sheet materials studied.

Sample No.	Materials	Contents (wt%)
1	Silica gel	30
	pulp	70
2	Silica gel	50
	pulp	50
3	Silica gel	70
	pulp	30
4	Silica gel	80
	pulp	20
5	Silica gel	70
	pulp	20
	Polyester fiber	10
6	Silica gel	70
	pulp	20
	Ceramic fiber	10

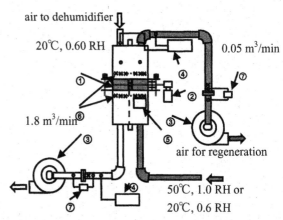

air to dehumidifier

20°C, 0.60 RH

0.05 m³/min

1.8 m³/min

air for regeneration

50°C, 1.0 RH or
20°C, 0.6 RH

Figure 1. Experiment apparatus used to evaluate of dehumidification performance.

①:Desiccant element, ②:Geared motor, ③:fan
④:Dew point meter, ⑤:400W-Heater,
⑥:Thermocouple, ⑦Differential pressure gauge.

mm height) to determine the rate of water vapor adsorption. The measurement was performed at a relative humidity (RH) of 0.65 at 25°C. The change in weight with time was measured under blowing air flowing at 51 m³/h.

Evaluation of Dehumidification Performance

The performance of the dehumidifier elements was evaluated using the apparatus illustrated in Figure 1. A 270 mm diameter and 20 mm thick element was partitioned in the circumferential direction into a part for dehumidification and a part for regeneration. The regeneration was conducted through desorption by heating with a 400 W heater as shown in Figure 1. The ratio of dehumidification area to regeneration area was 3:1. The rotating velocity of the element was optionally controlled with a motor. Each partition had a fan to maintain the airflow, and the flow rate was controlled by measuring the pressure of the nozzle set to the suction side of the fan using a differential manometer (DG-960, Fukuda Co., Ltd., Japan). The regeneration air was adjusted to RH = 1.0 at 50°C using a humidifier and a heater. In key places, the absolute humidity of the air was determined using a dew point meter (2001DF, Edge Tech Co., Ltd., Japan). The temperature and the relative humidity in the apparatus were monitored using a DW-4, (Ohnishi Netsugaku Co., Ltd., Japan).

The dehumidification performance, $D(H_2O\text{-kg/day})$, was calculated from the absolute humidity, defined as mass of water vapor per unit mass of dry air of inlet and outlet and the volumetric flow rate of air that passed through the element by consulting a psychrometric chart. D is expressed as follows:

$$D = 1440(X_i - X_o)(Q / \gamma) \qquad (1)$$

where, $X_i(H_2O\text{-kg/kg-dry air})$ is the absolute humidity of the inlet air, $X_o(H_2O\text{-kg/kg-dry air})$ is the absolute humidity of the outlet air, $Q(m^3/min)$ is the volumetric flow rate of the air, and γ (m³/kg-dry air) is the specific volume of the air that passed through the element.

RESULTS AND DISCUSSION

Water Vapor Adsorption Isotherms and BET Surface Areas

Figure 2 shows the water vapor adsorption isotherms for samples No. 1, No. 2, No. 3, No. 5, and No. 6. The adsorption isotherms for the silica gel and for

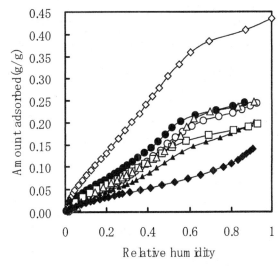

Figure 2. Water vapor adsorption isotherms of the dehumidifier sheets.
▲:No. 1, ○:No. 2, ●:No. 3, □:No. 5, △:No. 6, ◇:silica gel, ◆:pulp

the pulp alone are also shown. Almost no water adsorption was observed for the ceramic fiber or the polyester fiber alone. The amount of water adsorbed increases with increasing silica content in the samples with pulp. Water adsorption by sample No. 5 containing polyester fiber is lower than that by sample No. 3, although the content of silica gel is the same. This may be because the polyester fuses during the drying process in the papermaking machine to block the surface of the silica gel (i.e., the drying temperature is about 100°C, and the softening point of the polyester is ~70 °C). Sample No. 6 shows similar water adsorption to sample No. 3, indicating that the ceramic fiber does not significantly affect water vapor adsorption.

The isotherm for a dehumidifier sheet composed of both silica gel and pulp is influenced by both materials. The water adsorption can be calculated using the rule of mixtures, assuming that the water adsorption of the individual components is proportional to the total water concentration. The calculated values at 0.80 RH are compared with the measured values in Table II. This Table also contains the measured water adsorption for element S (produced by Seibu Giken), which has been the most suitable dehumidifier element in Japan.[5] The results in Table II indicate that the pulp may affect the adsorption performance of the silica gel.

Table II. Amount of water vapor adsorbed by dehumidifier sheets at 0.80 RH.

Sample No.	Materials	Contents (wt%)	①Calculated values of water adsorbed (g/g-sheet 0.80 RH)	②Measured values of water adsorbed (g/g-sheet 0.80 RH)	Ratio ②／①
—	Silica gel	100	0.40	0.40	1.0
	pulp	0			
—	Silica gel	0	0.107	0.107	1.0
	pulp	100			
1	Silica gel	30	0.195	0.175	0.897
	pulp	70			
2	Silica gel	50	0.254	0.222	0.874
	pulp	50			
3	Silica gel	70	0.312	0.235	0.753
	pulp	30			
5	Silica gel	70	0.301	0.181	0.601
	pulp	20			
	Polyester fiber	10			
6	Silica gel	70	0.301	0.230	0.764
	pulp	20			
	Ceramic fiber	10			
Element S	Silica gel, etc	—	—	0.230	—

Table III. BET surface area of dehumidifier sheets.

Sample No.	Materials	Contents (wt%)	Measured values of BET Surface area (m²/g-sheet)	①Calculated values of BET Surface area (m²/g-silica gel)	Ratio ①／783
—	Silica gel	100	783	783	1.0
	pulp	0			
3	Silica gel	70	246	351	0.448
	pulp	30			
4	Silica gel	80	476	595	0.760
	pulp	20			
5	Silica gel	70	320	457	0.584
	pulp	20			
	Polyester fiber	10			
6	Silica gel	70	452	646	0.825
	pulp	20			
	Ceramic fiber	10			
Element S	Silica gel, etc.	—	467	—	—

Table III shows the BET surface area of sample No. 3, No. 4, No. 5, and No. 6 calculated per gram of silica gel, assuming that the total surface area is governed by the silica gel. The ratio of the calculated surface areas to the silica 783 m²/g) is also shown in Table III. Sample No. 3, which contains 70 wt% silica gel and shows good performance (Figure 2), has a relatively low surface area of 246 m²/g. By comparison, Sample No. 4, which contains 80 wt% silica gel, has a surface area of 476 m²/g. This is 1.9 times higher than the surface area of sample No. 3, indicating that the effect of the pulp is quite large. Sample No. 5, has a surface area of 320 m²/g, which indicates that the polyester fiber present reduces the surface area. In samples No. 3 and No. 5, the effect of pulp and pulp plus polyester on the surface area is found to be larger than the effect on the amount of water adsorption. Based on the surface area of sample No. 6, the ceramic fiber appears to have little affect on silica gel surface coverage.

The pulp affects both water adsorption and surface area, with the effect on the surface area being larger. Although pulp covers the surface of the silica gel and may partially block the pores, it is presumed that water permeates into pores via the hydrophilic functional groups in the pulp.

Water Vapor Adsorption Rate
The rate of water adsorption was measured for samples No. 3 and No. 6, which showed the highest amount water vapor adsorption among the samples studied. The rate of water adsorption of element S was also measured for comparison. The results are illustrated in Figure 3. The curve for sample No. 6 is

almost the same as that for element S. However, the amount of water adsorbed by sample No. 3 is lower than that by element S, especially in the intermediate region from 2 to 8 min. This shows that the rate of water adsorption rate of the sample No. 3 is lower than that of the element S. The total water adsorption by sample No. 3 and No. 6 and of element S are almost the same; however, the BET surface area of sample No. 3 is half that of element S. Therefore, the water adsorption rate of the sample No. 3 is considered to be low due to a reduction of effective surface area of the silica gel by presence of the pulp.

Dehumidification Performance
    To assess the practical dehumidification performance of the sample No. 3 and sample No. 6 in comparison to element S, 270 mm diameter $\times$ 20 mm thick dehumidifier elements were fabricated. Air (0.60 RH at 20℃) was dehumidified, and air at 1.0 RH and 50℃ was regenerated. The dehumidification performance of element S, and of sample, No. 3, and sample No. 6 was determined to be 6.1, 5.3 kg-H_2O/day respectively. At 0.60 RH and 20℃, the dehumidification performance of element S sample No. 6 was determined to be3.3 and 3.1 kg-H_2O/day, respectively. The dehumidification performance of sample No. 6 is almost the same as that of element S.
    The dehumidification performance of sample No. 6 with ceramic fiber and pulp was found to be superior to that of sample No. 3. As mentioned, ceramic fiber does not influence the water vapor adsorption characteristics of the silica gel; however, dehumidifier sheets prepared with ceramic fiber, and without pulp cannot hold silica gel satisfactorily, and are extremely difficult to cut. The results indicate that pulp necessary to make sheets containing silica gel.

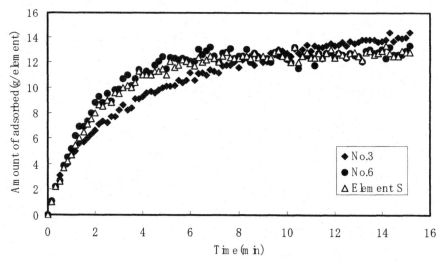

Figure 3. Water vapor adsorption rates for Sample No.3, No.6 and element S.

CONCLUSION
Dehumidifier sheets for air conditioners and dehumidifiers were prepared, and their dehumidification performance was investigated. The effect of dehumidifier sheet composition on water vapor adsorption and the rate of water vapor adsorption were examined. It was found that pulp lowers the amount of water vapor adsorption, although it is a good support for silica gel. In contrast, ceramic fiber does not influence the water vapor adsorption, but is unable to support the silica gel. As a result, pulp was proved to be necessary to produce a dehumidifier sheet containing silica gel.

REFERENCES
[1]Y. Tashiro, M.Kubo, Y. Katsumi, T. Meguro and K. Komeya, "Dynamical Dehumidification Performance of Colgate-Type Elements," *J. Mat. Sci. Japan*, **38** [4] 158-165 (2001).
[2]Y. Tashiro, M. Kubo, Y. Katsumi, T. Meguro and K. Komeya, "Experimental Evaluation and Simulation for Dehumidification Performance," *J. Mat. Sci. Japan,* **38** [4] 166-173 (2001).
[3]Y. Tashiro, M. Kubo, Y. Katsumi, T. Meguro and K. Komeya, "Assessment of Adsorption-Desorption Characteristic of Adsorbents," J. Mater. Sci., **39** 1-5 (2004).
[4]W. Jin, A. Kodama, M. Goto and T. Hirose, *J. Chem. Eng. Japan,* **24**[6] 894-900 (1998).
[5]T. Kuma and T. Hirose, "Performance of Honeycomb Rotor Dehumidifiers," *J. Chem. Eng. Japan*, **29**[2] 376-378 (1996).
[6]A. Kodama, M. Goto, T. Hirose, and T. Kuma, "Experimental Study of Operation for a Honeycomb Absorber,", *J. Chem. Eng. Japan*, **26** [5] 530-535 (1993).
[7]S. Brunauer, P. H. Emmett and E. Teller, *J. Am. Chem. Soc.*, **60** 309 (1938).

Novel Processing

DEVELOPMENT OF NEW MATERIALS BY MECHANICAL ALLOYING

José M Torralba and Elisa Ruiz-Navas
Dept. of Materials Science and Eng
Universidad Carlos III de Madrid
Av. Universidad 30, E-28911,
Leganés, Spain

João B. Fogagnolo
Materials Engineering Department
Universidade Federal de São Carlos
Brazil

ABSTRACT
   The synthesis of materials by high-energy ball milling or mechanical alloying (MA) of powders was first developed by John Benjamin and his co-workers to produce fine and uniform dispersions of oxide particles in nickel-base superalloys. Nowadays MA is being used to produce numerous materials and alloys. The aim of this work is to demonstrate that MA allows for the production of composite powders and materials with better structural properties than alternative routes. Alloys AA6061 and AA2014, reinforced with different particles, have been produced by MA and a conventional milling process. All of the powders produced were cold pressed and subsequently extruded with an extrusion ratio of 25/1 at 500°C. The powders and extruded bars were characterised by means of microstructural analysis, microhardness testing, and tensile testing.
   This work shows that MA can be used to produce composite powders with enhanced properties. With finer and more uniformly distributed the reinforcing particles, the MA extruded composites show improve mechanical properties.

INTRODUCTION
   Materials synthesis by high-energy ball milling was first developed by John Benjamin and his co-workers.[1] They produces fine and uniform dispersions of oxide particles (i.e., Al_2O_3, Y_2O_3 and ThO_2) in nickel-base superalloys that could not be made by conventional powder metallurgy (PM) methods. The process of mechanical alloying (MA) consists of the repeated welding, fracturing, and welding of a mixture of powder particles in a high-energy ball mill. The main event involves powder particles being trapped between colliding balls during milling. Those particles undergo deformation and/or fracture, depending on the mechanical behaviour of the powder. Presently, mechanical alloying is being used to produce numerous materials and alloys including supersaturated solid solutions, amorphous materials, intermetallic compounds and metal matrix composites.[2] Costa and co-workers[3] produced AA 2014 aluminium alloy from elemental powders (i.e., aluminium, copper, silicon, and magnesium) mixed in the required proportions. Lu and co-workers[4], in the same way, produced an Al-based alloy reinforced with SiC.

In the present work, mechanical alloying is used to produce metal matrix composite powders. The aim of the work is to demonstrate that MA allows for the production of materials with better structural properties than alternatives routes. Two different approaches were used: 1) Compare a AlN reinforced AA6061 produced by MA and by a conventional mix process; and 2) Compare TiC reinforced AA2014 produced using four different routes, including lab mixing, mixing in a planetary ball mill, high-energy mixing in an atrittor mill, and 'one step' mixing in a high-energy ball mill (attritor).

EXPERIMENTAL APPROACH

AA6061-based composites were made from 75 μm average particle size AA6061 (The Aluminium Powder Co. Ltd, U.K.), and with 8 μm average particle aluminium nitride (Advanced Refractory Technologies, Inc., U.K.). Composites with three different concentrations of AlN reinforcement were produced: 0, 5 and 15 wt.%. AA2014-based composites were made from mechanically alloyed powder AA2014 (optimised in a previous work[3]) mixed in different ways with 8.4 wt.% of <5 μm TiC (Goodfellow).

To produce the AA6061-based composites, the powders were mixed in either a low energy ball mill, or a high-energy centrifugal ball mill. To produce the 2014-based composites, three different mixing processes were used: a lab mixer (process 1); a low energy ball mill (process 2); or a high-energy attritor mill (process 3). There was also a process 4, called a 'one step' process, in which all of the raw materials were high energy milled together. Table I summarizes the different processes.

All of the powders produced were cold pressed at 200-300 MPa to form compacts. Zinc stearate was used as a die wall lubricant during pressing. Green powder compacts 80% of the theoretical density were produced by pressing. To avoid excessive grain growth due to the high level of stored elastic strain energy[5], the mechanically alloyed powders were annealed at 400 °C for 120 minutes prior to the extrusion. Without canning and degassing, the samples were then extruded with an extrusion ratio of 25/1 at 500°C. Graphite was used as a lubricant and a protective agent. Bars 5 mm in diameter and 98% of the theoretical density were produced by extrusion.

Metallographic analysis and Vickers micro-hardness tests were performed on the samples produced. Vickers indentation tests were used to determine hardness. Tensile tests were used to determine yield stress (YS) and ultimate tensile strength (UTS). Metallographic analysis was performed on the extruded samples in the

Table I. Materials and mixing/milling processes used to produce Al alloy-based composites.

Base alloy	Process 1	Process 2	Process 3	Process 4	Reinforcement (wt. %)
AA6061	---	Low energy ball mill	High-energy centrifugal ball mill	---	0, 5, 15 AlN
AA2014	Lab mixer		High-energy attritor mill	One step process in high-energy attritor mill	8.4 TiC

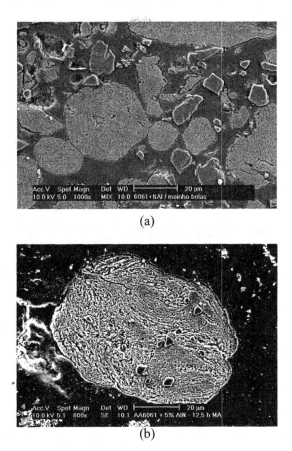

(a)

(b)

Figure 1. Microstructure of AA6061 aluminium reinforced
with 5 wt.% aluminium nitride produced by: a) conventional
low-energy milling; and b) mechanical alloying.

plane perpendicular to the extrusion direction. Keller's etchant[6] was used to
expose the microstructure of the extruded composite.

RESULTS

Figure 1a shows the microstructure of AA6061 aluminium reinforced with
5 wt.% aluminium nitride after low energy ball milling. Individual AA6061 and
AlN particles are observed after mixing. Figure 1b shows the microstructure of
AA6061 reinforced with 5 wt.% of aluminium nitride after 10 hours of high-
energy ball milling. Equiaxed particles are observed after mechanical alloying,
and aluminium nitride particles are distributed throughout the matrix.
Additionally, there is a random welding orientation.

Figure 2 shows the morphology of the AA2014-based composite powders
formed in this work. Figure 2a shows that TiC reinforced 2014 aluminium alloy
has Cu layers that are perpendicularly aligned to the action of the milling balls

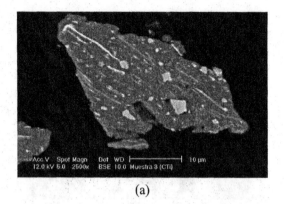

(a)

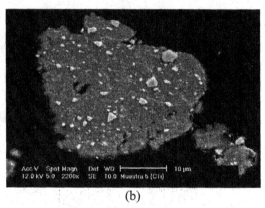

(b)

Figure 2. Microstructure of TiC reinforced AA2014 powder
produced by mechanical milling for: a) 5 hours; and b) 10 hours.

after milling for 5 h. However, after 10 h of milling, elemental Cu is no longer
observed. Also, after 10 hours of high-energy milling, the reinforcing TiC
particles are better distributed, and their size is reduced.

Figure 3a shows the microhardness of the composites particles produced in
this study. Composite particles produced after milling 10 h with either 5 or 15
wt.% AlN are compared with the as-received AA6061 aluminium powder in
Figure 3a. Figure 3b compares the microhardness of the as-fabricated AA2014
with that of the 8.4 wt.% TiC reinforced AA2014 powder milled for 5, 7 and 10 h.

Figures 4 and 5 show extruded the 5 and 15 wt.% of AlN reinforced AA6061
alloy composites produced from powders prepared by low energy ball milling,
and by high-energy ball mixing for 10 h. There is a homogeneous distribution of
the reinforcement particles in the alloy matrix; however, some agglomerates are
observed in the low energy milled composites. In the mechanically alloyed
composites, the reinforcement particles are smaller and better dispersed.

Figure 6 shows the yield strength (YS), the ultimate tensile strength (UTS)

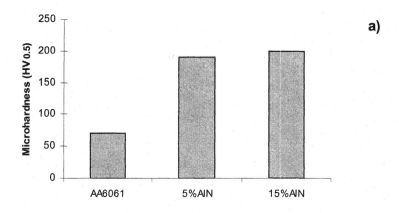

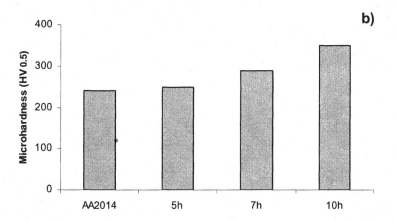

Figure 3. Vickers microhardness of the: a) AA 6061 aluminum alloy powder and the mechanical alloyed composite particles reinforced with 5 and 15 wt.% AlN; and b) the AA2014 aluminum powder and the mechanical alloyed composite particles reinforced with 8.4 wt.% TiC after 5 ,7, and 10 hours of mechanical milling.

and the Vickers hardness of the extruded composites, produced from powders prepared using the four different MA routes explained in the experimental section. YS and UTS usually decrease in conventional composites with the reinforcements[7], however, for the one step process, an increase in hardness, YS and UTS is observed.

DISCUSSION
    Benjamin and Volin[8] proposed the phenomenological mechanisms for mechanical alloying of ductile/ductile components. First, the particles are subjected to a microforging process, where the morphology changes from a

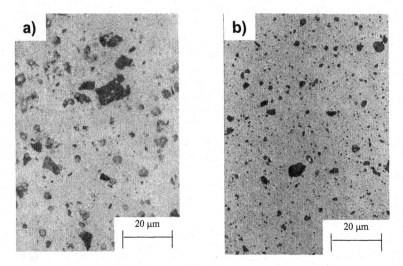

Figure 4. Microstructure of extruded 5 wt.% AlN reinforced AA6061 powder produced by: a) low energy ball milling; and b) mechanical alloying.

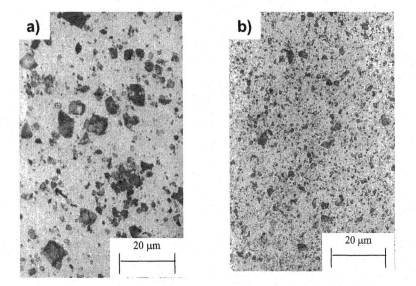

Figure 5. Microstructure of extruded 15 wt.% AlN reinforced AA6061 powder produced by: a) low energy ball milling; and b) mechanical alloying.

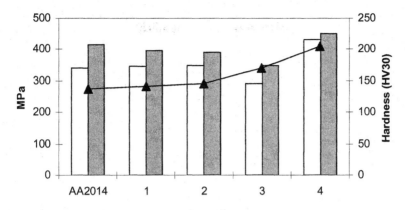

Figure 6. Mechanical properties of TiC reinforced AA2014 composites produced by the four different processes (according to Table I).

sphere to a flattened particle. The second stage is characterised by the predominance of the welding of the flattened particles. In the third stage, when the particles are hardened due to the high degree of deformation and welding, fracture is predominant. Random welding characterises the fourth stage. In the final stage, the microstructure remains the same, although crystalline refinement can occur. Recently, a new alternative model has been proposed for ductile-ductile components, based on macroscopic properties.[9] The mechanical alloying of ductile/brittle components also has been proposed by Benjamin.[10] Slightly modified Benjamin models also have been proposed for ductile-brittle components.[11-13] Based on these later models, the composite particles prepared in this study are fully processed, as they are characterized by the steady state equiaxial particle morphology[14]. Additionally, this work shows that the introduction of reinforcement particles has a strong influence on the hardness of the powder.

It has to be pointed out that the mechanical alloying process increases the particle hardness dramatically, and that the dependence of powder properties on the milling process and milling time is stronger than the dependence on the reinforcement.[15] This phenomenon can be seen in Figure 3 where the 15 wt.% AlN reinforcement increases the powder hardness in comparison to the 5% AlN. Additionally, the 8.4 wt% TiC reinforced AA2014 powder hardness increases significantly with increasing milling time from 5 to 10 hours. The phenomenon is further reinforced by comparing the hardness of the AA2014 aluminium alloy matrix after 10 hours of mechanical alloying to and the powder reinforced with 8.4wt.% of TiC obtained after 5 hours of milling. The hardness of both are similar despite the presence of the reinforcement. However, by comparing the hardness of the matrix alloy to the 8.4 wt.% TiC reinforced composite powder produced after 10 h of mechanical alloying, the influence of the reinforcement is evident.

The hard ceramic reinforcement particles absorb the milling energy by fracturing instead of deforming, and they also increase the deformation of the softer particles. The ceramic reinforcements are also trapped between the aluminium particles, which work as bonding points to help increase fracture and welding in the final powder.

By using a powder metallurgy (PM) route to prepare composite materials, it is possible to widely vary the concentration of the reinforcement phase without the segregation that is expected with a conventional casting process. This work shows how very different reinforcement materials (i.e., AlN and TiC) can be processed in similar way to achieve good processability and mechanical properties. However, depending on the size of the reinforcement, differences in density, flow, shape, and size between the matrix and reinforcement can produce agglomerates in the final product. With conventional mixing techniques, the smaller the reinforcement size, the greater the tendency to produce agglomerates. Agglomeration is the primary reason for low performance in this class of materials.[17-20] The extrusion process tends to minimize this problem.[16] Mechanical alloying can be used to improve the distribution of the reinforcement phase throughout the matrix. In this context, the one-step mechanical alloying processing route offers a processing advantage, as well as economic advantage of preparing the alloy and the composite powder in one simultaneous batch. Figures 2 and 3 show that each composite particle has a homogeneous distribution of the reinforcement phase after the mechanical alloying process. It is believed that the high degree of deformation, the high density of dislocations, and the oxide particle reinforcement in the matrix increase the hardness of the composite particles in this study.

Improvements in the mechanical properties of the MA prepared composites are believed to be due to the refinement of the grain size, and the better dispersion of the reinforcing particles throughout the aluminium matrix. For AA2014 reinforced with TiC, this improvement is only significant in the 'one-step' process. But it should be pointed out that the 2014 base alloy used in this work that was prepared by mechanical alloying has a UTS of ~430 MPa, which is much higher than the typical value obtained for conventional SiC-Al composites (100 MPa), and that is close to the value for 7xxx aluminium series alloys obtained by ingot metallurgy.[19]

CONCLUSIONS

The mechanical alloying processes can produce hard particle reinforced aluminium matrix composite particles, with equiaxial morphology and with a random orientation of welding. The hardness of these composite powders can reach three times the value of the as-fabricated aluminium powder. After extrusion, mechanically alloyed composites have a finer microstructure, with a good dispersion of the reinforcement particles.

It is believed that the hardness and UTS of the extruded materials increases due to the finer, well dispersed reinforcement particles.

The one-step process (i.e., alloying and reinforcing simultaneously) produces composite materials with enhanced mechanical properties.

Mechanical alloying has been demonstrated as an effective way to manufacture advanced performance powders suitable for forming metal matrix composites.

ACKNOWLEDGEMENTS
The authors would like to thank CAPES – Brasil (Fundação Coordenação de Aperfeiçoamento de Pessoal de Nível Superior) for the financial support and the Grant CICYT MAT-2000-0442-C02-01.

REFERENCES
[1] J.S. Benjamin, "Dispersion Strengthened Superalloys by Mechanical Alloying," *Metallurgical Transactions*, 1 2943 (1970).
[2] L. Lu, and M. Lai, "Mechanical Alloying", 1-64, Kluwer Academic Publishers, 1998.
[3] C.E. Costa, W.C. Zapata, J.M. Torralba, J. M. Ruiz-Prieto, and V. Amigó, "P/M MMC's Based Aluminium Reinforced with Ni_3Al Intermetallic Made by Mechanical Alloying Route," *Materials Science Forum*, 217-222 1859-1864 (1996).
[4] L. Lu, M.O. Lai and C.W. Ng, "Enhanced Mechanical Properties of an Al Based Metal Matrix Composite Prepared Using Mechanical Alloying," *Materials Science and Engineering A*, A252 203-211 (1998).
[5] J.B. Fogagnolo, M.H. Robert, and J.M. Torralba, "The Effects of Mechanical Alloying on the Extrusion Process of AA 6061 Alloy Reinforced with Si_3N_4," Proceedings (CD Rom) of the 15th Brazilian Congress of Mechanical Engineering, Sao Paulo, Brazil, 1999.
[6] G. Petzow, "Metallographic Etching", American Society for Metals, Ohio, U.S.A., 1978.
[7] H.F. Lee, F. Boey, and K.A. Khor, "High Deformation Consolidation Powder Metallurgy Processes for an Al-Li Alloy Composite," *Materials Science and Engineering A*, A189 173-180 (1994).
[8] J.S. Benjamin and T.E. Volin, "The Mechanism of Mechanical Alloying," *Metallurgical Transactions*, 5 1929-1934(1974).
[9] D.R. Amador, PhD Thesis, Universidad Carlos III de Madrid, Madrid, Spain, 2003.
[10] J.S. Benjamin, "Mechanical Alloying," *Science America*, 234 40-48 (1976).
[11] M.J. Tan and X. Zhang, "Powder Metal Matrix Composites: Selection and Processing", *Materials Science and Engineering A*, A244 80-85 (1998).
[12] E.M. Ruiz-Navas, C.E. da Costa, F. Velasco and J.M. Torralba, "Mechanical Alloying: a Way to Obtain Metallic Powders and Composite Materials," *Rev. Metalurgia-Madrid,* 36 279-286 (2000).
[13] J.B. Fogagnolo, F.J. Velasco, M.H. Robert and J.M. Torralba, "Effect of Mechanical Alloying on the Morphology, Microstructure and Properties of Aluminium Matrix Composite Powders," *Materials Science and Engineering A*, 342 131-143 (2003).
[14] B.J.M. Aikin and T.H. Courtney, "The Kinetics of Composite Particle Formation During Mechanical Alloying", *Metallurgical Transactions A*, 24A 647-657 (1993).
[15] D.R. Amador and J.M. Torralba, "Morphological and Microstructural Characterization of Low-alloying Fe Powder Obtained by Mechanical Attrition," 1203-1207, in *New Developments in Powder Technology,* Ed. J.M. Torralba, 2001.
[16] B.C. Ko, G.S. Park, and Y.C. Yoo, "The Effects of SiC Particle Volume Fraction on the Microstructure and Hot Workability of SiCp/AA 2024 composites," *Journal of Materials Processing Technology,* 95 210 – 215 (1999).

[17]K.A. Khor, Z.H. Yuan, and F. Boye, 499-507 in "Processing of Submicron SiC Reinforced Al-Li Composites by Mechanical Milling," in Processing and Fabrication of Advanced Materials VI – Eds. T. S. Srivatsan and J. J. Moore, The Minerals, Metals & Materials Society, 1996.

[18]J.J. Lewandowski and C. Liu, 117-137 "Microstructural Effects on the Fracture Micromechanism in 7XXX Al P/M-SiC Particulate Metal Matrix Composites," in Processing and Properties for Powder Metallurgy Composites, Eds. P. Kumar, K. Vedula, and A. Ritter, The Metallurgical Society, 1988.

[19]A.J. Mourisco, 37-42 in "Processing and Properties of Composites Al/SiC Produced by PM," in Proceedings of the PM World Congress, Granada, **Vol. 5,** 1998.

[20]J.B. Fogagnolo, E.M. Ruiz-Navas, M.H. Robert, and J.M. Torralba, "6061 Al Reinforced with Silicon Nitride Particles Processed by Mechanical Milling," *Scripta Materialia,* **47** 243-248 (2002).

MAGNETIC PROPERTIES OF NI-FERRITE PRODUCED BY HIGH ENERGY MILLING

M.E. Rabanal, A. Várez, B. Levenfeld, and J.M. Torralba
Department of Materials Science and Engineering
Universidad Carlos III de Madrid,
Avda de la Universidad n° 30, E-28911
Leganés (Spain)

ABSTRACT
The structural and magnetic evolution of nickel ferrite ($NiFe_2O_4$) powder as consequence of high-energy milling has been evaluated. The Ni-ferrite polycrystalline powder was prepared by solid-state reaction at 1400°C. Structural changes of the samples were examined by x-ray diffraction. The crystallite size and internal strain were evaluated using the Williamson-Hall and Scherrer methods. Nanometer size powders were obtained after short times. The Curie transition temperature, determined by a thermomagnetometry (TGM) drops from 591°C to 548°C after 30 h of milling as a consequence of decreasing the crystallite size. Magnetization curves showed a broadening of the hysteresis loop as milling time increases. Strong changes in the saturation magnetization, from 52.7 emu/g to 14.8 emu/g after 14 h, also correlates with milling time.

INTRODUCTION
Ferrites constitute a very large and important group of ceramics that exhibit spontaneous magnetization and most of the other properties attributed to ferromagnetic materials. An important property that distinguishes ferrites from other ferromagnetic materials is that the spontaneous magnetization in ferrites is the result of two oppositely magnetised sublattices, each of which is ferromagneticly ordered.[1] Ferrimagnetic materials, owing to their ease of preparation and their good magnetic characteristics, are frequently used in high-frequency devices.[2] An important property of these materials is high specific resistivity, which greatly reduces eddy current losses at high frequencies.[3] In contrast, the application of ferrites for corrosion protection applications requires a very small specific resistivity, *i.e.,* less than 10 Ωcm. It may be possible to control the stoichiometry of a ferrite by doping and heat-treating to achieve this requirement.[4]

The synthesis of ultrafine magnetic particles has been intensively investigated in recent years because of the potential applications in high-density magnetic recording and magnetic fluids.[5] Various preparation techniques, such as sol-gel pyrolysis[6], hydrothermal[7], and mechanical grinding[8-10] are used to produce nanoparticles. Recently, a mechanochemical route, which has been used to prepare various nanocrystalline spinel ferrites[8], has been reported as a powerful tool to synthesize amorphous alloys[11], nanocrystalline metals and alloys[12], and ceramic materials.[13] Nanosized magnetic particles have properties that are drastically different from those of the corresponding bulk materials.[14]

In this work, we describe the influence of milling on structural changes and the magnetic properties of Ni-ferrite. Experiments were conducted to determine the optimum milling time to obtain uniform ferrite nanoparticles with good magnetic properties.

EXPERIENTAL PROCEDURES

Nickel ferrite ($NiFe_2O_4$) was prepared in polycrystalline form by solid-state reaction of a stoichiometric mixture of $Ni_3(CO_3)Ni(OH)_4 4H_2O$ and α-Fe_2O_3, powders having a purity of 99%. The mixture was heated at 1200°C in air for 24 h, followed by slow cooling. Afterwards, the powder was compacted and sintered at 1400°C for 48 h, followed by quenching in air.

A high-energy centrifugal ball mill (Fritsch Pulverisatte-6) was used to grind the Ni-ferrite at 650 rpm in air without any additives. To reduce contamination, agate was used for the milling container and balls. The influence of different ball-to-powder weight ratios (RM_{b-p}) was evaluated (10:1 and 20:1 which are designated as Sample 4 and Sample 5, respectively). A constant 10 mm ball diameter was used. The direction of rotation was reversed every 5 minutes and samples were extracted from the mill at different times.

Structural changes of Ni-ferrite milled powder were analysed by powder X-ray diffraction (XRD) using a Philips X'Pert automatic diffractometer with Cu Kα radiation (λ=1.5418 Å) operating at 40 KV and 40 mA. Data were recorded between 10° and 70° 2θ with a 0.04° step size. A counting time of 13 s was used for the starting Ni-ferrite powder while 1 s was used for the milled powders. The unit cell parameters were obtained by a pattern machine refinement by means Rietveld method[15] using of the FULLPROF Programme.[16]

Crystallite size (called TDC, in Å) and internal strain (ε, in %) with milling time were calculated using the Williamson-Hall[17] method (W-H) and Scherrer Formula[18] (S-F). In the first method, the width (β_{exp}) of every peak was measured as the *integral breadth* considering only the $\lambda_{\alpha1}$=1.54060 Å contribution. A correction for instrumental broadening (β_{inst}) was calculated from a standard silicon (NISC) reference material. The peak integral breadth due to the sample (β_{sample}) was calculated according to:

$$\beta_{sample}^2 = \beta_{exp}^2 - \beta_{inst}^2 \qquad (1)$$

Afterwards, crystallite size and internal strain were determined by fitting to the Williamson-Hall (W-H) plot:

$$\beta_{sample} \cos\theta = \frac{(K \times \lambda_{\alpha1})}{L} + 2\varepsilon\sin\theta \qquad (2)$$

where L is the coherent scattering length ("crystallite size" in Å); K is a constant (1.0<K<1.3 when integral breadths are used) that we have assumed to be 1 in all cases; β_{sample} is the integral width of the sample (in radians); and ε is the inhomogeneous internal strain in %.

In the case of the Scherrer formula, the width peak was determined at *full width at half-maximum (FWHM)*, considering only the $\lambda_{\alpha1}$=1.54060 Å contribution. The breadth peak (β_{sample}) due to the sample was calculated according to:

$$L = \frac{(K \times \lambda_{\alpha 1})}{(\beta_{sample} \times \cos\theta)} \quad (3)$$

The evolution of magnetic interaction with milling time was evaluated by means the Curie Temperature (T_C). A Perkin-Elmer TGA7 thermogravimetric analyser (TGA) equipped with a small permanent magnet was used for the thermomagnetometry (TGM) measurements.[19] Initially, a thermogravimetry analysis (TGA) run was made to verify that there was no weight loss. To keep the magnetic field gradient constant during the analysis, the magnet was kept in the same position as much as possible. Data were collected during heating from 50 to 500°C at 10°C/min. under flowing air. TGM with XRD is very helpful to determine the magnetic phases and the different values of T_C.

Magnetization measurements were performed using a SQUID magnetometer operating at 5 K in a magnetic field between -50 and 50 Oe.

RESULT AND DISCUSSION

The XRD pattern for the $NiFe_2O_4$ starting material was indexed according to a cubic unit cell (S.G.=Fd3m, N° 227) with parameters a=0.3379(2) Å, in concordance with previously reported values.[20,21] In Figures 1a and 1b, XRD patterns for Samples 4 and 5 after different milling times are shown, respectively. The starting Ni-ferrite powder has sharp diffraction peaks, indicating high crystallinity. With increased milling time, mechanical grinding produces a broadening of the diffraction peaks. This is believed to be due to crystallite size reduction and the accumulation of internal strain. At lower milling times (t < 178 h), all of the diffraction peaks are due to Ni-ferrite phase. After 178 h (Figure 1a) a new diffraction peak is observed at about 2θ=28°, which corresponds to the most intense diffraction peak for SiO_2. The presence of this phase is attributed to contamination. For Sample 5 with only a higher RM_{b-p}

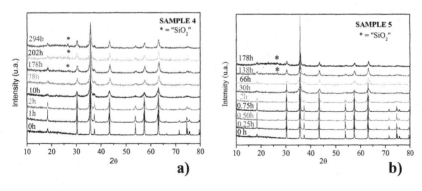

Figure 1. XRD patterns of: 1a) Sample 4 after different milling times showing a new peak (*) after 178 hours (indicating contamination from the milling equipment; and b) Sample 5 after different milling times showing a new diffraction peak (*) corresponding to SiO_2 at 138 h.

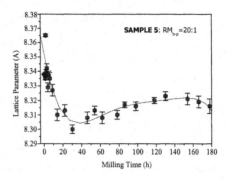

Figure 2. Evolution of the lattice parameter with milling in Sample 5, as determined by the Rietveld method.

(20:1), the SiO_2 peak appears after 138 h (Figure 1b). The higher RM_{b-p} decreases in milling time, indicating that the milling process is more efficient.

The lattice parameter of Ni-ferrite as a function of milling time was calculated using the Rietveld method (Figure 2). At short milling times (t<30 h) the lattice parameter decreases from 8.34 Å to 8.30 Å. At longer milling times the lattice parameter increases smoothly to 8.32 Å. This change could be related to a modifications at the octahedral and tetrahedral sites of Fe^{3+} and Ni^{2+}, producing changes in length between the cations, and distortion between the octahedral and tetrahedral polyhedrals.[22,23] These structural modifications will produce important changes in the magnetic properties of Ni-ferrite milled powders, which can be measured and analyzed.

Soft ferrites systems show very different behaviour during milling.[24] In Mg-ferrite milled under similar conditions and, for short times, new diffraction peaks corresponding to α-Fe_2O_3 are observed due to the decomposition of Mg-ferrite, and there is a related decrease in the lattice paramete.[25]

The crystallite size of the coherent domains can be estimated using the W-H method and S-F. Figure 3a shows the evolution of crystallite size and microstrain with milling time. For Sample 4, at milling times less than 22 h, crystallite size decreases rapidly from 4910 Å to 121 Å, while microstrain increases. At intermediate times, the crystallite size remains constant at 10 nm. The crystallite size increases to 22 nm after 294 h. No correlation to a family of crystallographic planes is found, which is consistent with the lack of anisotropic internal strain usually found in metallic samples.[12] This behavior is explained by the formation of defects such as dislocations and/or stacking faults during the milling process. We also determined the crystallite size using the S-F. Figure 3b compares the evolution of crystallite size with milling time for Sample 4 and Sample 5. Independent of RM_{b-p}, at low milling times (t <25 h), a strong decrease in crystallite size always occurs to produce nanometer scale powder. At intermediate times we observe a small increment of crystallite size. This increment is more important in order to higher RM_{b-p}.

In Figure 4, the evolution of crystallite size with the milling time determined by the W-H method and by the S-F are compared. The crystallite sizes calculated

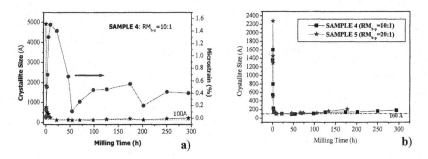

Figure 3. a) Evolution of crystallite size, TDC (Å), and microstrain (%) with milling time in Sample 4, as determined by the W-H method, and b) Evolution of TDC with milling time in Sample 4 ($RM_{b-p}=10:1$) and Sample 5 ($RM_{b-p}=20:1$), as determined by the S-F.

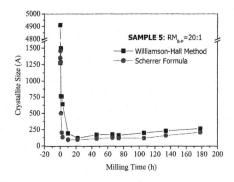

Figure 4. Crystallite size with milling time estimated using the W-H method and the S-F.

by the S-F are smaller than those determined by the W-H method. This difference could be due to peak broadening associated with internal strain that is not considered in the S-F.

TGA analysis performed on the milled samples before TGM analysis, confirmed a very small weight loss at low temperature. The weight loss is higher with milling time and RM_{b-p}, but it is always lower than 8%. The weight loss could be due to water or CO_2 adsorbed on the powder surface due to the high energy milling process.

We determined that the drop in the TGM curves is associated, exclusively, with the magnetic transition (from the ferrimagnetic to the paramagnetic state). The magnetic transition temperature for the starting of Ni-ferrite is around 591°C. However, the reported transition temperature for Ni-ferrite is 585°C. It is known that the synthesis process, mechanical treatments, and/or thermal treatments can modify the magnetic properties of soft ferrites.[26] In the TGM curves for all of the milled samples, there is only a single thermal event that corresponds to the Ni-ferrite magnetic transition. This indicates that there are no other magnetic phases, *i.e.*, magnetite or α-Fe_2O_3 present.

In Figure 5 the decrease in Curie temperature with milling time is shown. For Sample 4, at short times (t <50 h), the T_c ($\cong$538°C) shows a strong decrease. Longer milling times do not modify the magnetic transition. For Sample 5, the strong decrease in T_C takes place at shorter milling times than for Sample 4. This is a consequence of the milling process being more effective. The changes in Curie temperature could be a consequence of the evolution of crystallite size with time, or/and a modification in cationic distribution from inverse spinel to mixed spinel during the grinding process. Magnetic materials in the nanometer size range have a higher surface area than bulk materials.[27] This produces a weak magnetic interactions and a decrease in T_C. Since the milling process creates defects or/and leads to amorphization of the crystal structure, the magnetic properties will be modified.

Figure 6 shows the hysteresis loop for Sample 4 after different milling times. At short times, the hysteresis curves are typical of soft magnetic materials, with high values of saturation magnetisation (m_S) and low coercivity, H_C. The starting material has a m_S 52.7 emu/g and 0.3 Oe, corresponding to a cation distribution of $(Fe_{0.96}Ni_{0.04})(Fe_{1.04}Ni_{0.96})O_4$ consistent with the inverse spinel structure. Figure 6 shows the evolution of hysteresis loop with milling time. At longer milling times the shapes of hysteresis curves widen due to a decrease in ms and an increase in the coercivity field, H_C. Therefore, in Sample 4, a significant difference can be detected from 52.7 emu/g for the un-milled sample to 28 emu/g after milling for 294 h. The changes in the saturation magnetization are a consequence of mechanically induced cation redistribution between tetrahedral (A) and octahedral (B) sites. Similar behaviour is observed in Sample 5. The larger modifications are consequence of the higher RM_{b-p}. Also, an increase in the coercivity is observed from 0.3 Oe for the starting material to 3000 Oe after milling for 14 h. This increase is attributed to the high internal stresses[28] and/or a decreasing of magnetic anisotropy.[29] Also, high values of coercivity are associated with particles smaller than the single-domain size.[30]

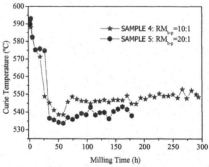

Figure 5. Evolution of transition temperature (°C) with the milling time. At short milling times, a strong decrease in T_C is observing in both samples. Sample 5, with a higher RM_{b-p}, shows a larger decrease

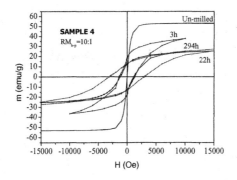

Figure 6. Widening of the hysteresis loop for Sample 4 with milling time, showing the effect of the milling process on the shape of the demagnetisation curve.

CONCLUSION

Mechanical grinding of $NiFe_2O_4$ powder decreases the crystallite size and increases the internal strain in the polycrystalline powder. The crystallite size, calculated using two different methods, reaches the nanometer size range in a short period of time during milling. Changes in crystallite size produce changes in the magnetic properties. The saturation magnetisation decreases with decreasing crystallite size with milling time, possibly due to changes in the crystal structure. In this way, mechanical grinding increases the coercivity of the samples. TGM analysis is very helpful in determining the magnetic transition temperature, which decreases with milling time. The results demonstrate that mechanical grinding is an adequate route to modify the magnetic properties of soft magnetic materials.

REFERENCES
[1]M. Sugimoto, "The Past, Present and Future of Ferrites," *J. Am. Ceram. Soc.*, **82** [2] 269-280 (1999).
[2]M. Pardavi-Horvath "Microware Applications of Soft Ferrites," *Journal of Magnetism and Magnetic Materials* **215-216** 171-183 (2000).
[3]E.C. Snelling, *"Soft Ferrites, Properties and Applications,"* 2nd Edition, Butterworths, London, 1998.
[3]J. Tasaki, T. Izushi, "Effect of Additives on Magnetization of Ferrites," *Proceeding of the International Conference*, September-October 1980, Japan
[4]V.E. Fertman, *"Magnetic Fluids Guidebook: Properties and Applications,"* Ed. Hemisfere, New York, 1990.
[5]J. Ding, W.F. Miao, E. Pirault, R. Street, and P.G. McCormick, *J. Magnetism and Magnetic Materials*, **177-181** 933-934 (1998).
[6]K.J. Davies, Wells S, K. OGrady, S. Morup, "The Observation Of Multiaxial Anisotropy In Ultrafine Cobalt Ferrite Particles Used In Magnetic Fluids," *J. Magn. Magn. Mater.* **149** 14-18 (1995).
[7]J. Ding, W.M. Miao, P.G. McCormick, and R. Street, "Hexaferrite Magnetic-Materials Prepared By Mechanical Alloying," *J. Magnetism And Magnetic Materials* **150** (3) 417-420 (1995).

[8]J. Ding, P.G. McCormick, and R. Street "Formation of Spinel Mn-Ferrite During Mechanical Alloying," *J. Magnetism and Magnetic Materials* **171** 309-314 (1997).

[9]V. Sepelak, A. Buchal, and K.D. Becker "Nanocrystalline Structure of the Metastable Ball-milled Inverse Spinel-Ferrites," *Materials Science Forum* **278-281** 862-867 (1998).

[10]I.W. Modder and H Bakker, "Changes in the Structural and Magnetic Properties of GdIr2 with the Milling Process," *Physica Review B* **58** (21) 14479-14483 (1998).

[11]C.C. Koch, "Synthesis of Nanostructured Materials by Mechanical Milling: Problems and Opportunities" *NanoStructured Materials* **9** (1-8) 13-22 (1997).

[12]S.F. Moustafa and M.B. Morsi, "The formation of Mg Ferrite by Mechanochemical Alloying and Sintering," *Materials Letters* **34** 241-247 (1998).

[13]C. Suryanarayana, "Structure and Properties of Nanocrystalline Materials," *Bull. Mater. Sci.*, **17** [4] 307-346 (1994).

[14]H.M. Rietveld, "A Profile Refinement Method for Nuclear and Magnetic Structure," *J. Appl. Crys.*, **2** 65 (1969).

[15]J. Rodríguez-Carbajal, "Fullprof Program: Rietveld, Profile Matching & Integrated Intensity Refinement of X-Ray and/or Neutron Data," Vers.3.5d. 1998.

[16]G.K. Williamson and W.H. Hall," X-Ray Line Broadening From Filed Aluminium And Wolfram" *Acta Metallurgica.* **1** [1] 22-31 (1953).

[17]B.D. Cullity *"Elements of X-Ray Diffraction"*. Ed. Addison-Wesley (1978).

[18]R.G. Rupard and P.K. Gallagher. "The Thermal Decomposition of Coprecipitates and Physical Mixtures of Magnesium-Iron Oxalates," *Thermochimica Acta* **272** 11-26 (1996).

[19]Raul Valenzuela *"Magnetic Ceramic,"* Cambridge University Press, 1994

[20]R.A. McCurrie *"Ferromagnetic Materials: Structure and Properties,"* Academic Press, 1994.

[21]B. Guillot, B.B. Domenichini, and P. Perriat "Effect of the Preparation Method and Grinding Time of Some Mixed Valency Ferrite on Their Cationic Distribution and Thermal Stability Toward Oxygen," *Solid State Ionics* **84** 303-312 (1996).

[22]Y.T. Pavlyukhin, Y.Y. Medikov, and V.V. Boldyrev, "Magnetic and Chemical Properties of Mechanically Activated Zinc and Nickel Ferrites," *Mat. Res. Bull.*, **18** 1317-1327 (1983).

[23]S. Music, D. Balzar, S. Popovic, M. Gotic, I. Czakó-Nagy, S. Dalipi, "Formation and Characterization of NiFe$_2$O$_4$," *Croatica Chemica Acta CCACAA* **70** [2] 719-734 (1997).

[24]M.E. Rabanal, A. Várez, B. Levenfeld, and J.M. Torralba, "Magnetic Properties of Mg-Ferrites after Milling Process," *J. Materials Processing Technology*, **143-144** 470-474 (2003).

[25]Y. Shi, J. Ding, and H. Yin "CoFe$_2$O$_4$ Nanoparticles Prepared by the Mechanochemical Method," *J. of Alloys and Compounds* **308** 290-295 (2000).

[26]C.N. Chinnasamy, A. Narayanasamy, N. Ponpandian, R. J. Joseyphus, B. Jeyadevan, K. Tohji, and K. Chattopadhyay "Grain Size Effect on the Néel Temperature and Magnetic Properties of Nanocrystalline NiFe$_2$O$_4$ Spinel," *J. Magn. Magn. Mater.* **238** 281-287 (2002).

[27]J. Ding, Y. Li, L.F. Chen, C.R. Deng, Y. Shi, Y.S. Chow, and T.B. Gang, Microstructure and Soft Magnetic Properties of Nanocrystalline Fe-Si Powders," *J. Alloys and Compounds* **314** 262-267 (2001).

[28]G. Buttino, M. Poppi, "Dependence on the temperature of magnetic anisotropies in Fe-based alloys of Finemet," *J. Magn. Magn. Mat.* **170** [1-2] 211-218 (1997).

[29]G. Mendoza-Suárez, J.A. Matutes-Aquino, J.I. Escalante-García, H. Mancha-Molinar, D. Ríos-Jara, and K.K. Ojal "Magnetic Properties and Microstructure of Ba-Ferrite Powders Prepared by Ball Milling," *Journal of Alloys and Compounds* **223** 55-62 (2001).

CENTRIFUGAL PRESSURE ASSSITED DIFFUSION BONDING OF CERAMICS

Y. Kinemuchi and K. Watari
National Institute of Advanced
Industrial Science and Technology
2266-98 Anagahora
Shimoshidami, Moriyama
Nagoya 463-8560, Japan

S. Uchimura
Shinto V-Cerax Ltd.
3-1 Honohara,
Toyokawa 442-8505, Japan

ABSTRACT

Centrifugal pressure assisted diffusion bonding of ceramics has been demonstrated. In this work, an advanced centrifuge has been developed that enables one to produce an acceleration rate of 100 m/s^2 at 1000°C. Because of the centrifugal acceleration, material transfer is enhanced at the joining interface. This results in a high tensile strength joint at high centrifugal pressure. Furthermore, because hot-pressing dies are not required in the process, it is possible to produce joints free of contamination caused by solid contact. These advantages may lead to the application of this technique for clean joining of ceramics to metals.

INTRODUCTION

Interfacial matching is necessary in order to develop good bonding. In diffusion bonding this matching can be enhanced by external pressure. External pressure enhances material transfer via plastic deformation, diffusion, and creep, resulting in the elimination of interfacial defects. In practice, the pressure applied is normally selected to be smaller than the yield stress in order to avoid macroscopic deformation; that is, a pressure ranging from several MPa to 100 MPa is applied. Such pressure is thought to be sufficient to remove voids at the interface, and to provide oxide breakdown through plastic deformation at the bonding temperature.[1]

Currently, the main focus of joining research is directed towards commercializing devices such as layered actuators[2] or solid oxide fuel cells.[3] In these devices, interfaces are subjected to severe repetitive stress, which results in failure caused by interfacial delamination, crack formation, or breakdown of the hermetic seal. One solution to this problem is to improve adhesion strength, which can be accomplished by diffusion bonding as a result of interface matching. One method of diffusion bonding involves conventionally clamping or inserting specimens into dies, and subsequently pressurizing the joint mechanically while heat is applied. However, it is hard to apply this process to complex shape devices. Furthermore, contamination caused by solid contact with the pressing media inherently occurs.

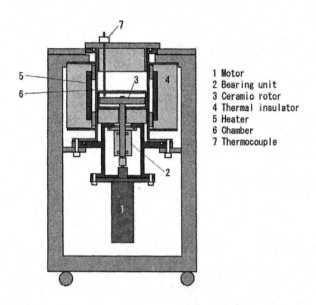

1 Motor
2 Bearing unit
3 Ceramic rotor
4 Thermal insulator
5 Heater
6 Chamber
7 Thermocouple

Figure 1. Schematic of the experimental, centrifugal joining apparatus.

One option is to employ a non-contact pressing process, for example, by centrifugal, magnetic, or electrical means. Among these promising measures, centrifugal pressure is thought to be the most practical, because it applies to a wide variety of materials. The present paper reports on a centrifugal pressing apparatus and its potential use for fine joining.

EXPERIMENT
Apparatus
An illustration of the experimental joining apparatus is shown in Figure 1. The centrifugal force is generated by rotating a ceramic rotor made of Si_3N_4 or SiC. These ceramics were selected to increase the resonance frequency (for heating), and because they maintain mechanical strength at high temperature. The rotor is held in place by a ball bearing that is kept hermetically sealed by means of a magnetic fluid. The motor driving the ceramic rotor rotates at the selected speed, which is controlled by an inverter circuit. The dimensions of the rotor are 90 mm in radius and 20 mm in thickness. The specimen is placed in the rotor and subjected to centrifugal force caused by high-speed rotation at a rotational radius of 80 mm. A maximum rotational speed of 20,000 rpm is currently feasible, which is soon to be increased to 50,000 rpm. This rotational unit is installed in a stainless steel chamber that is heated by a resistance heater or by radio frequency. The temperature of the specimen can be raised to 1000°C. The chamber can be evacuated by a rotary pump to a vacuum of 10 Pa. Temperature and the rotational speed are controlled and monitored as a function of time using a computer.

Diffusion Bonding Procedure

Aluminum oxide (Nikkato Co., SSA-S grade, 99.6% purity) was joined to stainless steel (JIS SUS304) using an interlayer of aluminum (99.9% purity). Here, an aluminum plate with a thickness of 0.5 mm was used. The selection of these materials was based on published work by Nicholas et al.[4] The surface of the Al_2O_3 and the stainless steel to be joined had a roughness of several micrometers, i.e., a ground but unpolished surface. Before joining, the materials were cleaned in ethanol. Then they were pressed in a stack under a pressure of 100 MPa by cold isostatic pressing (CIP) to increase the contact area at the interfaces. Prior to CIP, the stacked materials were wrapped in graphite sheet or adhesive plastic tape. During diffusion bonding, the chamber was evacuated by means of a rotary pump to a pressure of less than 100 Pa. The temperature was increased to 650°C at a heating rate of 5°C/min, and held for 2 h. Samples were furnace cooled at about 7°C/min down to 500°C. The centrifugal pressure was controlled by controlling the rotational speed, and the contact area of the joint. Each condition examined is summarized in Table I.

The tensile strength of the Al_2O_3/Al/stainless steel joints was measured at a head speed of 0.5 mm/min. The strength of five test pieces was evaluated.

An electron-probe microanalyzer (EPMA, JEOL JXA8900L) was used to characterize the diffusion layers. The thickness of the diffusion layer was determined from backscattered electron images.

RESULTS AND DISCUSSION

Interface

EMPA results of the interface joined under centrifugal pressure are shown in Figure 2. A clear diffusion layer is observed at the Al/stainless steel interface (Figure 2a). In contrast, no diffusion layer is observed at the Al/Al_2O_3 interface (Figure 2b). The diffusion layer in Figure 2a forms at a location that was originally metallic Al before joining, and that consisted of two layers. The composition of the layer located near the Al interlayer and next to the stainless

Table I. Diffusion boding conditions for Al_2O_3/Al/stainless steel

rotation number [rpm]	acceleration*) [km/s^2]	contact area [mm^2]	mass of Al_2O_3 [g]	mass of Al [g]	P_c at Al_2O_3 / Al [MPa]	P_c at Al / SUS**) [MPa]
5 000	21.9	24.0	0.74	0.32	0.7	1.0
7 000	43.0	24.0	0.74	0.32	1.3	1.9
9 000	71.0	10.9	0.48	0.15	3.1	4.1
10 000	87.7	10.9	0.48	0.15	3.9	5.1

*) rotational radius 80 mm

**) Stainless steel of JIS SUS304 is abbreviated as SUS.

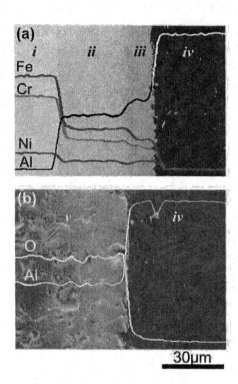

Figure 2. EPMA analysis of a joined interface: a) the interface between the stainless steel and the Al; b) the interface between the Al_2O_3 and the Al. Each region is designated by a Roman numeral that corresponds to the following composition: i - stainless steel; ii - $(Fe_{0.68}Cr_{0.21}Ni_{0.11})Al_{3.5}$; iii - $(Fe_{0.68}Cr_{0.21}Ni_{0.11})Al_{3.5}$; iv - Al; and v - Al_2O_3.

steel plate is $(Fe_{0.68}Cr_{0.21}Ni_{0.11})Al_{3.5}$ and $(Fe_{0.65}Cr_{0.21}Ni_{0.14})Al_{2.4}$, respectively. These compositions are almost the same as those reported by Nicholas et al.[4] The constitutive intermetallic phases can be reasonably explained from the phase diagram of the Fe-Al system;[5] that is, $FeAl_3$ is a stable intermetallic phase with the highest quantity of Al. The next most Al-rich compound is Fe_2Al_5. The observed compositions are very close to those of Al-rich intermetallic compounds, a logical similarity considering that the layers were formed by the diffusion of stainless steel into Al. Concerning the Al/Al_2O_3 interface, no clear compounds or diffusion layers have been reported so far, but these analyses are beyond the scope of the present study.

It is noted that the centrifugal force produces a pressure gradient along the rotational radius, whereas hot isostatic pressing or hot pressing forms materials more homogeneously. This pressure gradient is one of the driving forces for diffusion. Because the thickness of the diffused layer strongly influences the strength of the diffusion bonded joint,[4] we characterized the thickness of the Al/stainless steel interface joined under different centrifugal pressures. Figure 3 shows backscattered electron images of the polished intersection. The thickness of the diffusion bond layer was found to be ≈50 μm, independent of pressure. Parabolic growth kinetics of this layer has been reported when the diffusion is governed by a compositional gradient.[4]

$$X^2 = At \exp\left(-\frac{Q}{RT}\right) \qquad \qquad (1)$$

Here, X is the thickness of reaction layer, t is the bonding time, T is the bonding temperature, A is a constant (1.5 m^2/s), Q is the activation energy (227 kJ/mol), and R is the gas constant. Applying our experimental conditions of 7200 s and 923 K (650°C) to the relationship predicts a reaction layer thickness of 40 μm. Hence, it can be said that diffusion under centrifugal force is effectively governed by the pressure gradient. The pressure gradient estimated from both the geometry and the mass of the specimen. is in the range of 0.5–2.5 x 10^3 Pa/μm; that is, the pressure difference within a diffusion bonded layer of 50 μm is likely to be 0.3–1.3 x 10^5 Pa.

Adhesion Strength
    The tensile strength of a joint produced by centrifugal diffusion bonding is shown in Figure 4 as a function of centrifugal pressure. The pressure corresponds to the pressure at the Al/stainless steel interface. Increasing pressure clearly enhances the strength of the joint. Bonding was not achieved without centrifugal pressure.
    After centrifugal bonding with a pressure of 5 MPa, the strength of the joint is ~40 MPa. This value is comparable to that obtained by conventional diffusion

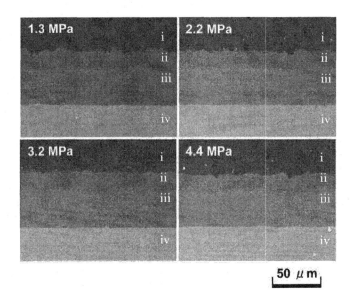

Figure 3 The reaction layer formed at the diffusion bonded Al/stainless steel interface formed using different centrifugal pressures. Regions i-iv correspond to the phases described in Figure 2.

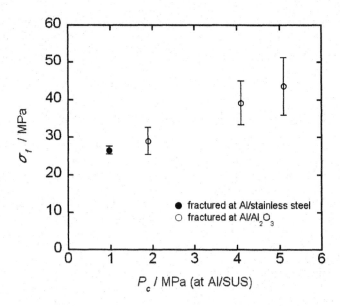

Figure 4. The tensile strength of a pressure diffusion bonded Al$_2$O$_3$/Al/stainless steel joint (σ_f) as a function of centrifugal pressure (P_{CF}).

bonding with the present set of materials.[4] This strength also corresponds to the yield stress of annealed Al.[6] A rapid deformation mismatch at the interface caused by plastic deformation of the Al may lead to formation of voids that act as origins of fractures.

CONCLUSIONS

Centrifugal pressure has been applied to diffusion bond Al$_2$O$_3$ to stainless steel using an Al interlayer. Interface analysis by EPMA reveals that a clear intermetallic layer is formed near the interface between the stainless steel and the Al. On the other hand, no distinguishable reaction layer or diffusion layer is observed at the interface between the Al and the Al$_2$O$_3$. The constitutive phases in the joint are identical to those formed in conventional diffusion bonding, indicating a similar joining mechanism. The pressure gradient, (0.5–2.5 x 10^3 Pa/μm), does not significantly influence the diffusion kinetics, which is confirmed by the growth of the intermetallic layer. The tensile strength of the joint increases with centrifugal pressure, and a strength of ~40 MPa is achieved. These results clearly indicate the potential of centrifugal bonding as a technique for clean and fine joining, which is necessary for ceramic-device fabrication.

ACKNOWLEDGMENT
This work was financially supported by the New Energy and Industrial Technology Development Organization (NEDO).

REFERENCES
[1] O.M. Akselsen, "Review Diffusion Bonding of Ceramics," *J. Mater. Sci.*, **27** 569-79 (1992).
[2] S. Saito and T. Inoi, "Laminated Electroceramic Parts and its Fabrication Process," Japanese Patent, P3006518, Nov. 15, 1996.
[3] K.L. Ley, M. Krumpelt, R. Kumar, J.H. Meiser, and J. Bloom, "Glass-Ceramic Sealants for Solid Oxide Fuel Cells: Part I. Physical Properties," *J. Mater. Res.*, **11** [6] 1489-1493 (1993).
[4] M.G. Nicholas, and R.M. Crispin, "Diffusion Bonding Stainless Steel to Alumina using Aluminium Interlayers," *J. Mater. Sci.*, **17** 3347-60 (1982).
[5] T. Lyman, "Metals Handbook 8th edition, **vol. 8**"; pp. 260 American Society for Metals, Ohio, 1973.
[6] K. Suganuma and N. Kawakami, "Novel Processing of Brazing Aluminium to Aluminium and to Austenitic Stainless Steel," *Material Science and Technology*, **9** 349-56 (1993).

FORMATION OF INTERFACIAL PHASES AT A SiC/Cu JOINT FRICTION-BONDED WITH Ti

Akio Nishimoto and Katsuya Akamatsu
Materials Science & Engineering Dept
Kansai University
3-3-35 Yamate-cho, Suita
Osaka 564-8680, Japan

Makoto Takahashi and Kenji Ikeuchi
Joining and Welding Research Institute
Osaka University
11-1 Mihogaoka, Ibaraki
Osaka 567-0047, Japan

Masaaki Ando
Akita National College of Technology
1-1 Iijima Bunkyo-cho
Akita 011-0923, Japan

Masatoshi Aritoshi
Hyogo Prefectural Institute of Industrial
Research
3-1-12 Yukihira-cho, Suma
Kobe 654-0037, Japan

ABSTRACT
The interfacial microstructure of a SiC/Cu joint friction-bonded with a Ti intermediate layer has been characterized with TEM to determine the process by which interfacial layers of mainly TiC and a Cu-alloy (adjacent to the SiC matrix) form. The formation of these layers is greatly suppressed by water quenching just after the forging stage. In the quenched joint, almost the entire interface is comprised of a Ti-rich layer a few nm thick. TEM observations of the interface after reheating quenched joints at 1073 K and 773 K indicate that the reaction to form the TiC and the Cu-alloy layers can proceed at 1073 K but not at 773 K. These results suggest that the formation of the TiC and Cu-alloy layers starts during the forging stage, and that the layers grow at temperatures higher than 773 K during the cooling process.

INTRODUCTION
As previously reported[1,2], we found that the interfacial microstructure in a silicon carbide/copper joint that is friction-bonded with a Ti intermediate layer can be characterized by a TiC layer and a Cu-alloy layer that is adjacent the SiC matrix. The Cu-(Si) alloy layer (reaction product) has been observed in diffusion-bonded or brazed joints by several authors[3-8], although the layer is much thinner in our friction-bonded joint. However, no clear explanation has been given for why the Cu-alloy layer forms between the TiC layer and the silicon carbide. To get a better understanding of how the TiC and the Cu-alloy layers form, the influence of thermal quenching (just after the bonding operation) on the interfacial microstructure has been investigated.

EXPERIMENTAL DETAILS

The ceramic specimen to be bonded was a silicon carbide (SiC) produced through pressureless sintering at 2273 K. A few % Al_2O_3 and Er_2O_3 were used to aid sintering. Oxygen-free copper (Cu) annealed for 3.6 ks at 773 K was used as the metal specimen. The SiC and Cu specimens consisted of 16 mm diameter rods. An intermediate layer of 20 μm thick commercial titanium (Ti) foil was also used.

The friction-bonding parameters employed were as follows: rotation speed, $N = 40^{-1}$, friction time, $t_1 = 3$, 10 s, friction pressure, $P_1 = 20$ MPa, forge time, $t_2 = 2$ or 6 s, and forge pressure, $P_2 = 30$ MPa. To determine the process by which the reaction layers form, the joint was quenched in iced water just after the forging stage for $t_2 = 2$ and 6 s. A heat treatment of quenched joints was carried out for 3.6 ks at 773 K or 1073 K in a vacuum furnace at ~10^{-2} Pa. For the reference joint that was not quenched, it was cooled slowly in a heat insulator after the forging stage.

EXPERIMENTAL RESULTS

Interfacial Microstructure After Quenching

In the SiC/Cu joint friction-bonded with a Ti intermediate layer, a 10-50 nm thick TiC layer (depending on the friction time and the distance from the center axis of the specimen) was formed over almost the entire interface area. Additionally, a 10 nm thick Cu-alloy layer was formed between the TiC layer and the SiC matrix when the joint was cooled slowly in a heat insulator (Figure 1).[1,2]

When the joint was quenched in iced water after friction bonding for a friction time of $t_1 = 3$ s and a forge time of $t_2 = 2$ s, the joint broke at the interface. However, when the friction time was increased to 10 s, the joint withstood the quenching stress. A TEM micrograph of the interface formed by quenching after friction bonding for $t_1 = 10$ s and $t_2 = 2$ s is shown in Figure 2. No obvious reaction layer is present at the interface, but a very thin layer a few nm thick with a bright contrast is observed. No obvious diffraction pattern is detected from this layer by selected area diffraction (SAD) or NBD. Energy dispersvie x-ray (EDX) analysis, however, indicates that the layer is rich in Ti, as shown in Figure 3. These results suggest that the TiC and Cu-alloy reaction layers observed in a slowly-cooled joint were formed during the cooling process after the 2 s forging stage; i.e., only a Ti-rich layer a few nm thick exists between the Cu-Ti mixed region and the SiC after the friction stage and the 2 s forging

Figure 1. TEM micrograph of a SiC/Cu joint interface bonded with a Ti intermediate layer.

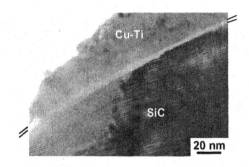

Figure 2. TEM micrograph of a quenched SiC/Cu joint interface bonded with a Ti intermediate layer. ($t_1 = 10$ s, $t_2 = 2$ s)

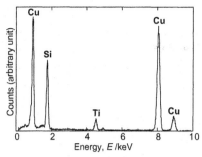

Figure 3. EDX spectrum of the Ti-rich layer at the interface in Fig. 2.

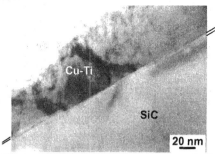

Figure 4. TEM micrograph of a quenched SiC/Cu joint interface bonded with a Ti intermediate layer. ($t_1 = 10$ s, $t_2 = 6$ s)

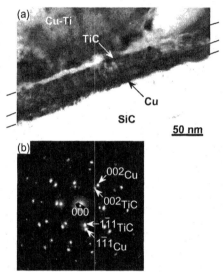

Figure 5. TEM micrographs of a quenched SiC/Cu joint interface bonded with a Ti intermediate layer ($t_1=10$ s, $t_2=6$ s): a) bright field image; and b) NBD pattern from the [110] pole of the TiC and Cu layers.

stage. This means that the relative motion of the SiC and Cu surfaces, at a rotation speed of $N = 40^{-1}$, occurs in the thin Ti-rich layer during the friction stage.

To investigate the formation of the reaction layer during the forging stage, a joint was quenched in iced water after forging for 6 s. No obvious reaction layer is present over most of the interface of the quenched joint (Figure 4); however, reaction layers about 40 nm thick are observed in small areas (Figure 5). A NBD pattern from this reaction layer is shown in Figure 5b. The diffraction pattern from the reaction layers identifies the presence of TiC and Cu-alloy, similar to a slowly cooled joint. The thickness of these layers is smaller than that observed in a slowly cooled joint. These results suggest that the formation of the TiC and Cu-alloy reaction layers starts in small areas of the interface during the forging stage (within 6 s).

Effect Of Heat Treatment On Interfacial Microstructure Of Quenched Joint

To determine the temperature at which the TiC and Cu-alloy layers form, a quenched joint was heat-treated at 773 K or 1073 K for 3.6 ks. Figure 6 shows TEM micrographs of a quenched joint after holding at 1073 K. A SAD pattern from the interface (Figure 6e) identifies TiC and Cu-alloy. Dark-field images shown in Figure 6c and 6d indicate that the TiC and Cu alloy layers are formed at the interface similar to a slowly cooled joint. The TiC and Cu-alloy layers have preferred orientation relationships with the SiC, which can be expressed by the following equations.

$$(0001)_{6H\text{-SiC}}//(1\bar{1}1)_{TiC}, [1\bar{2}10]_{6H\text{-SiC}}//[110]_{TiC} \qquad (1)$$

$$(0001)_{6H\text{-SiC}}//(1\bar{1}1)_{Cu}, [1\bar{2}10]_{6H\text{-SiC}}//[110]_{Cu} \qquad (2)$$

These orientation relationships are observed in brazed joints with a filler of Cu-Ti alloy and Cu-Ag-Ti alloy, and are reported in several papers.[9-12] The following relationships are observed in a friction-bonded joint that is cooled slowly[2]:

$$(10\bar{1}2)_{6H\text{-SiC}}//(1\bar{1}1)_{TiC}, [\bar{1}2\bar{1}0]_{6H\text{-SiC}}//[110]_{TiC} \qquad (3)$$

$$(10\bar{1}2)_{6H\text{-SiC}}//(1\bar{1}1)_{Cu}, [\bar{1}2\bar{1}0]_{6H\text{-SiC}}//[110]_{Cu} \qquad (4)$$

The close-packed planes and directions of the SiC, Cu, and TiC crystals align in a parallel structure similar to that of the close-packed plane that appears periodically on the {10$\bar{1}$2} plane of 6H-SiC. Diffraction from graphite is also observed (Figure 6e). The dark-field image from the graphite (Figure 6b) indicates that the graphite precipitates in the Cu-alloy layer adjacent the TiC layer. These results suggest that a reaction of SiC with Cu and Ti proceeds to form the TiC and Cu-alloy layer at 1073 K. The TiC and Cu-alloy layers observed in the joint that was heat-treated at 1073 K subsequent to quenching are much thicker than those observed in the friction-bonded joint cooled slowly. This result can be explained by the very long hold time at 1073 K compared with the slowly cooled joint. The formation of graphite particles implies that the dissolution of SiC into the Cu-alloy layer supplies a greater amount of C for consumption by reaction with the Ti supplied through the TiC layer.

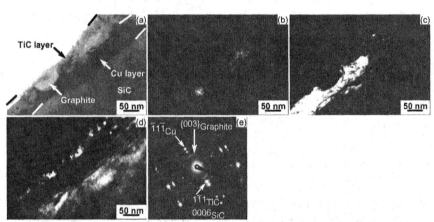

Figure 6. TEM micrographs of a SiC/Cu joint bonded with a Ti intermediate layer that was quenched and then heat-treated at 1073 K for 3.6 ks: a) bright field image; b) dark field image of graphite; c) dark field image of Cu; d) dark field image of TiC and SiC; and e) SAD pattern from the TiC layer, graphite layer, Cu layer, and SiC matrix, where pole figures of [112] TiC, [112] Cu, and [1$\bar{1}$00] 6H-SiC are observed.

In contrast, when a quenched joint is held at 773 K, neither the TiC nor the Cu-alloy layer form (Figure 7); a Ti-rich layer similar to that observed in a quenched joint is observed. In addition, small particles 20 nm in size are observed at the joint interface as (Figure 8). A NBD pattern from this particle (Figure 8b) suggests it consists of TiC. Thus, TiC precipitates in the form of a particle rather than a layer at a temperature of 773 K. These results suggest that the formation of the TiC and the Cu alloy layers takes place at temperatures higher than 773 K.

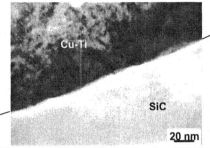

Figure 7. TEM micrograph of a SiC/Cu joint interface bonded with a Ti intermediate layer after quenching and heat-treating for 3.6 ks at 773 K.

DISCUSSION

The TEM observations described in the experimental results section suggest that the SiC matrix makes contact with the Cu-Ti mixed region only through a Ti-rich layer that is a few nm thick during the friction stage. The formation of the TiC and Cu-alloy layers starts during the forging stage within 2-6 s after the application of the forging pressure. After the forging stage ($t_2 = 6$ s), the TiC and Cu-alloy layers form only in small areas at the joint interface; the dominant microstructure is still a Ti-rich layer a few nm thick. During the slow cooling process at temperatures higher than 773 K, the TiC and Cu-alloy grow and form an interfacial layer.

As to the formation or the growth of TiC and Cu-alloy layers, the dissolution of SiC into Cu to form graphite and Cu-Si solid solution is thermodynamically permissible at temperatures higher than 773 K.[13] The reaction of C with Ti makes the dissolution reaction of SiC more favorable thermodynamically, because of the decrease in the chemical potential of Cu. Thus, the formation of the TiC and Cu-alloy layers at the joint interface are thermodynamically permissible.

The determination that the TiC and Cu-alloy layers form mainly during the cooling process after the forging stage indicates that the reaction to form these layers proceeds through solid-state diffusion. The diffusion path corresponding to the observed interfacial microstructure can be discussed on the basis of a chemical

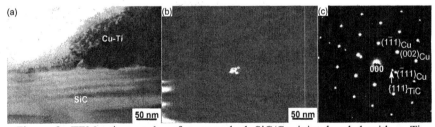

Figure 8. TEM micrographs of a quenched SiC/Cu joint bonded with a Ti intermediate layer, and heat-treated for 3.6 ks at 773 K: a) bright field image; b) dark field image; and(c) SAD pattern of the TiC layer and SiC matrix, where a pole figure of [110] Cu, and a spot from TiC can be observed.

potential diagram, since the driving force for diffusion is the gradient in the chemical potential. Although, four elements Si, C, Cu, and Ti participate in the interfacial reaction, it is rather difficult to present a quaternary chemical potential diagram. Therefore, the diffusion path was plotted on the chemical potential diagram for the Cu-Ti-C ternary system, considering that the main reaction products at the interface are Cu_4Ti, TiC, and Cu-alloy. Since the SiC specimen used in the present investigation contains 0.05 % C, the activity of the C in the SiC can be assumed to be 1 (chemical potential = 0). Unfortunately, a ternary phase diagram for the Cu-Ti-C system is not available. Thus, we constructed a chemical potential diagram based on

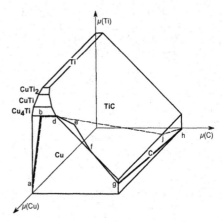

Figure 9. Estimated three-dimensional chemical potential diagram for the Cu-Ti-C system at 1073 K.

the thermodynamic data for the Ti-Cu[14] and Ti-C[15] binary systems, as shown in Figure 9 (in the Cu-C system, the mutual solubility of Cu and C is very small, and no intermetallic compound forms). Although intermetallic compounds of Cu_3Ti_2 and Cu_4Ti_3 are also form in the Cu-Ti system, they are neglected in the potential diagram shown in Figure 9 due to the lack of thermodynamic data, and because of the absence of these compounds at the interface. In Figure 9 the Cu-Ti mixed region lies on the intersection of Cu and Cu_4Ti faces, since in this region Cu_4Ti particles precipitate in the Cu matrix. The Cu-Ti mixed region is followed by the TiC layer, the Cu-alloy layer, and the SiC matrix. The diffusion path corresponding to this interfacial microstructure can be schematically illustrated by line a-b-d-e-f-g-h. In this diffusion path, indicated by line e-f, Cu seems to diffuse against the chemical potential gradient. However, the Cu in the Cu-alloy layer corresponding to line f-g can lower the chemical potential more, considering the alloy composition; e.g. the Si content in the Cu-alloy layer is estimated to be 8%, compared to 5% in the Cu-Ti mixed region. Assuming Raoult's law, this difference in the Si content lowers the chemical potential of Cu in the Cu-alloy layer by 0.28 kJ/mol in comparison to that of the Cu in the Cu-Ti mixed region. Thus, the formation of the Cu-alloy layer between the TiC layer and the SiC matrix can probably be explained thermodynamically considering the effect of the Si content on the chemical potential of Cu.

CONCLUSIONS

To discuss the formation of the TiC and Cu-alloy layers observed in a SiC/Cu joint that is friction-bonded with a Ti intermediate layer, the influence of quenching (just after the forging stage) on the interfacial microstructure was investigated. The results obtained are summarized as follows:

1) In the quenched joint, a Ti-rich layer a few nm thick is present over almost the entire interface. The formation of the TiC and Cu-alloy layers starts in small regions of the interface during the forging stage as the forge time is increased from 2-6 s.

2) When a quenched joint is held for 3.6 ks at 1073 K, the TiC and Cu-alloy

layers are much thicker than those observed in a slowly cooled joint. Additionally, graphite precipitates are also observed in the Cu-alloy layer adjacent to the TiC layer. In contrast, when the quenched joint is held at 773 K, no TiC or Cu-alloy layers form; however, very small TiC particles are detected at the interface.

3) The TiC and Cu-alloy layers have a preferred orientation relationship with SiC, which can be expressed by the following equations.

$$(0001)_{6H-SiC}//(\bar{1}11)_{TiC}, [1\bar{2}10]_{6H-SiC}//[1\bar{1}0]_{TiC} \qquad (1)$$

$$(0001)_{6H-SiC}//(\bar{1}11)_{Cu}, [1\bar{2}10]_{6H-SiC}//[1\bar{1}0]_{Cu} \qquad (2)$$

4) These results suggest that the TiC and Cu-alloy starts to form during the forging stage, and then grow to form interfacial layers during cooling above 773 K.

REFERENCES
[1]A. Nishimoto, M. Ando, M. Takahashi, M. Aritoshi, and K. Ikeuchi, "Friction Bonding of Silicon Carbide to Oxygen-Free Copper with an Intermediate Layer of Reactive Metal," *Materials Transactions, JIM*, **41** [12] 1636-45 (2000).
[2]A. Nishimoto, M. Ando, M. Takahashi, M. Aritoshi, K. Akamatsu, and K. Ikeuchi, "Interfacial Microstructure of SiC/Cu Joint Friction-Bonded with Ti Intermediate Layer," *J. Japan Institute of Metals*, **67** [10] 538-46 (2003).
[3]Y. Nakao, K. Nishimoto, K. Saida, and K. Katada, "The Role of Elements in an Insert Metal for Bonding of Si_3N_4 to Refractory Metals Using Cu-Co-Cr-Ni-W Alloy as an Insert Metal," *Quarterly J. Japan Welding Society*, **5** [1] 54-59 (1987).
[4]S. Kato, T. Yano, and T. Iseki, "Interfacial Structures Between Ag-Cu-Ti Alloy and Sintered SiC with Various Additives," *J. Ceram. Soc. Jpn.*, **101** [3] 325-30 (1993).
[5]F. Tamai and M. Naka, "Microstructure and Strength of SiC/SiC Joints brazed with Cu-Ti Alloys," *Quarterly J. Japan Welding Society*, **14** [2] 333-37 (1996).
[6]H.K. Lee and J.Y. Lee, "A Study of the Wetting, Microstructure and Bond Strength in Brazing SiC by Cu-X (X=Ti,V,Nb,Cr) Alloys," *J. Mater. Sci.*, **31** 4133-40 (1996).
[7]T. Nishino, S. Urai, and M. Naka, "Interface Microstructure and Strength of SiC/SiC Joint Brazed with Cu-Ti Alloys," *Engineering Fracture Mechanics*, **40** [4/5] 829-36 (1991).
[8]W. Tillman, E. Lugscheider, R. Xu, and J.E. Indacochea, "Kinetic and Microstructural Aspects of the Reaction Layer at Ceramic/Metal Braze Joints," *J. Mater Sci.*, **31** 444-52 (1996).
[9]T. Yano, H. Suematsu, and T. Iseki, "High-Resolution Electron Microscopy of a SiC/SiC Joint Brazed by a Ag-Cu-Ti Alloy," *J. Mater Sci.*, **23** 3362-66 (1988).
[10]C. Iwamoto, H. Ichinose and S.I. Tanaka: "Atomic Observations at the Reactive Wetting Front on SiC," *Philosophical Magazine A*, **79** [1] 85-95 (1999).
[11]Q.H. Zhao, J.D. Parsons, H.S. Chen, A.K. Chaddha, J. Wu, G.B. Kruaval, and D. Downham, Single Crystal Titanium Carbide, Epitaxially Grown on Zincblend and Wurtzite Structures of Silicon Carbide, *Mater. Res. Bull.*, **30** 761-69 (1995).
[12]K. Suganuma and K. Nogi, "Interface Structure Formed by Characteristics Reaction between α-SiC Single Crystal and Liquid Cu," *J. Japan Institute of Metals*, **59** [12] 1292-98 (1995).
[13]A. Nishimoto, M. Ando, M. Takahashi, M. Aritoshi, and K. Ikeuchi, "Interfacial Reaction Layers in SiC/Cu Joint Friction-Bonded with Nb Intermediate

Layer," *Materials Transactions, JIM,* **40** [9] 953-56 (1999).

[14]M. Arita, R. Kinaka, and M. Someno, "Application of the Metal-Hydrogen Equilibration for Determining Thermodynamic Properties in the Ti-Cu System," *Metallurgical Transactions A,* **10A** 529-34 (1979).

[15]O. Knacke, O. Kubaschewski, and K. Hesselmann, "*Thermochemical Properties of Inorganic Substances,*" 2104, Springer-Verlag, Berlin, **II**, 1991.

AN AEROSOL DEPOSITION METHOD AND ITS APPLICATION TO MAKE MEMS DEVICES

Jun Akedo
National Institute of Advanced Industrial Science and Technology,
Namiki 1-2-1 Tsukuba,
Ibaraki, 305-8564 Japan

ABSTRACT

An aerosol deposition method (ADM) for the shock-consolidation of fine ceramic powder to form dense and hard layers is reported. Submicron ceramic particles were accelerated in gas flowing through a nozzle at up to several hundred m/s. During the interaction with the substrate, the particles formed thick (10 ~ 100 µm), dense, uniform and hard ceramics layers. A high-speed optical micro scanner MEMS device was successfully fabricated using this method. This paper describes the particular features of ADM, and comments on the deposition mechanism.

INTRODUCTION

Ceramic integration technology is required in many applications, such as, in the fabrication of micro electro mechanical systems (MEMS), display devices, fuel cells, and RF components. A ceramic layer thickness over 1 µm is required in these applications. However, thick layers produced by conventional thin or thick film methods usually have cracks, and may easily peel from the substrate.[1] Because of thermally induced diffusion, it can also be difficult to control material compositions at the interface using these methods. Additionally, fabrication can be time-consuming. It is very important to have a high-speed deposition rate, a low process temperature, and fine patterning.

For example, an effective and reliable technology is required to fabricate piezoelectric actuators for optical scanners, micro motors,[1,2] scanning force microscopy,[3] micro-pumps and ultrasonic mixers,[4,5] and micro manipulators for medical applications in MEMS. This technology also will be very important in the near future to produce ink-jet printer heads to reduce printing time, and to produce flapper-actuators to control high speed positioning of recording heads for high density storage drive.[6] Piezoelectric or electrostrictive materials with large strain and high-speed response in these applications often require dense and thick micro-patterned films on Si or stainless steel with a thickness of more than 5 µm.[2,7,8]

Figure 1 summarizes the conventional methods used to fabricate PZT films on substrates, and the applications of these films. There are many reports on fabricating 0.08 µm to 5 µm thick lead-zirconate-titanate ($Pb(Zr_{0.52},Ti_{0.48})O_3$)

Characterization & Control of Interfaces for High Quality Advanced Materials   245

(PZT) films by sol-gel,[9,10] sputtering,[11,12] metal organic chemical vapor deposition (MO-CVD),[13] pulse laser ablation,[14] electron beam evaporation,[15] and ion-beam deposition.[16] In these methods, dense PZT film can be formed and oriented on a Pt/Ti/SiO$_2$/Si substrate. However, it is difficult to make PZT films thicker than 1 ~ 3 μm using these methods. Hydrothermal synthesis[17,18] has the advantage of a low process temperature at 150°C, and eliminates the poling

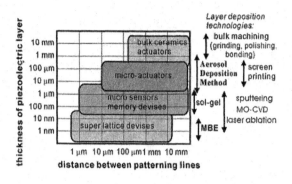

Figure 1. Methods to Fabricate PZT thick films

procedure; however, the surface roughness and the density of the films produced are poor. PZT thick films fabricated by screen-printing[19] have low density, and PZT/Pt/Si structures can be damaged during long firing times at temperatures higher than 800°C. An improved screen-printing method with a low-temperature sintering and a high resistance electrode has been reported,[20] but the piezoelectric properties of the films derived by this method were not reported.

For sputtered and sol-gel derived PZT thin films, a post deposition rapid thermal annealing treatment[21] was introduced to reduce the damage to the substrate or structure, and to improve the electrical properties; however, for a film over 1 ~ 3 μm thick, this process is not so effective. The etching of thick ceramic films by plasma etching,[22] inductively coupled plasma etching,[23,24] or reactive ion etching[25] is also difficult. For bulk PZT adhered to a Si membrane, it is difficult to ensure adequate mechanical and electrical coupling between films, and it is difficult to assemble complex structures. Thus, we can conclude, that fine patterning of over 1 μm thick PZT films on Si-based substrates by conventional methods is still difficult.

For this reason, we introduced a new deposition technique, called the Aerosol Deposition Method (ADM), which uses impact adhesion of ultra-fine particles to form and micro-pattern thick, functional ceramic layers.

EXPERIMETAL PROCEDURE
Deposition Apparatus

The ADM apparatus is made up of two vacuum chambers connected by a gas pipe. The first chamber is the deposition chamber used to form and pattern the films. It contains the spray nozzle, the substrate, the substrate heating system, and a mask alignment system. This chamber is evacuated during the deposition process using a rotary vacuum pump and a mechanical booster pump. The second chamber is the aerosol chamber to generate ceramic aerosol. It has the carrier gas system, and a vibration system to mix the powder with the carrier gas. The fine ceramic powder contained in the aerosol chamber is delivered to the deposition chamber using a pressure difference between the two chambers. The fine ceramic powder flows through a micro orifice nozzle and deposits onto the substrate behind a patterning masks. Particle velocities are controlled by the gas flow consumption, which is controlled by a mass flow controller. Details of the

ADM apparatus are described elsewhere.[26,27]

During ADM, as fine particles collide with the substrate, part of their kinetic energy is converted into energy that helps bond them to the substrate and to each other. However, the actual deposition mechanism has not been completely determined yet.[26,27]

In our experiments, the starting powder used was a commercially available raw-powder. To improve the deposition rate the powder was dry-milled. Based on SEM observations, the particle size of the powder ranges from 0.08 – 1 μm. The size of the particles in the aerosol coming out of the nozzle is 0.3 ~ 7 μm, as measured by optical scattering (using PCS-2000, PALAS Co.). This suggests that partial aggregation of the starting powder particles takes place during ADM. To reduce mechanical damage to the ceramic layers and to the substrate during deposition, large aggregated particles with a diameter >2 μm were discarded using an aerodynamic filter. The typical deposition parameters used for ADM are described elsewhere.[26,27]

RESULTS AND DISCUSSION

Coating of Ceramics Layers (α-Al$_2$O$_3$ and PZT) at Room Temperature

Different materials were successfully deposited on metal, glass, and Si substrates. Dense ceramic layers of α-Al$_2$O$_3$ and PZT with thickness of 1-100 μm were formed at room temperature on glass and metal substrates. The deposition rate for α-Al$_2$O$_3$ ranges from 1 to 3 μm/min over a deposition area 10 x 10 mm^2. The measured micro Vickers hardness, Hv, (DUH-W201, Shimazu Co.) of the deposited layers is shown in Table I. For the measurements, >5 μm thick samples were deposited at room temperature on soda lime glass substrates (Hv = 5 GPa). The hardness measurement was made for 10 s with 50 gf. The penetration depth of the indenter into the film sample is <1 μm; therefore, the influence of the underlying substrate on the measurement can be ignored. For comparison, the hardness of the bulk material prepared by conventional sintering is also presented. The results indicate that the deposited layers exhibit mechanical properties similar to those of the bulk material. This means that hard solidification occurs in oxides as well as non-oxides during the ADM.[28]

The tensile strength of the interface between the deposited layer and the stainless steel or fused silica substrate was measured using a tensile testing machine. As a result, the resin layer that was bonded between the film and the tensile testing holder (diameter: 5 mm) broke. The tensile strength is estimated

Table I. Vickers hardness of ADM as-deposited layers.[28]

material	Micro Hardness, (Hv)		Crystallite size, (nm)
	Layer	Bulk*	
Oxides			
Al$_2$O$_3$	1000 ~ 1600	1900±100	13
PZT	530	350±50	18
(Ni,Zn)Fe$_2$O$_3$	750	1040±80	5 ~ 20
Non-oxide			
AlN	1470	1180±90	
MgB$_2$	700	-	5 ~ 10

* Bulk samples were prepared from the same starting powder used to make the ADM films, and then sintered conventionally at >1200°C.

to be higher than 50 MPa. During the ADM deposition process, the substrate temperature never exceeded 50°C.

The crystal structure of the deposited layer was characterized using X-ray diffraction (XRD). The deposited layer has a randomly oriented polycrystalline structure, and the XRD pattern is similar to that for the starting-powder. However, a broadening of the peak in the pattern and a slight peak shift at higher angles is observed. The difference between the starting powder and the deposited layer may be due to a smaller crystallite size in the ADM layer, or due to non-uniform distortion during deposition.[27,29] According to TEM observations (Figure 2), the particle size of the starting α-Al_2O_3 powder is about 200 – 500 nm. Selective area diffraction (SAD) indicates that the particles have a single crystal structure. In contrast, the dense, as-deposited layer consists of randomly oriented crystallites that are <15 nm, which is an order of magnitude smaller than that of the starting powder. SAD patterns show that the layer is crystallized (i.e., it is not an amorphous phase). The microstructure results for PZT and other materials are described in detail elsewhere.[29]

To determine the density of a >100 μm thick film produced by ADM, the film volume was measured using a three-dimensional stylus profiler (Tokyo Seimitsu Surfcom 480A). The film weight was obtained using a precise electrical weight balance (Shimazu Co. AEM-5200: precision: ± 1 μg.). The density of the PZT layer is estimated to be 7.76 g/cm³, which is more than 95% of the theoretical density.[26]

Electrical Properties of Deposited Layers

ADM as-deposited layers generally have high electrical breakdown strength and resistivity. The electrical breakdown strength of a α-Al_2O_3 layer is shown in the Figure 4. For a 1.2 μm thick film, the breakdown strength is over 300 V (2 MV/cm). The resistivity of the deposited film is about 10^{15} Ωcm at 100 kV/cm, and decreases to 10^{14} Ωcm at 2 MV/cm. High velocity particle impacts over 500 m/s reduce the electrical breakdown strength.

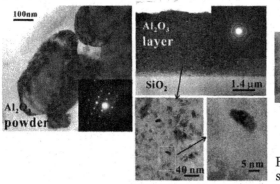

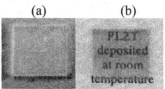

Figure 3. Influence of the starting powder on the transparency of the deposited layer.[33] Starting powder: a) with particles < 100 nm; and b) without particles < 100 nm.

Figure 2. TEM images of the α-Al_2O_3 starting powders and the layers deposited at room temperature on SiO_2 by ADM.

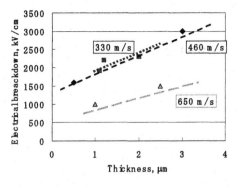

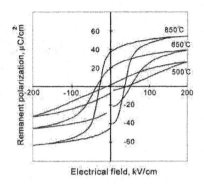

Figure 4. Electrical breakdown strength of the ADM α-Al₂O₃ layer vs. impact particle velocity.

Figure 5. P-E hysteresis of the ADM PZT layer vs. annealing temperature.[30]

This may be due to mechanical defects that may be introduced in the layer during particle impact, like shot penning.   The ferroelectric properties of the PZT layer are shown in Figure 5.   The as-deposited layers do not exhibit ferroelectricity under and externally applied field of up to 200 kV/cm, until after heat-treating. The remnant polarization (P_r) markedly increases at temperatures higher than 500°C, and reaches 20 and 38 μC/cm^2 at 600 and 850°C, respectively.[30]   The piezoelectric constant (d_{31}) of a layer annealed at 600°C is estimated to be ~100 pm/V.[31]   These properties are comparable with those of a PZT layer deposited by conventional thin film methods.[9-18]

Influence of Primary Powder Particle Size and Velocity on Deposition and Film Properties
    The influence of pre-processing the primary powder on ADM deposition rate and film properties was investigated.   Dry ball milling the primary ferroelectric powder for 5 h increases rate of deposition during ADM dramatically, by up to 30 times (up to 73 μm/min) for a 5 mm^2 area, as shown in Figure 6. However, the density and electrical properties of the thick film degrade.

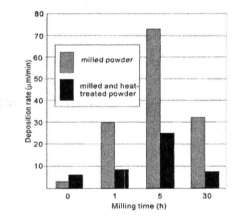

Figure 6. PZT film deposition rate by ADM as a function of raw powder dry milling time.[33]

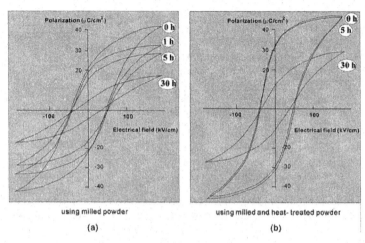

Figure 7. Hysteresis loops for PZT films deposited on a stainless steel substrate using: a) dry milled powder; and b) dry milled powder heat-treated at 800°C for 4 h in air.[33]

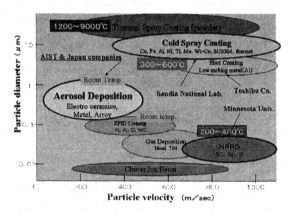

Figure 8. The particle diameter-velocity space for aerosol deposition and other spray coating methods

Heat-treating 5 h dry milled powder at 800°C for 4 h increases the deposition rate 7 - 10 times relative to the original starting powder, and does not degrade the electrical properties of a 600°C annealed layer (Pr = 20 to 32 $\mu C/cm^2$ at Ec=45 kV/cm as shown in Figure 7).

Remarkably, if the starting powder has particles <100 nm, the deposited layer looks like a pressed powder. On the other hand, after heat-treating the starting powder to grow the particle size, the film density and the optical transmittance significantly improved (Figure 3).[32,33] These results show that the particle size of the starting powder is very important to produce a dense film by ADM.

Previous studies indicate that deposition rate and film properties strongly depend on particle size and velocity.[26,32,33] Figure 8 compares ADM particle sizes and velocities with those for other coating methods. The velocity of the particles in the jet flow in the ADM was measured by time-of-flight method, in which part of the particle flow was mechanically cut from the total flow and deposited onto a moving substrate. The estimated particle velocity varies from 150 to 500 m/s. The velocity of the particles in ADM, even for a ceramic coating, is smaller than that for conventional spray coating methods.[35-39] In contrast, the cold spray method[38] for metal coating requires extremely high particle velocities over 500 m/sec to produce a high density layer.

Application to a MEMS Device

    An optical micro scanner driven by a PZT thick film [40] was fabricated by the ADM and conventional MEMS processing. An SEM image of the device fabricated is shown in Figure 9. PZT layers with a thickness of 6μm were deposited on 50 μm thick Si cantilevers. The four thin beams were not destroyed or deformed by the impact of the PZT jet. Simple masking kept the PZT powder off the mirror. These results are significant and important to reduce the production costs for specialty MEMS; the total number of the process steps required to fabricate this structure was reduced from over 20 steps using photo resist and etching, to less than 10 steps. The 100 μm thick mirror structure is guaranteed to be flat within 1/8 λ during operation in air. The thick torsional beam is also guarantied to modulate at only a specific vibrational mode. The frequency response of the scanner is shown in Figure 10. The resonant

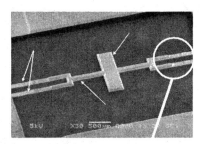

(Scanning pattern of laser beam)

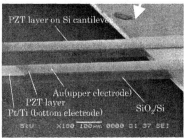

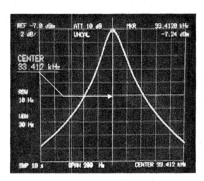

Figure 9. High-speed optical micro scanner Direct deposition of PZT thick film (6 μm) on thin Silicon membrane (50 mm), which was fabricated by conventional bulk micro machining.

Figure 10. Scanning image and dynamic property of micro-scanner fabricated by ADM [40]

frequency in the torsional mode is 33.4 kHz, with a large scan angle (peak to peak value) of more than 30°. Additionally, a high Q factor of 600 is also measured. This means that the scanner has the properties necessary for a high-resolution laser display. The performance and the production cost of the micro optical scanner fabricated by the ADM are superior to that of a conventionally prepared device.

CONCLUSION
Dense and thick ceramic layers having a nanocrystalline structure were fabricated at room temperature using an aerosol deposition method. A reduction in crystallite size (relative to the size of the starting powder) is observed after depositing ceramic materials. The deposition rate for the ADM at room temperature is 3 to 10 μm/min. The hardness and electrical breakdown strength of the layers are close to those of bulk material formed by traditional processing. ADM provides an easy way to fabricate nanostructured ceramics layers, and has the potential to fabricate MEMS devices.

ACKNOWLEDGEMENTS
This research was partially supported by NEDO projects of "Fundamental technology development for energy conservation" and "Nano structure forming for advanced ceramic integration technology in Nano technology program of Japan".

REFERENCE
[1]Muralt, "Ferroelectric Thin Films for Microsensors and Actuators: A Review," *Journal of Micromechanics and Microengineering*, **10** 136-46 (2000).
[2]M. Ikeda, H. Totani, A. Akiba, H. Goto, M. Matsumoto, and T. Yada, "PZT Thin-Film Actuator Driven Micro Optical Scanning Sensor by 3D Integration of Optical and Mechanical Devices," *Proc. IEEE MEMS Workshop '99,* 435-40 (1999).
[3]S. Akaminea, T.R. Albrecht, M.J. Zdeblicka, and C.F. Quate, "A Planar Process for Microfabrication of a Scanning Tunneling Microscope", *Sensors & Actuators A,* **21-23** 964-70 (1990).
[4]T. Massood, ed., Microactuators, Kluwer Academic Publishers, 1998.
[5]A. Manz, N. Graber, and H.M. Widmer, "Miniaturized Total Chemical Analysis Systems: A Novel Concept for Chemical Sensing," *Sensors & Actuators B, 1* 244-48 (1990).
[6]S. Koganezawa, Y. Uematsu, T. Yamada, H. Nakano, J. Inoue, and T. Suzuki, "Dual-stage Actuator System for Magnetic Disk Drives Using a Shear Mode Piezoelectric Microactuator," Digest of the Asia-Pacific Magnetic Recording Conference, 1998, *Singapore (1998) FB-02-1,2.*
[7]J. Akedo, "Deposition Method Using an Ultrafine Particle Beam and its Application to Microfabrication," *Oyo Buturi,* **68** 44-47 (1999) [in Japanese] .
[8]J.-R. Cheng, W. Zhu, N. Li, and L. E. Cross, "Electrical Properties of Sol-Gel-Derived $Pb(Zr_{0.52}Ti_{0.48})O_3$ Thin Films on a $PbTiO_3$-Coated Stainless Steel Substrate," *Applied Physics Letters,* **81** 4805-07 (2002).
[9]H.D. Chen, K.R. Udayakumar, C.J. Gaskey, L.E. Cross, J.J. Bernstein, and L.C. Niles, "Fabrication and Electrical Properties of Lead Zirconate Titanate Thick Films," *Journal of the American Ceramic Society,* **79** 2189-92 (1996).
[10]P. Luginbuhl, G-A. Racine, Ph. Lerch, B. Romanowicz, K.G. Brooks, N.F. de Rooij, Ph. Renaud, and N. Setter, "Piezoelectric Cantilever Beams Actuated by

PZT Sol-Gel Thin Film," *Sensors & Actuators A*, **54** 530-35 (1996).

[11]S.Watanabe, T. Fujiu, and T. Fujii, "Effect of Poling on Piezoelectric Properties of Lead Zirconate Titanate Thin Films Formed by Sputtering," *Applied Physics Letters*, **66** 1481-83 (1995).

[12]I. Kanno, S. Fujii, T. Kamada, and R. Takayama, "Piezoelectric Properties of C-Axis Oriented $Pb(Zr, Ti)O_3$ thin Films," *Applied Physics. Letters*, **70** 1378-80(1997).

[13]Y. Sakashita, T. Ono, H. Segawa, K. Tominaga, and M. Okada, "Preparation and Electrical Properties of MOCVD-Deposited PZT Thin Films," *Journal of Applied Physics*, **69** 8352-57 (1991).

[14]H. Kidoh, T. Ogawa, A. Morimoto, and T. Shimizu, "Ferroelectric Properties of Lead-Zirconate-Titanate Films Prepared by Laser Ablation," *Applied Physics Letters*, **58**, 2910-12 (1991).

[15]M. Oikawa and K. Toda, "Preparation of $Pb(Zr,Ti)O3$ Thin Films by an Electron Beam Evaporation Technique," *Applied Physics Letters*, **29**, 491-93 (1976).

[16] R.N. Castellano and L.G. Feinstein, "Ion-Beam Deposition of Thin Films of Ferroelectric Lead Zirconate Titanate (PZT)," *Journal of Applied Physics* **50**, 4406-11 (1979).

[17] Y. Ohba, M. Miyauchi, T. Tsurumi, and M. Daimon, "Analysis of Bending Displacement of Lead Zirconate Titanate Thin Film Synthesized by Hydrothermal Method," *Japanese Journal of Applied Physics*, **32**, 4095-98 (1993).

[18] T. Morita, T. Kanda, M. Kurosawa, and T. Higuchi: "Single Process to Deposit Lead Zirconate Titanate (PZT) Thin Film by a Hydrothermal Method," *Japanese Journal of Applied Physics*, **36** 2998-99, (1997).

[19]H.D. Chen, K.R. Udayakumar, L.E. Cross, J.J. Bernstein and L.C. Niles, "Dielectric, Ferroelectric, and Piezoelectric Properties of Lead Zirconate Titanate Thick Films on Silicon Substrate," *Journal of Applied. Physics*, **77** 3349-53 (1995).

[20]Y. Akiyama, K. Yamanaka, E. Fujisawa and Y. Kowata, "Development of Lead Zirconate Titanate Family Thick Films on Various Substrates," *Japanese Journal of Applied Physics*, **38** 5524-27 (1999).

[21]C.V.R. Vasant Kumar, M. Sayer, R. Pascual, D.T. Amm, and Z. Wu, "Lead Zirconate Titanate Films by Rapid Thermal Processing," *Applied Physics Letters*, **58**, 1161-63 (1991).

[22]M.R. Poor and C.B. Fledderman, "Measurements of Etch Rate and Film Stoichiometry Variations During Plasma Etching of Lead-Lanthanum-Zirco-nium-Titanate Thin Films," *Journal Applied Physics*, **70** 3385-87 (1991).

[23]C.W. Chung, "Reactive Ion Etching of $Pb(Zr_xTi_{1-x})O3$ Thin Films in an Inductively Coupled Plasma," *Journal of Vacuum Science Technology B*, **16** 1894-00 (1998).

[24]X. Li, T. Abe, and M. Esashi, "Deep Reactive Ion Etching of Pyrex Glass," Proc. IEEE MEMS'2000, Miyazaki, (2000) 271-76.

[25] K. Saito, J. H. Choi, T. Fukuda, and M. Ohue, "Reactive Ion Etching of Sputtered PbZr1-xTixO3 Thin Films," *Japanese Journal of Applied Physics*, **31** L1260-62 (1992).

[26]J. Akedo and M. Lebedev, "Aerosol Deposition Method (ADM): A Novel Method of PZT Thick Films Producing for Microactuators," *Recent Research and Development in Materials Science*, **2** (Research Signpost, India) 51-77(2001).

[27]J. Akedo and M. Kiyohara, "Nanostructuring and Shock Compaction Using Fine Particle Beam-Aeorsol Deposition for Forming of Nanocrystal Layer and Powder Techology," *Journal of the Society of Powder Technology Japan*, **40**,

192-01 (2003).

[28]J. Akedo, M. Lebedev, A. Iwata, H. Ogiso and S. Nakano, "Aerosol Deposition Method (ADM) for Nano-crystal Ceramics Coating Without Firing," *Materials Research Society Symposium Proc.*, **778** (U8.10/W7.10), 289-94 (2003).

[29]J. Akedo and M. Lebedev, "Microstructure and Electrical Properties of Lead Zirconate Titanate (Pb(Zr52/Ti48)03) Thick Film Deposited with Aerosol Deposition Method," *Japanese Journal of Applied Physics,* **38** 5397-01 (1999).

[30]J. Akedo and M. Lebedev, "Effects of Annealing and Poling Conditions on Piezoelectric Properties of Pb(Zr0.52,Ti0.48)O3 Thick Films Formed by Aerosol Deposition Method," *Journal of Crystal Growth,* **235** 397-02 (2002).

[31]J. Akedo and M. Lebedev, "Piezoelectric Properties and Poling Effect of Pb(Ti,Zr)O3 Thick Films Prepared for Microactuators by Aerosol Deposition Method," *Applied Physics Letters,* **77** 1710-12 (2000).

[32]J. Akedo and M. Lebedev, "Influence of Carrier Gas Conditions on Electrical and Optical Properties of Pb(Zr, Ti)O3 Thin Films Prepared by Aerosol Deposition Method," *Japanese Journal of Applied Physics,* **40** 5528-32 (2001).

[33]J. Akedo and M. Lebedev, "Powder Preparation for Lead Zirconate Titanate Thick Films in Aerosol Deposition Method," *Japanese Journal of Applied Physics*, **41** 6980-84 (2002).

[34]M. Levedev, J. Akedo, K. Mori and T. Eiju, "Simple Self-Selective Method of Velocity Measurement for Particles in Impact-Based Deposition," *Journal of Vacuum Science & Technology A,* **18**, 563-65 (2000).

[35]R.C. Dykhuizen, M.F. Smith, D.L. Gilmore, R.A. Neiser, X. Jiang and S. Sampath, "Impact of High Velocity Cold Spray Particle", *Journal of Thermal Spray Technology,* **8** 559-64 (1999).

[36]T. Ide, Y. Mori, I. Konda, N. Ikawa and H. Yagi, "A Film Formation Method using Hypervelocity Microparticle Impact by Electrostatic Acceleration (2nd Report)-Preparation of Diamond-like Carbon Film," *Journal of the Japanese Society for Precursor Engineering,* **57**, 143-48(1991) .

[37]N.P. Rao, N. Tymiak, J. Blum, A. Neuman, H.J. Lee, S.L. Girshick, P.H. McMurry, and J. Heberlein, "Hypersonic Plasma Particle Deposition of Nano-structured Silicon and Silicon Carbide," *Journal of Aerosol Science,* 707-20 (1998).

[38]C. Hayahi, "Ultrafine Particles," *Journal of Vacuum Science Technology A,* **5** 1375-84 (1987).

[39]E. Barborini, P. Piseri, A. Podesta and P. Milani, "Cluster Beam Microfabrication of Pattern of Three-Dimensional Nanostructured Objects," *Applied Physics Letters,* **77** 1059-61 (2000).

[40]N. Asai, R. Matsuda, M. Watanabe, H. Takayama, S. Yamada, A. Mase, M. Shikida, K. Sato, M. Lebedev and J. Akedo, "A Novel High Resolution Optical Scanner Actuated by Aerosol Deposited PZT Films," Proc. of IEEE Micro Electro Mech. Systems, Kyoto (2003) 247-50.

MORPHOLOGY AND PERFORMANCE OF A Ni-YSZ CERMET ANODE FOR SOLID OXIDE FUEL CELLS

T. Fukui and K. Murata
Nano Particle Technology Center,
Hosokawa Powder Technology Research
Institute
1-9 Shoudai Tajika, Hirakata
Osaka 573-1132, Japan

H. Abe, M. Naito, and K. Nogi
Joining and Welding Research Institute
(JWRI), Osaka University
11-1 Mihogaoka, Ibaragi, Osaka
567-0047, Japan

ABSTRACT

NiO-Y_2O_3 stabilized ZrO_2 (YSZ) composite powders for solid oxide fuel cells (SOFCs) were processed using a novel processing technique. The composite powders consisted of submicron size NiO particles with finer YSZ particles. Ni-YSZ cermet anodes produced from the composite powders had a uniformly dispersed porous structure with Ni and YSZ grains smaller than several hundred nano-meters. The YSZ content in the cermet anode affects the anode performance. The cermet anode with 19.9 mol% YSZ has the most homogeneous porous structure with almost no Ni agglomerates. The cermet anodes produced have improved electrical performance at a lower operating temperature. ($\leqq 800°C$).

INTRODUCTION

Physical, chemical, and electrical properties of functional materials depend on the morphology and composition of starting powders.[1-4] Therefore, the properties of the functional materials can be improved by controlling the morphology and composition of the starting powders. In solid oxide fuel cells (SOFCs), starting powders with controlled morphology and composition have been used to improve the electrical performance.[1-7] In particular, the performance of SOFC electrodes is affected by morphology, composition, and structure; the electrochemical activity for power generation depends on the electrochemical reaction area created by the electrode grains, and electric and/or ionic conductivity depends on how the electrode grains interconnect. For instance, the long-term stability of a Ni-YSZ cermet anode during high temperature operation (1000°C) is highly influenced by its morphology, and it has been improved with a NiO-YSZ composite powder consisting of NiO grains covered with fine YSZ particles.[3,8] The electrochemical activity of the Ni-YSZ cermet anode is also influenced by the YSZ content in the starting NiO-YSZ composite powder.[6]

In this study, NiO-YSZ composite powders with different YSZ contents were produced using a novel processing technique called mechano-chemical bonding (MCB) technique. MCB is a novel method to create chemical bonds between particles in a dry powder process without any binder. It can be used to control the morphology and composition of multifunctional composite powders for various applications. During

MCB, mechanical, chemical, and thermal interactions produce strong bonds between particles.

In this paper, the relationship between the morphology and performance of Ni-YSZ cermet anodes fabricated from MCB-treated NiO-YSZ composite powder will be discussed, and the anode performance will be evaluated in comparison with a Ni-YSZ cermet anodes fabricated from NiO-YSZ powder produced by conventional powder mixing.

EXPERIMENTAL

Nickel oxide (NiO, 99.9%, average particle size: 0.79 μm, Yamanaka chemical Co.) and 8-mol% Y_2O_3 stabilized ZrO_2 (YSZ, average particle size: 0.08 μm, Tosoh Co., TZ-8Y) were used as the starting materials. The NiO and YSZ powders were processed using mechano-chemical bonding to make multifunctional composite particles. NiO and YSZ powders were treated for 6 min using MCB (Special Type, Hosokawa micron Co. Ltd.). In this experiment, The NiO and YSZ molar ratios studied were 77.0:23.0, 80.1:19.9 and 83.2:16.8. The NiO and YSZ powders were also treated by conventional dry ball milling for 30 min. The NiO-YSZ powders produced by these operations were characterized using scanning electron microscopy (SEM, Model S-3500N, Hitachi Ltd, Japan) with energy dispersive analysis of X-rays (EDAX, Model EX-200, HORIBA Ltd., Japan). Particle size distribution was measured by laser diffraction and scattering (MICROTRAC, Model HRA9320-X100, NIKKISO Co. Ltd., Japan). Powder samples were dispersed in distilled water for the particle size measurement.

The NiO-YSZ powders were mixed with organic binder, and were printed onto one side of a 13 mm diameter by 0.2 mm thick YSZ electrolyte pellet. The printed body was fired at 1350°C in air to produce the NiO-YSZ anode. Then, the $(La,Sr)MnO_3$(LSM)-YSZ powder selected as a cathode material[1,6] was printed onto the other side of the YSZ electrolyte pellet, and the pellet was fired at 1200°C. Pt wire was used as the reference electrode. The anode polarization between the anode and the reference electrode was measured using the current interruption technique with a current density of up to 1 A/cm^2. The single cell produced was operated at 800 and 700°C in H_2 – 3% H_2O for the anode, and in air for the cathode. The NiO-YSZ anode was reduced at 800°C in H_2 – 3% H_2O to make the Ni-YSZ cermet anode. The morphology of the Ni-YSZ cermet anode was characterized using SEM with EDAX after the cell test was finished.

RESULTS AND DISCUSSION

Figure 1 shows a SEM photograph of the NiO-YSZ composite powder with 23.0 mol% YSZ. It is obvious from this figure that the NiO-YSZ composite powder consists of fine particles coated with finer, ~100 nm particles. SEM-EDAX spot analysis confirms that the fine particles are NiO, and that the finer, 100 nm size particles are YSZ. SEM observation also confirms that other

Figure 1. SEM photograph of NiO-YSZ composite powder with 23.0 mol% YSZ prepared by MCB.

Figure 2. SEM photographs of Ni-YSZ cermet anode produced from NiO-YSZ composite powder with 19.9 mol% YSZ prepared using MCB: a) high magnification; and b) low magnification image.

composite powders with 19.9 and 16.8 mol% YSZ have almost the same morphology as the composite powder with 23.0 mol% YSZ. The average particles size of the NiO starting powder and of the composite powder was measured by laser diffraction and scattering to be 0.79 μm and 0.42 μm, respectively.

Figure 2 shows SEM photographs of the surface of a Ni-YSZ cermet anode fabricated with the NiO-YSZ composite powder with 19.9 mol% YSZ that was produced by MCB.

Figure 3 shows a SEM photograph of a Ni-YSZ cermet anode produced from NiO-YSZ powder prepared by conventional dry ball mixing. The Ni-YSZ cermet anodes fabricated with the different YSZ content powders produced by MCB have submicron size grains and a relatively homogeneous porous structure. The Ni-YSZ cermet anode with 19.9 mol% YSZ, especially, shows a uniformly porous structure with Ni and YSZ grains less than several hundred nano-meters (see Figure 2). The anodes with 16.8 and 23.0 mol% YSZ show some Ni agglomeration. By comparison, the cermet anode produced from the conventionally prepared NiO-YSZ powder has micron size grains, and an apparently less homogeneous structure. SEM-EDAX element analysis confirms that the micron size grains are nickel. It is believed that NiO agglomeration may be a problem in the NiO-YSZ powder prepared by conventional ball milling. The uniform and finer dispersion structure of the cermet anode fabricated from MCB powder is attributed to the

Figure 3. SEM photograph of a Ni-YSZ cermet anode produced from NiO-YSZ powder with 19.9 mol% prepared by conventional dry mixing.

starting morphology of the composite powder that consisted of fine NiO particles with 100 nm size YSZ particles. Normally, the NiO grains in a conventional NiO-YSZ mixture sinter easily during the anode formation process, because NiO densification occurs at a temperature lower than the anode fabrication temperature (1300-1500°C). Fine and highly dispersed YSZ grains prevent NiO sintering in the anode structure during the anode formation process.

Figure 4 shows the polarization of the Ni-YSZ cermet anodes at 800°C as a function of the current density. Lower polarization at the same current density indicates a higher performance anode. The polarization of the Ni-YSZ cermet anode fabricated from the NiO-YSZ composite powder produced by MCB is lower than that of the anode fabricated from powder produced by conventional mixing. The performance of the anode actually is improved with MCB in comparison with conventional mixing. The polarization at 800°C and 700°C of MCB treated cermet anodes as a function of YSZ content is shown in Figure 5. The Ni-YSZ cermet anode with 19.9 mol% YSZ has the best performance. Additionally, anode performance improves at the lower operating temperature of 700°C. The electrochemical reaction for power generation occurs on the three-phase boundary (TPB) created by the Ni grains, YSZ grains, and the pores; the electrochemical activity increases with an increase of the TPB length.[9,10] Finer Ni and YSZ grains in the cermet anode can act to increase the TPB length, and improve anode performance. The MCB cermet anode has fine (nano-size) Ni and YSZ grains in the porous structure. It is believed that this excellent structure leads to high performance at lower operating temperature ($\leq$800°C).

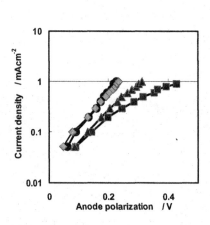

Figure 4. Anode polarization at 800°C as a function of current density for an anode fabricated with: MCB powder with a YSZ content of 23.0 mol%($\blacklozenge$), 19.9 mol% ($\bullet$) and 16.8 mol% ($\blacktriangle$); and conventionally processed powder with 23.0 mol% YSZ($\blacksquare$).

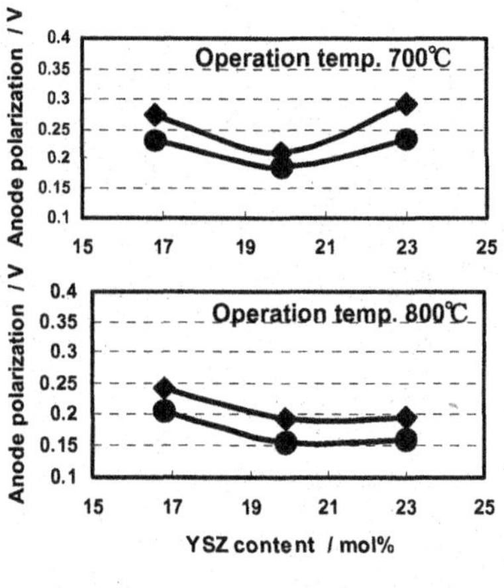

Figure 5. Anode polarization as a function of YSZ content at a current density of: 0.3 A/cm^2($\bullet$); and 0.5 A/cm^2($\blacklozenge$).

CONCLUSIONS
(1) NiO-YSZ composite powders with different concentrations of YSZ were successfully fabricated by mechano-chemical bonding (MCB). The composite powders processed by MCB consist of submicron sized NiO particles with finer YSZ particles.
(2) Ni-YSZ cermet anodes for SOFCs produced from the NiO-YSZ composite powder show Ni and YSZ grains less than several hundred nano-meters with a homogeneous porous structure, and almost no Ni agglomerates.
(3) The MCB powder produces a Ni-YSZ cermet anode for SOFCs that has improved electrical performance at lower operating temperature. The YSZ content in the cermet anode affects electrical performance. The Ni-YSZ anode with 19.9 mol% YSZ shows the best performance.

ACKNOWLAGEMENT
This work was supported by the project to create university-based businesses and develop practical application of NEDO Japan.

REFERENCES
[1]G.L. Messing, S-C. Zhang, and G.V. Jayanthi, "Ceramic Powder Synthesis by Spray Pyrolysis," *Journal of the American Ceramic Society*, **76** 2707-26 (1993).
[2]T. Fukui, T. Oobuchi, Y. Ikuhara, S. Ohara, and K. Kodera, "Synthesis of (La,Sr)MnO₃-YSZ Composite Particles by Spray Pyrolysis," *Journal of the American Ceramic Society,* **80** 261-63 (1997).
[3]T. Fukui, S. Ohara, and K. Mukai, "Long-Term Stability of Ni-YSZ Anode with a New Microstructure Prepared from Composite Powder", *Electrochem. Solid-State Lett.* **1**, 3, 120-22 (1998).
[4]R. Maric, S. Ohara, T. Fukui, T. Inagaki, and J. Fujita, "High-Performance Ni-SDC Cermet Anode for Solid Oxide Fuel Cells at Medium Operating Temperature," *Electrochem. Solid-State Lett.,* **1**, 5, 201-203 (1998).
[5]R. Maric, S. Ohara, T. Fukui, H. Yoshida, M. Nishimura, T. Inagaki, and K. Miura, "Solid Oxide Fuel Cells with Doped Lanthanum Gallate Electrolyte and LaSrCoO₃ Cathode, and Ni-Samaria-Doped Ceria Cermet Anode," *Journal of the Electrochemical Society,,* **146**, 2006-10 (1999).
[6]T. Fukui, S. Ohara and M. Naito, and K. Nogi, "Morphology and Performance of SOFC Anode Fabricated from NiO/YSZ Composite Particles," *Journal of Chemical Engineering of Japan* **34** 964-66 (2001).
[7]T. Fukui, S. Ohara, M. Naito, and K. Nogi, "Morphology Control of the Electrode for Solid Oxide Fuel Cells by Using Nanoparticles," *Journal of Nanoparticle Research*, **3** 171-74 (2001).
[8]T. Fukui, S. Ohara, M. Naito, and K. Nogi, "Performance and Stability of SOFC anode Fabricated from NiO-YSZ Composite Particles," *Journal of Power Sources,* **110** 91-95 (2002).
[9]N. Q. Minh, "Ceramic Fuel Cell," *Journal of the American Ceramic Society,* **76** 563-68 (1993).
[10]N. Nakagawa, K. Nakajima, M. Sato, and K. Kato, "Contribution of the Internal Active Three-Phase Zone of Ni-Zirconia Cermet Anodes on the Electrode Performance of SOFCs," *Journal of the Electrochemical Society*, **146** 1290-95 (1999).

FUEL CELL TECHNOLOGY IN THAILAND

P. Aungkavattana, S. Charojrochkul,
H. Mahaudom, A. Kittiwanichawat,
W. Wattana, and M. Henson
National Metal and Materials
Technology Center
114 Thailand Science Park
Pathumthani 12120, Thailand

S. Kuharuangrong
Suranaree University of Technology
Nakhon
Ratchasima 30000, Thailand

S. Assabumrungrat and S. Srichai
Chulalongkorn University
Patumwan
Bangkok 10330, Thailand

J. Charoensuk, W. Khan-ngern,
P. Khamphakdi, and N. Nakayothin
King Mongkut's Institute of
Technology
Ladkrabang
Bangkok 10520, Thailand

ABSTRACT
An overview of fuel cell research and development in Thailand is presented, with emphasis on a solid oxide fuel cell (SOFC) project. The focus of the project is to develop technology to generate electricity from electrochemical reactions between ethanol and oxygen gas in air using solid oxide fuel cells. Research work on cell development and characterization (including performance testing) of solid oxide fuel cells, and on the development of a prototype electric generator from solid oxide fuel cells is presented. The results from this project will contribute to a better understanding and the development of technology to generate power from solid oxide fuel cells. The learning process will lead to commercialization of a clean energy source as an alternative to fossil fuels to generate electricity in Thailand in the future.

INTRODUCTION
In recent years, fuel cell technology has been investigated for its potential use in transport and power generation applications in many countries in the world. Fuel cells have the capacity to replace various conventional technologies with cleaner and more efficient energy systems.[1] A fuel cell, in principle, performs like a battery, but never needs recharging. As long as fuel is provided, the fuel cell will produce energy in the form of electricity and heat.[2] Hydrogen is the fuel that is fed into the fuel cell. It can be obtained from any hydrogen source, such as natural gas, methanol, ethanol, or ammonia.

Fuel cells have many distinct advantages. Dramatic reductions in urban air pollution, and a decrease in oil imports can be realized with the extensive use of fuel cells.[3] If just 10 % of American automobiles were powered by fuel cells, one million tons of air pollutants, and 60 million tons of greenhouse carbon dioxide gas would be eliminated annually.[4] Fuel cells are now utilized in buses, boats, and airplanes, and all of the major automobile manufacturers are currently in the process of developing or testing fuel cell vehicles.[4-5] Fuel cells are providing power to hospitals, police stations, and banks, and in Tokyo, 25-kilowatt units are already on line.[5] Remarkably, fuel cells are also in operation at landfills and wastewater treatment facilities, converting the methane gas produced there into electrical power.[6] Micro fuel cells, used in computers, palm pilots, cellular phones and hearing aids, run on methanol.[6]

Fuel cells are categorized by their electrolytes. The alkaline, phosphoric acid, and solid polymer types are all past the fundamental development stage, and are being tested or used in commercial applications, though not necessarily on a fully cost-competitive basis. Polymer electrolyte membrane fuel cells (PEMFCs) are suitable for application as a mobile power source primarily for the automobile industry.[4] Molten carbonate fuel cells (MCFCs) and solid oxide fuel cells (SOFCs) offer high-temperature operation and remarkable generating efficiencies when combined with gas and/or steam turbines. These high temperature types of fuel cells still require research and testing to address a series of fundamental problems, however, the technique is still evolving, and excellent technical progress is being made. Multikilowatts of solid oxide fuel cells have been operated for thousand of hours, and have shown excellent performance.[7]

Solid oxide fuel cells (SOFCs) are of particular interest. This type of fuel cell is promising for industrial use and large-scale electrical generating stations. These cells have the highest efficiencies of any conventional fuel cell design, they can be used with a wide variety of fuels, and they can deliver a quiet, uninterruptible source of power.[7,8] Generating efficiencies can reach as high as 60% to 85% with co-generation, and the electrical output can be 100 kW.[8] Composed of ceramic materials, SOFCs can operate at the high temperatures (up to approximately 1000°C) needed to produce the exhaust gas temperature necessary for co-generation applications.[9] If the capture and use of waste heat occurs, overall fuel use efficiencies could reach 80-85%.[9]

The high temperature SOFC is an electrochemical conversion device composed of three ceramic layers (Figure 1) . A dense, solid electrolyte layer is

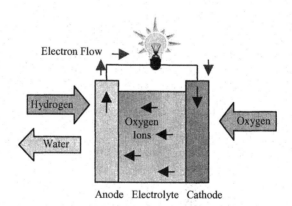

Figure 1. A schematic of the principle of operation of a solid oxide fuel cell.

usually sandwiched between a porous anode layer[10] and a porous cathode layer.[11] Oxygen from the air undergoes chemical reduction at the cathode layer, forming the anode where they react with hydrogen fuel. The fuel is then oxidized, which releases electrons to an external circuit, thereby creating electricity.

FUEL CELL RESEARCH IN THAILAND

Interests in fuel cell research and development in Thailand started in fiscal year (FY) 1999, when a government agency called the National Energy Policy Office (NEPO) provided funding (5 millions bahts, which is equivalent to $ 0.12 million) to the King Mongkut University of Technology Thonburi (KMUTT) to do research on molten carbonate fuel cells (MCFCs). Researchers at the National Metal and Materials Technology Center (MTEC) joined in developing materials for the electrolyte and the electrodes for MCFCs. In FY 2000, due to their expertise, and their in-house facilities for ceramic fabrication using tape casting and screen-printing techniques, MTEC started to focus on SOFCs. A feasibility study on SOFCs was first conducted, and a 3-year project on planar SOFCs was funded in FY 2001 with the budget of about 11 millions bahts ($0.3 millions). After a year on planar SOFCs, MTEC decided to promote R&D on tubular SOFCs with a collaboration with Advanced Ceramic Limited (ACL) in the United Kingdom (UK) in FY 2002. The project goal is to construct a tubular fuel cell prototype to generate electricity using ethanol. A progress update on this project will be discussed.

SOLID OXIDE FUEL CELL DEVELOPMENT IN THAILAND

The objective of this project is to study and develop technology to generate electricity from electrochemical reactions between ethanol and the oxygen gas in air using a planar solid oxide fuel cell. The research work is divided into two parts: 1) Cell development and characterization (including performance testing) of solid oxide fuel cells; and 2) Development of an electricity generator prototype from solid oxide fuel cells. Both parts of the project run in parallel by Thai researchers from various institutes. The results from this project will contribute to a better understanding of SOFC materials and fabrication technology. The learning process will lead to the commercialization of an energy source as an alternative to fossil fuels to generate electricity in Thailand in the future.

Planar SOFC Materials and Fabrication

The materials and ceramic fabrication techniques used to make the SOFCs are summarized in Table I.

Anode material synthesis and forming: The anode was made from a nickel-based cermet composition synthesized with approximately 50 vol% each of 8 mol% yttria stabilized zirconia (YSZ) and NiO powder. Electrodes were formed

Table I. Summary of the planar SOFC materials and fabrication techniques

Components	Materials	Fabrication Technique
anode	Ni/YSZ and Ni/doped CeO_2	tape casting
electrolyte	YSZ and doped CeO_2	tape casting
cathode	LSM, Doped-$LaSrMnO_3$	tape casting/screen printing
interconnector	stainless steel	machining

by tape-casting green ceramic tapes of approximately 150-200 microns thick. The green tapes were debindered, and then sintered in a tube furnace at 1200°C in a controlled atmosphere (5%H$_2$ - 95%N$_2$).

Cathode material synthesis and forming: The cathodes were synthesized from La$_2$O$_3$, SrCO$_3$, and Mn$_2$O$_3$ (LSM) powders using the mixed-oxide route. This included ball-milling for 6 hrs followed by calcining at 1200°C for 2 hrs. The LSM powders were fully reacted after calcination as shown in the XRD pattern in Figure 2. Similar to the anode, LSM cathodes were formed into a ceramic green tape by tape casting. The LSM-based materials were doped with Fe or Co to improve the electrical conductivity. The LSM formulations examined are shown in Table II. The tape cast LSM specimens were sintered at 1450°C to produce a fully dense microstructure (Figure 3).

Electrolyte Material Synthesis and Forming: To fabricate the electrolyte, yttria-stabilized zirconia (8YSZ, Daiichi powder) was prepared as a slurry (organic/water-based system), and tape-cast in the same manner as the electrodes. The green tapes were de-bindered, and then sintered in air at 1500°C for 2 hours as shown in Figure 4.

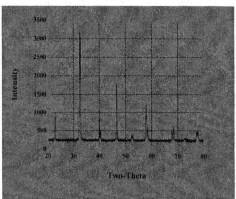

Figure 2. XRD pattern for the SOFC cathode (La$_{0.82}$Sr$_{0.16}$MnO$_3$) after calcination 1200°C 2h.

Table II. The LSM formulations examined for the cathode.

Formula	La$_2$O$_3$ (mol%)	SrCO$_3$ (mol%)	Mn$_2$O$_3$ (mol%)	Co$_3$O$_4$ (mol%)	Fe$_2$O$_3$ (mol%)
LSM16C2	0.42	0.16	0.40	0.02	-
LSM16C4	0.42	0.16	0.30	0.12	-
LSM16F2	0.42	0.16	0.40	-	0.02
LSM16F4	0.42	0.16	0.30	-	0.12

These basic materials are generally used to make conventional SOFCs. The present design is an electrolyte-supported tube in which the dimensions of the YSZ tube are 4 mm OD, and 20 cm in length (Figure 5). To make an electric generator, individual cells were connected in series. For a 6-cell generator operated at about 850-900°C in hydrogen, the performance power was about 7.34 watts. Test runs of single tubular SOFCs in hydrogen (with nitrogen as carrier gas), and with steam reformed ethanol at 600 and 700°C also were completed (Figures 6). The results show that the performance of the cell with steam-reformed ethanol at 700°C is not as high as in hydrogen. The Model MTEC-01 cell stack reactor (Figure 7), which has 29 tubular SOFCs connected in series, was test run in hydrogen. The total power, when operated at 850°C in hydrogen, was only about 10 watts. It is suspected that a few cells in the stack were not at peak performance, and may have reduced the total power of the stack. Further experiments are now underway to improve the stack performance.

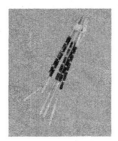

Figure 5. Tubular, co-sintered SOFCs.

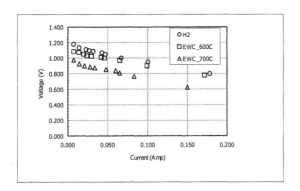

Figure 6. The current-voltage (I-V) performance curve for a tubular SOFC (single cell) operated under different conditions: H2 = in 7% H_2 in Ar at 750°C; EWC_600°C = with Ethanol + H_2O reformed at 600°C; and EWC_700°C = with Ethanol + H_2O reformed at 700°C.

Figure 7. The model MTEC-01 reactor, which has 29 tubular SOFCs connected in series.

CONCLUSION
This paper provides an overview and the status of the R&D on fuel cell development in Thailand. The development of fuel cell technology has come a long way. Further improvements through R&D and feedback from testing under realistic operation conditions will help bring the technology to full maturity.

ACKNOWLEDGEMENT
The author would like to thank her colleagues at the National Metal and Materials Technology Center (MTEC), Thailand Science Park, who contributed to this paper. The technical supports from researchers at Chulalongkorn University, KMUTL, KMUTT and Suranaree University of Technology are gratefully acknowledged. This work is financially support by MTEC RD&E through the funding no. MT-B-45-CER-07-127-I (B21 C10127).

REFERENCES
[1]J.H. Hirschenhofer, D.B. Stauffer, R.R. Engleman, and M.G. Klett, "Fuel Cell Handbook," 4th ed. DOE/FETC-99/1076. Edited by U.S. Department of Energy, Federal Energy Technology Center, (1998).
[2]N.Q. Minh, and T. Takahashi, "Science and Technology of Ceramic Fuel Cells," *Elsevier*, 5 (1995).
[3]New Sunshine Program: R&D Program on Energy and Environment Technologies, Edited by New Sunshine Program Promotion Headquarters, Agency of Industrial Science and Technology, Ministry of International Trade and Industry, Tokyo, (1998).
[4]Tom Koppel, "Powering The Future, The Ballard Fuel Cell and the Race to Change the World," John Wiley & Sons Canada, Ltd. (1999).
[5]Peter Hoffmann, "Tomorrow's Energy, Hydrogen, Fuel Cells and the Prospects for a Cleaner Planet," The MIT Press, U.S.A. (2002).
[6]C. Hebling and U. Groos, "Miniature Fuel Cell Systems for Consumer Applications," International Conference The Fuel Cell World Proceedings, Lucerne, Switzerland, 41-48, (2002).
[7]K. Foger and B. Godfrey, "System and Demonstration Program at Ceramic Fuel Cells Ltd in Australia," **167**, Proceedings of the 4th European Solid Oxide Fuel Cell Forum. Edited by A.J. McAvoy. Lucerne, Switzerland, 10-14 July (2000).

[8]M. Pastula, R. Boersma, D. Prediger, M. Perry, A. Horvath, J. Devitt, and D. Ghosh, "Development of Low Temperature SOFC Systems for Remote Power Applications," ibid., 123.

[9]A. Schuler, T. Zähringer, B. Doggwiler, and A. Rüegge, "Sulzer Hexis SOFC Running on Home Heating Oil," ibid., 107.

[10]G.M. Christie, P. Nammensma, and J.P.P. Huigsmans, "Status of Anode Supported-Thin Electrolyte Ceramic SOFC Component Development at ECN," ibid., 3.

[11]Tietz, H.P. Buchkremer, and D. Stover, "Components Manufacturing for Solid Oxide Fuel Cells," *Solid State Ionics* **152-153,** 373-381 (2002).

ELECTROPHORETIC DEPOSITION OF A HIGH PERFORMANCE La(Sr)Ga(Mg)O₃ ELECTROLYTE FILM FOR A LOW TEMPERATURE SOLID OXIDE FUEL CELL

Motohide Matsuda, Osamu Ohara, and
Michihiro Miyake
Department of Environmental
Chemistry and Materials
Faculty of Environmental Science and
Technology
Okayama University
Tsushima-Naka
Okayama 700-8530, Japan

Kenji Murata and Takehisa Fukui
Hosokawa Powder Technology
Research Institute
Hirakata
Osaka 573-1132, Japan

Satoshi Ohara
Institute of Multidisciplinary Research
for Advanced Materials
Tohoku University
Katahira, Aoba-ku
Sendai 980-8577, Japan

ABSTRACT

Self-supported, dense La(Sr)Ga(Mg)O₃ electrolyte films <100 μm thick were successfully fabricated by electrophoretic deposition. The La(Sr)Ga(Mg)O₃ ceramic electrolyte films, with oxide ion conductivities higher than those of well-known Y_2O_3-stabilized ZrO_2, exhibited low ohmic resistance at elevated temperatures. To investigate the cell performance of the La(Sr)Ga(Mg)O₃ electrolyte films thus fabricated, Ni-$(Y_{0.2}Ce_{0.8})O_{2-\delta}$ powders were prepared for the anode by spray pyrolysis, and La(Sr)CoO₃₋δ powders were prepared for the cathode by spray-freeze-drying. A maximum output density around 0.5 W/cm² was obtained for a single cell consisting of Ni-$(Y_{0.2}Ce_{0.8})O_{2-\delta}$ | 50 μm-thick La(Sr)Ga(Mg)O₃ electrolyte film | La(Sr)CoO₃₋δ using humidified (3% H₂O) hydrogen as the fuel.

INTRODUCTION

Solid oxide fuel cells (SOFCs) have been widely regarded as relatively benign, and highly efficient devices to convert chemical energy directly into electrical energy. However, current SOFCs with an Y_2O_3-stabilized ZrO_2 (YSZ) electrolyte face serious problems, such as the degradation of the physiochemical properties of the cell-component materials at the high operating temperature around 1000°C. Therefore, it is desirable to reduce the operating temperature.

It has recently been found that the perovskite type of La(Sr)Ga(Mg)O$_3$ (LSGM) material shows excellent oxide ion conduction properties.[1,2] The oxide ion conductivities of LSGM at 800°C are comparable to those of YSZ at 1000°C. Therefore, LSGM becomes a strong candidate as an alternative electrolyte for SOFCs operated at or below 800°C. Fukui et al.[3] have succeeded in fabricating 130 μm-thick LSGM electrolyte films by tape casting. They reported that maximum output densities around 0.41 W/cm^2 can be attained at 700°C from SOFCs with the LSGM electrolyte. Recently, electrophoretic deposition (EPD), which is a film-fabrication technique based on colloidal processing, has been applied to form thinner LSGM electrolytes with low ohmic resistance.[4-7] EPD has some technological advantages over other film-fabrication techniques including: 1) the experimental system is simple; and 2) uniform thickness ceramic films can be fabricated in complex forms and large sizes.

This paper describes the microstructure and electrical properties of LSGM films fabricated using EPD. In addition, the cell performance of 50 μm-thick LSGM films at 700°C is reported. For cell performance tests, mixed Ni-(Y$_{0.2}$Ce$_{0.8}$)O$_{2-\delta}$ (Ni-YDC) powders with high electrode activity were prepared for the anode through spray pyrolysis of solutions containing the respective compounds. In this paper, the preparation of high performance Ni-YDC cermet anode powders is also described.

EXPERIMENTAL

Electrophoretic Fabrication and Characterization of LSGM Films

For the EPD experiments, LSGM powders with atomic ratio of La : Sr : Ga : Mg = 9 : 1 : 8 : 2 were prepared according to a previous report[8] using lanthanum acetate, strontium acetate, magnesium acetate, and gallium nitrate as starting materials.

The LSGM powders prepared were ultrasonically dispersed in acetylacetone containing iodine. The concentration of the LSGM powder in the suspensions was about 6.7 g/L. Two electrode plates separated by a distance of 1 cm were placed in the suspensions containing the LSGM powders. One was graphite on which the LSGM powders accumulated, and the other was Pt. EPDs of the LSGM powders were carried out with a constant current density of 14 mA/cm^2. The films deposited on the graphite substrates were heated up to 800°C, held for 9 h, and then sintered at 1500°C for 10 h. Heating at 800°C was indispensable to eliminate the graphite substrate and to form flat films.

Microstructure was characterized using scanning electron microscopy (SEM). Electrical resistance was measured using the AC impedance method over frequencies ranging from 5 Hz to 13 MHz. For the resistance measurements, Pt paste was used to produce the electrodes on the LSGM films.

Preparation of Ni-YDC Cermet Anode Powders

Mixed Ni-YDC powders with Ni : YDC = 1 : 1 by volume were prepared by spray pyrolysis of solutions containing Ni^{2+}, Ce^{3+} and Y^{3+} ions. This was and effective way to prepare high performance anode powders for intermediate temperature SOFCs.[9-13] For the spray pyrolysis, an apparatus consisting of a mist generator with an ultrasonic vibrator operated at 1.7 MHz was used to produce the spray mist. The mist was transported to the furnace using air as the carrier gas. Pyrolysis occurred in an electric furnace with four heating zones in series The temperatures of respective heating zones, 1, 2, 3, and 4, were kept at 200, 400, 800, and 1000°C, respectively. The details of the preparation of the mixed anode powders are described elsewhere.[9-13] The as-prepared powders were composed of mixture of NiO and YDC. The NiO was completely reduced to Ni during SOFC testing.

Preparation of Cathode Powders

For the $La_{0.6}Sr_{0.4}CoO_{3-\delta}$ (LSC) cathode, the precursor powders were prepared by spray-freeze-drying a mixed aqueous solution that included the respective compounds according to the stoichiometric composition. These precursor powders were heated at 900°C for 10 h in air, pulverized and then re-heated at 1100°C for 10 h to obtain pure LSC. The products were sufficiently pulverized prior to the use.

Construction and Test of SOFC

The Ni-YDC cermet powders were mixed with ethyl carbitol, painted on the surface of the LSGM films, and heated at 1250°C for 2 h. The LSC powders were mixed with ethyl carbitol, painted on the surface of the LSGM films with the Ni-YDC anode and heated at 1000°C for 2 h. The thickness of the anode and cathode layer was about 50 μm and 100 μm, respectively.

The electrical power generation characteristics of the SOFC were measured at 700°C using humidified (3% H_2O) hydrogen gas as the fuel, and air as an oxidant. The measurements were made after completely reducing the NiO in the anode to Ni.

RESULTS AND DISCUSSION

Since it has been reported that an abnormal phase transition is induced during EPD of $BaTiO_3$ with the perovskite structure using non-aqueous solutions as the suspension medium,[14] the crystalline phases in the as-deposited and sintered films were characterized using X-ray diffraction (XRD) with Cu-Kα radiation. Figure 1 shows XRD patterns of the LSGM powders used for the EPD, and the electrophoretically fabricated films. All the peaks observed on the two XRD patterns can be indexed on the basis of a monoclinic cell with the space group, I2/a. No change in crystalline phase was observed in the present work.

Figure 2 shows a SEM photograph of the electrophoretically deposited LSGM on a graphite substrate before sintering. No large cracks are observed, although individual particles of various sizes are recognizable. The thickness of the deposited film, which is about 100 μm, was controlled by the EPD time.

Figure 3 shows an SEM photograph of an as-sintered surface, and of a fractured section of an EPD

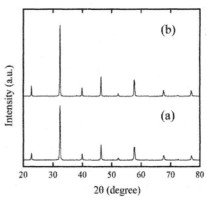

Figure 1. XRD patterns of: a) LSGM powders used in the present work; and b) electrophoretically fabricated films.

film after sintering. The EPD films were transformed into self-supported continuous bodies without the graphite substrate by sintering at 1500°C. The thickness of the EPD films decreases by about 20% during sintering. Pronounced grain growth is observed, and some small pores are present in the sintered films. The sintered films have relatively flat surfaces. In the present work, varying the EPD time produced LSGM electrolyte films with thickness of about 20-80 μm.

The temperature dependence of resistance for a 50 μm-thick sintered LSGM film with a 1.69 cm^2 cross

Figure 2. SEM photograph of an electrophoretically deposited LSGM film on a graphite substrate. (bar = 10 μm)

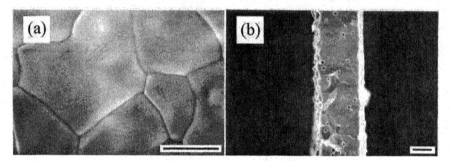

Figure 3. SEM photographs of: a) an as-sintered surface; and b) a fractured section of an EPD film after sintering. (bars = 10 μm)

section is shown in Figure 4. Resistance decreases with increasing temperature, and is about 50 and 15 milliohms at 730 and 900°C, respectively. The electrical conductivity estimated from the measured resistance is in fair agreement with oxide ion conductivity reported previously for bulk LSGM specimens.[15]

Therefore, it is clear that the low resistance results from reducing thickness of the LSGM electrolyte.

Figures 5 and 6 show SEM photographs of the Ni-YDC anode and LSC cathode powders, respectively. The

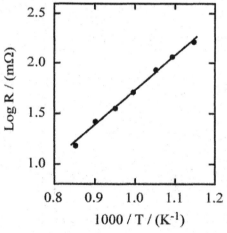

Figure 4. Temperature (T) dependence of resistance (R) of electrophoretically-fabricated 50 μm-thick LSGM films.

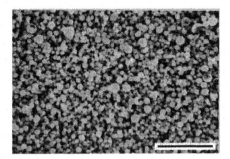

Figure 5.   SEM photograph of the mixed Ni-YDC powder used for the anode.   (bar = 10 μm)

Figure 6.   SEM photograph of the LSC powder used for the cathode. (bar = 10μm)

anode powders are uniform and equiaxial.   According to previous reports,[9,12] good connections are expected between Ni-Ni, Ni-YDC and YDC-YDC grains in the anode.   The mixed cermet anode powders reportedly show high electrode activities.[9-13]   The cathode powders are not as uniform as the anode powders but are fine, with average size of about 2 μm.

In Figure 7, the current-voltage (I-V) and current-power (I-P) characteristics at 700°C for a SOFC with a 50 μm-thick LSGM film as the electrolyte are shown.   A maximum output power density around 0.5 W/cm^2 was attained. The cell performance was superior to that of SOFCs constructed with thicker LSGM electrolytes with the high performance mixed cermet anode powders, including Sm instead of Y.[3]   The enhanced cell performance is, therefore, considered to be due to the reduction of the ohmic loss in the electrolyte.   An

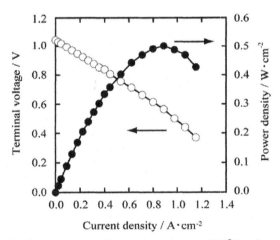

Current density / A·cm^{-2}

Figure 7. Power generation properties at 700°C of a SOFC constructed with a 50 μm-thick LSGM electrolyte film, with Ni-YDC for the anode, and with LSC for the cathode.

open-circuit voltage of 1.044 V was obtained, which was slightly lower than the value expected theoretically. This may be due to some gas-leakage between the electrodes.

CONCLUSIONS

In the present work, self-supported dense LSGM electrolyte films with a thickness less than 100 µm were successfully fabricated by EPD of ceramic powders onto graphite substrates followed by sintering at 1500°C in air. The LSGM films fabricated exhibited low ohmic resistance at elevated temperatures. The resistance of a 50 µm thick LSGM film with a 1.69 cm^2 cross section was around 50 and 15 milliohms at 730 and 900°C, respectively. The low electrical resistance at elevated temperature resulted from the reduced thickness of the LSGM film. To investigate the cell performance of the LSGM film, a mixed Ni-YDC anode and LSC cathode were prepared and used. A single cell consisting of Ni-YDC | 50 µm-thick LSGM electrolyte film | LSC produced a maximum output power density around 0.5 W/cm^2 at 700°C using humidified (3% H_2O) hydrogen as the fuel.

ACKNOWLEDGMENTS

This work was supported, in part, by a Grant-in-Aid for Science Research (14550668) from Japan Society for the Promotion of Science and the Ministry of Education, Science, Sports and Culture, and by a Grant-in-Aid for the COE project, Giant Molecules and Complex Systems, 2003. The financial support of the Kazuchika Okura Memorial Foundation and the Uesuko Foundation for the Promotion of Science are greatly acknowledged. The authors thank Mr. T. Hosomi for carrying out some of the EPD experiments.

REFERENCES

[1] T. Ishihara, H. Matsuda, and Y. Takita, "Doped LaGaO$_3$ Perovskite Type Oxide as a New Oxide Ionic Conductor," *J. Am. Chem. Soc.*, **116** [9] 3801-03 (1994).

[2] M. Feng and J. B. Goodenough, "A Superior Oxide-Ion Electrolyte," *Eur. J. Solid State Inorg. Chem.*, **T31**, 663-72 (1994).

[3] T. Fukui, S. Ohara, K. Murata, H. Yoshida, K. Miura and T. Inagaki, *J. Power Sources*, **106**, 142-45 (2002).

[4] A. M. El-Sherik, P. Sarkar, and A. Petric, "Electrophoretic Deposition of Lanthanum Gallate Solid Electrolyte Coating on Cathode Substrates," 643-50 in *Proceedings of the Third International Symposium on Ionic and Mixed Conducting Ceramics*, Edited by T. A. Ramanarayanan, The Electrochemical Society, Pennington, NJ, 1998.

[5] T. Mathews, Nadine Rabu, J. R. Sellar, and B. C. Muddle, "Fabrication of La$_{1-x}$Sr$_x$Ga$_{1-y}$Mg$_y$O$_{3-(x+y)/2}$ Thin Films by Electrophoretic Deposition and its Conductivity Measurement," *Solid State Ionics*, **128**, 111-15 (2000).

[6] I. Zhitomirsky and A. Petric, "Electrophoretic Deposition of Ceramic Materials for Fuel Cell Applications," *J. Eur. Ceram. Soc.*, **20**, 2055-61 (2000).

[7] M. Matsuda, S. Ohara, K. Murata, S. Ohara, T. Fukui, and M. Miyake, "Electrophoretic Fabrication and Cell Performance of Dense Sr- and Mg-Doped LaGaO$_3$-Based Electrolyte Films," *Electrochem. Solid-State Lett.*, **6** [7] A140-03 (2003).

[8] K. Huang, M. Feng, and J. B. Goodenough, "Sol-Gel Synthesis of a New Oxide-Ion Conductor Sr- and Mg-Doped LaGaO$_3$ Perovskite," *J. Am. Ceram.*

Soc., **79** [4] 1100-104 (1996).

[9]R. Maric, S. Ohara, T. Fukui, T. Inagaki, and J. Fujita, "High-Performance Ni-SDC Cermet Anode for Solid Oxide Fuel Cells at Medium Operating Temperature," *Electrochem. Solid-State Lett.,* **1** [5] 201-03 (1998).

[10]R. Maric, S. Ohara, T. Fukui, H. Yoshida, M. Nishimura, T. Inagaki, and K. Miura, "Solid Oxide Fuel Cells with Doped Lanthanum Gallate Electrolyte and LaSrCoO₃ Cathode and Ni-Samaria-Doped Ceria Cermet Anode," *J. Electrochem. Soc.,* **146**[6] 2006-10 (1999).

[11]X. Zhang, S. Ohara, R. Maric, K. Mukai, T. Fukui, H. Yoshida, M. Nishimura, T. Inagaki, and K. Miura, "Ni-SDC Cermet Anode for Medium-Temperature Solid Oxide Fuel Cell with Lanthanum Gallate Electrolyte," *J. Power Sources,* **83**, 170-77 (1999).

[12]S. Ohara, R. Maric, X. Zhang, K. Mukai, T. Fukui, H. Yoshida, T. Inagaki, and K. Miura, "High Performance Electrolyte for Reduced Temperature Solid Oxide Fuel Cells with Doped Lanthanum Gallate Electrolyte I. Ni-SDC Cermet Anode," *J. Power Sources,* **86**, 455-58 (2000).

[13]R. Maric, T. Fukui, S. Ohara, H. Yoshida, M. Nishimura, T. Inagaki, and K. Miura, "Powder Prepared by Spray Pyrolysis as an Electrode Material for Solid Oxide Fuel Cells," *J. Mater. Sci.,* **35**, 1397-404 (2000).

[14]K. Yamashita, M. Matsuda, Y. Inda, T. Umegaki, M. Ito, and T. Okura, "Dielectric Depression and Dispersion in Electrophoretically Fabricated BaTiO₃ Ceramic Films," *J. Am. Ceram. Soc.,* **80** [7] 1907-909 (1997).

[15]K. Huang, R. S. Tichy, and J. B. Goodenough, "Superior Perovskite Oxide-Ion Conductor; Strontium- and Magnesium-Doped LaGaO₃: I, Phase Relationships and Electrical Properties," *J. Am. Ceram. Soc.,* **81** [10] 2565-75 (1998).

FORMATION OF MgB$_2$ SUPERCONDUCTING PHASE FROM Mg AND B COMPOSITE PARTICLES PRODUCED BY MECHANICAL MIXING

H. Abe, M. Naito, and K. Nogi
Joining and Welding Research Institute
Osaka University
Ibaraki
Osaka 573-1132, Japan

S. Ohara
Tohoku University,
Sendai 980-8577, Japan

A. Kondo and T. Fukui
Hosokawa Powder Technology
Research Institute
Hirakata
Osaka 573-1132, Japan

M. Matsuda and M. Miyake
Department of Environmental
Chemistry and Materials
Okayama University
Okayama 456-8587, Japan

ABSTRACT
The influence of mechanical milling on the formation of MgB$_2$ was investigated. After mechanical milling Mg and B powders, B particles are embedded in Mg particles, and a 1 μm thick Mg and B composite layer forms on the Mg particles. After relatively low temperature annealing, a MgB$_2$ superconducting phase forms from this composite layer. The results indicate that the Mg and B composite surface layer may be microscopically mixed, and may have a highly disordered crystal structure, which would provide a strong driving force for the formation of MgB$_2$ at lower temperature than can be achieved conventionally.

INTRODUCTION
Following the discovery of superconductivity in MgB$_2$[1], attempts have been made to synthesize this material with the aim of improving superconducting properties. Although systematic studies to relate the effects of synthesis parameters with superconducting properties have been reported,[2-4] some of the practical aspects of forming MgB$_2$ have not been addressed, including the influence of the mixing or milling conditions of the constituent powders.

Mechanical milling has been known to be an effective technique for mixing, and to drive a wide range of chemical reactions.[5] In the latter case, mechanical energy is utilized instead of thermal energy to provide the activation energy for solid-state reaction. Mechanical milling can be designed as an intermediate step to promote reactions that could otherwise only be completed at high temperature.[6]

In this paper, we investigate the influence of the mechanical milling of Mg and B powders on the formation of superconducting MgB$_2$. This study focuses on are the structure of the Mg and B composite particles formed by mechanical milling, and the relevance to MgB_2 evolution during annealing at relatively low temperature.

EXPERIMENTS

The raw materials used in this study were 99.5% pure Mg powder with a sieve size of -330 mesh (Wako Pure Chemical Industries, Japan) and 97% pure amorphous B powder with an average particle size of 0.8 μm (Rare Metallic, Japan). Images of the powders are shown in Figure 1. Powder mixtures consisting of an Mg:B molar ratio of 1:2 were mechanically milled using an attrition type device (Mechanofusion System[7], Hosokawa Micron Corp., Japan). A milling vessel rotation speed of 2000 rpm was used for 5 min and 60 min under atmospheric pressure argon to mill the powders. The milled powder mixture was removed from the milling vessel for analysis, and for subsequent annealing. Annealing of the milled powder mixtures was conducted in a tube furnace at 773 K for 10 h under atmospheric pressure argon flowing at 1 l/min. The milled and annealed powder mixtures were characterized by X ray diffraction, XRD (Rigaku RINT2100) and dc magnetization (Quantum Design MPMS2). The reaction characteristics of the powder mixtures were evaluated by differential thermal analysis, DTA, using a heating rate of 10 K/min. Particle morphology was characterized using scanning electron microscopy, SEM (Hitachi S3500N). Element mapping of the milled powder mixture was performed with electron probe microanalysis, EPMA (JEOL-8600).

RESULTS AND DISCUSSION

Figure 1c and d show SEM images of typical particles found in a powder mixture milled for 5 min and 60 min, respectively. After mechanical milling for 5 min, amorphous B particles adhere to the surface of the Mg particles. In contrast, a considerable amount of amorphous B particles are embedded in the surface of

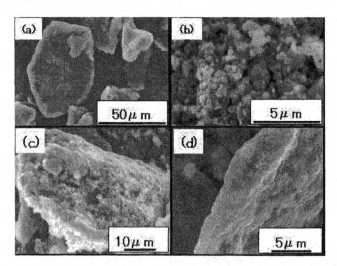

Figure 1. SEM images of the: a) Mg powder; and b) B powder, and of the Mg and B powder mixture after milling for: c) 5 min; and d) 60 min.

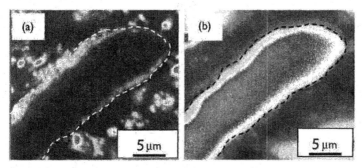

Figure 2. EPMA image of the cross section of a particle in the powder mixture after mechanical milling for 60 min showing an elemental map (in white) of: a) of B, and b) Mg.

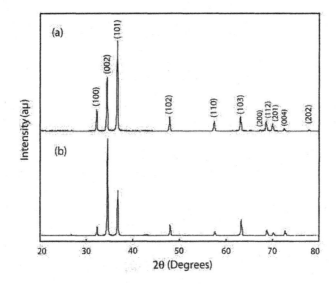

Figure 3. XRD pattern of the powder mixture after mechanical milling for: a) 5 min; and b) 60 min.

the Mg particles after mechanical milling for 60 min.

Figure 2 shows elemental maps of B and Mg in the cross sectional plane of a particle after milling the powder mixture for 60 min. The B embedded layer of the composite particle is about 1 μm thick.

Figure 3 shows XRD patterns of the powder mixtures milled for 5 min and 60 min. Only crystalline diffraction peaks for Mg are evident. The reflections and the intensities observed in the powder mixture milled for 5 min are identical to those observed in the starting Mg powder. In the XRD pattern of the powder mixture milled for 60 min, the [002] reflections of Mg are intensified. The [002] reflection corresponds to the slip plane of the Mg, indicating that the longer milling time results in plastic deformation of the Mg powder.

Figure 4 shows DTA curves for the powder mixture milled for 5 min and 60 min. The DTA curves show the exothermic reaction characteristic of MgB_2 synthesis. The DTA curve for the powder mixture milled for 5 min is similar to that for a powder mixture prepared conventionally using a mortar and pestle. Obviously, the formation of MgB_2 occurs at lower temperature after longer mechanical milling. It is noted that MgB_2 cannot form *in situ* during milling.[8]

Figure 5 shows the temperature dependence of field cooled (FC) magnetization for mechanically milled powder mixtures after annealing at 773 K for 10 h. In this measurement, the sample temperature is reduced to 4 K through the critical temperature, T_C, under a fixed magnetic field of 10 G. The magnitude of the diamagnetism for the powder mixture milled for 60 min is about 50 times larger than that for the powder milled for 5 min. Clearly, a higher concentration of superconducting phase is produced after the longer milling time. When annealing at 903 K, the magnetization curves for the two powder mixtures are almost the same, and the show a $T_C \sim 38K$.

Figure 6 shows the surface structure of the processed particles (Figure 1d) after annealing at 773 K for 10 h. The crystallites of close-packed hexagonal MgB_2 are clearly visible. Thus, MgB_2 is formed at the surface of the composite particle by annealing mechanically milled Mg and B powders at 773 K.

It has been shown already that solid-state reactions are enhanced by mechanical processing.[5,6,9] This study also shows that the longer mechanical milling of Mg and B powders raises the reactivity for the formation of MgB_2, and leads to the lower temperature synthesis of MgB_2. The enhanced reactivity is strongly related to the structure of the composite particles, in which B particles

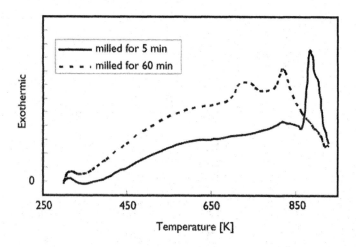

Figure 4. DTA curve for the Mg + B powder mixture after mechanical milling for 5 min and 60 min.

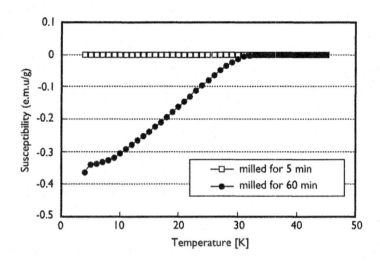

Figure 5. Temperature dependence of field cooled magnetization under 10 G for the mechanically mixed MG + B powders after annealing at 773 K for 10 h.

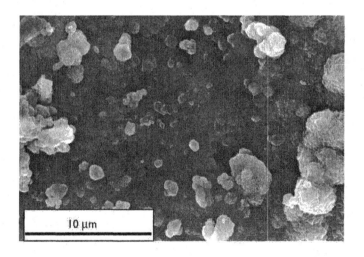

Figure 6. Surface morphology of the mechanically milled (60 min) powder mixture after annealing at 773 K for 10 h.

are embedded in the surface of Mg particles. This embedded region may have a highly disordered crystal structure and microscopic mixing, which will contribute to a strong driving force for the formation of MgB_2. Further study, especially microscopic structural analysis of the composite layer, is currently underway.

Many groups have been trying to draw MgB$_2$ wire using the powder in tube (PIT) method.[10,11] It is common to employ an ex situ PIT method, in which commercial MgB$_2$ powders are used as raw materials. However, there is potential to improve current density, J$_C$, with the in situ method, by controlling the grain size, and by doping with other elements. In the latter case, a sophisticated powder design will be needed. Mechanical milling will be useful to promote the synthesis of MgB$_2$ at lower temperature, and to produce doped powders that are homogeneous.

CONCLUSION

Large Mg particles (-330 mesh) and submicron amorphous B particles (average particle size, 0.8 μm) were mechanically milled for 60 min using an attrition type device. Milling produced B/Mg composite particles, in which B particles were embedded in Mg particle surfaces at a depth of about 1 μm. The superconducting MgB$_2$ phase forms from mechanically milled powder by annealing at relatively low temperature.

ACKOWLEGEMENT

This work was partially supported by *Kinki-chiho* innovation center.

REFFERENCE

[1]J. Nagamatsu, N. Nakagawa, T. Muranaka, Y. Zenitani, and J. Akimitsu, "Superconductivity at 39K in Magnesium Diboride," *Nature*, **410** 63-64 (2001).

[2]S. Lee, "Crystal Growth of MgB$_2$," *Physica C*, **385** 31-41 (2003).

[3]Y. Takano, H. Takeya, H. Fujii, H. Kumakura, T. Hatano, K. Togano, H. Kito, and H. Ihara, "Superconducting Properties of MgB$_2$ Bulk Materials Prepared by High-Pressure Sintering," *Applied Physics Letters*, **78** 2914-16 (2003).

[4]J.D. Jorgensen, D.G. Hinks, and S. Short, "Lattice Properties of MgB$_2$ versus Temperature and Pressure," *Physical Review B*, **63** 224522-1-5 (2003).

[5]L.L. Shaw, Z. Yang, and R. Ren, "Mechanically Enhanced Reactivity of Silicon for the Formation of Silicon Nitride Composites," *Journal of American Ceramic Society*, **81** 760-64 (1998).

[6]K.W. Liu, J.S. Zhang, J.G. Wang, and G.L. Chen, "Formation of Nanocomposites in the Ti-Al-Si System by Mechanical Alloying and Subsequent Heat Treatment," *Journal of Materials Research*, **13** 1198-03 (1998).

[7]M. Naito, A. Kondo, and T. Yokoyama, "Application of Comminution Techniques for the Surface Modification of Powder Materials," *ISIJ International*, **33** 915-24 (1995).

[8]H. Abe, M. Naito, K. Nogi, M. Matsuda, M. Miyake, S. Ohara, A. Kondo, and T. Fukui, "Low Temperature Formation of Superconducting MgB$_2$ Phase from Elements by Mechanical Milling," *Physica C*, **391** 211-16 (2003).

[9]A. Gulino, R.G. Egdell, and I. Fragala, "Mechanically Induced Phase Transformation and Surface Segregation in Bismuth-Doped Tetragonal Zirconia," *Journal of American Ceramic Society*, **81** 757-59 (1998).

[10]S. Jin, H. Mavoori, C. Bower, and R.B. van Dover, "High Critical Currents in Iron-Clad Superconducting MgB$_2$ Wires," *Nature*, **411** 563-65 (2001).

[11]T. Machi, S. Shimura, N. Koshizuka, and M. Murakami, "Fabrication of MgB$_2$ Superconducting Wire by *in situ* PIT Method," *Physica C*, **392-396** 1039-42 (2003).

Microstructure

MICROSTRUCTURE AND MECHANICAL PROPERTIES OF AlN CERAMICS WITH Y_2O_3

Seiji Iwasawa, Junichi Tatami, Katsutosi Komeya, and Takesishi Meguro
Yokohama National University
79-7 Tokiwadai, Hodogayaku
Yokahama, 240-8501, Japan

ABSTRACT
The effects of 1 to 3 wt% Y_2O_3 additions and sintering time on the microstructure and properties of AlN are reported. The bulk density of the sintered body is almost independent of soak time. The lattice constant, c, of AlN increases with increasing soak time, which means that purification of the AlN grains occurs during firing. Grain size and fracture toughness increase with increasing soak time. The fracture toughness of the grains, G_{IC}^{grain}, and of the grain boundary, $G_{IC}^{boundary}$, were calculated from experimental data and a stochastic model for the fracture toughness and fracture behavior of polycrystalline ceramics. G_{IC}^{grain} increases as the oxygen content in the AlN grains decreases. Furthermore, it is probable that $G_{IC}^{boundary}$, which decreases slightly with soak time, depends on the secondary phases present on the grain boundaries after sintering.

INTRODUCTION
AlN, which has high thermal conductivity and good electric resistivity, has been used for intergraded circuit substrates and packages, structural parts for semiconductor processing, and radiators for inverters.[1-4] Although dense and high thermal conductivity AlN ceramics can be produced using suitable additives like Y_2O_3, strength and fracture toughness must still be improved.[5,6] These properties strongly depend on the microstructure, including grain size and the grain interfaces. In particular, the AlN grains themselves are also an important microstructural factor because oxygen dissolves in the AlN grains during sintering forms point defects and stacking faults that can influence mechanical properties.[7] The objectives of this study are to evaluate the microstructure of AlN ceramics fabricated with Y_2O_3 using various firing conditions, to measure their mechanical properties, and determine the relationship between microstructure and mechanical properties.

EXPERIMENTAL PROCEDURE

High purity AlN (Tokuyama Co., Ltd.; F grade, 0.6 μm) and Y_2O_3 (Shinetu Chemical Co., Ltd,; RU grade, 0.6 μm) powders were used as the raw materials. Batch compositions of AlN with 1 and 3 wt% additions of Y_2O_3 were prepared. The powder mixture was ball milled in ethanol. A solution containing paraffin was added to the mixture. After the solvent was removed, the powder mixture was granulated by sieving through a 48 μm sieve. The mixture was then molded into cylinders 15 mmϕ by 7 mm tall, and into $40 \times 40 \times 5$ mm blocks by uniaxial pressing at 50 MPa, followed by cold isostatic pressing (CIP) at 200 MPa. Dewaxing was performed at 600°C for 2 h in air, and the powder compacts were fired at 1850°C for 2, 6, and 12 h in 0.6 MPa N_2. The density of the AlN sintered body was measured by the Archimedes method. The phases present and the lattice constant, c, of AlN were determined by X-ray diffraction (XRD) with CuKα radiation. Bend strength was measured on $3 \times 3 \times 30$ mm specimens using a three-point bend test with a lower span of 16 mm and a crosshead speed of 0.5 mm/min. Fracture toughness was measured by the surface crack in failure (SCF) method (ASTM C1421-99). To evaluate fracture behavior, the fracture surface was characterized using a scanning electron microscope (SEM).

RESULTS AND DISCUSSION

Tables I and II summarize the properties of the sintered AlN specimens. The bulk density is over 3.2 g/cm^3 in all of the samples, confirming that the specimens prepared and examined in this study are dense sintered bodies. XRD identified $Al_5Y_3O_{12}$ (YAG) and AlN in the AlN-Y_2O_3 (1 wt%)-series sintered body. The AlN-Y_2O_3 (3 wt%)-series sintered body contains $Al_5Y_3O_{12}$, $AlYO_3$ (YAL), and AlN, and the $Al_5Y_3O_{12}$ content in the specimen decreases after sintering for 12 h. The c axis of AlN lengthens from 0.49792 nm to 0.49803 nm in both series as sintering time increases. This means that oxygen impurity in the AlN grains is removed, resulting in higher purity AlN grains.[9] Therefore, it was shown that an increase in the soak time during sintering results in the purification of the AlN grains. The grain size of the AlN ceramics in both series increases with increasing soak time. Furthermore, it was found that the AlN ceramics with 1 wt% Y_2O_3 have larger grains than those with 3 wt% Y_2O_3. More specifically, the grain size of the AlN ceramics with 1 wt% Y_2O_3 is about twice as large as that with 3 wt% Y_2O_3.

Tables I and II also show the bend strength of each sample. The bend strength of the specimens with 1 wt% Y_2O_3 is about 350 MPa. In the AlN-Y_2O_3 (3 wt%)-series sintered body, the strength slightly increases from 420 to 460 to 480 MPa as the sintering time increases. Fracture toughness increases with increasing soak time in both systems. In the AlN-3 wt%Y_2O_3 system, it was found that the change in bend strength with soak time correlates with the change in fracture toughness. In the AlN-1 wt% Y_2O_3 system, although the bend strength is almost constant, this tendency probably results from the increase in the fracture toughness and the flaw size related to the grain size.

Figure 1 shows SEM photographs of the fracture surfaces of the sintered bodies. The observation was made in the area close to the tip of the induced crack used to measure fracture toughness, because the fracture behavior depends on the speed of the crack. In all sintered bodies, the fracture behavior is a mixture of transgranular and intergranular fracture. From the photographs, the fraction of transgranular fracture was measured for 100 grains in each specimen. As a result, it was found that the fraction of transgranular fracture, which varies from 20 to

Table I. Properties of AlN ceramics with 1 wt% Y_2O_3 sintered at 1850°C.

Soak time (h)	2	6	12
Bulk density, ρ_{bulk} (g/cm^3)	3.25	3.28	3.28
Crystal phase	AlN YAG	AlN YAG	AlN YAG
Lattice constant, c, of AlN (nm)	0.49792	0.49798	0.49802
Grain size, d (μm)	6.3	12.2	12.3
Bending strength, σ_f (MPa)	370±40	370±20	350±20
Fracture toughness, K_{IC} (MPa m$^{1/2}$)	3.0	3.2	3.4
Fraction of transgranular fracture, f (%)	32	30	32
Fracture toughness of grain, G_{IC}^{grain} (J/m^2)	43.4	54.3	58.4
Fracture toughness of grain boundary $G_{IC}^{boundary}$ (J/m^2)	7.7	7.2	8.0

Table II. Properties of AlN ceramics with 3 wt% Y_2O_3 sintered at 1850°C.

Soak time (h)	2	6	12
Bulk density ρ_{bulk} (g/cm^3)	3.32	3.33	3.33
Crystal phase	AlN YAL YAG	AlN YAL YAG	AlN YAL YAG
Lattice constant, c, of AlN (nm)	0.49798	0.49801	0.49803
Grain size, d (μm)	4.4	5.1	5.5
Bending strength, σ_f (MPa)	420±40	460±65	480±20
Fracture toughness, K_{IC} (MPa m$^{1/2}$)	3.0	3.2	3.5
Fraction of transgranular fracture, f (%)	22	36	22
Fracture toughness of grain G_{IC}^{grain} (J/m^2)	46.6	48.3	66.9
Fracture toughness of grain boundary $G_{IC}^{boundary}$ (J/m^2)	8.1	9.8	10.7

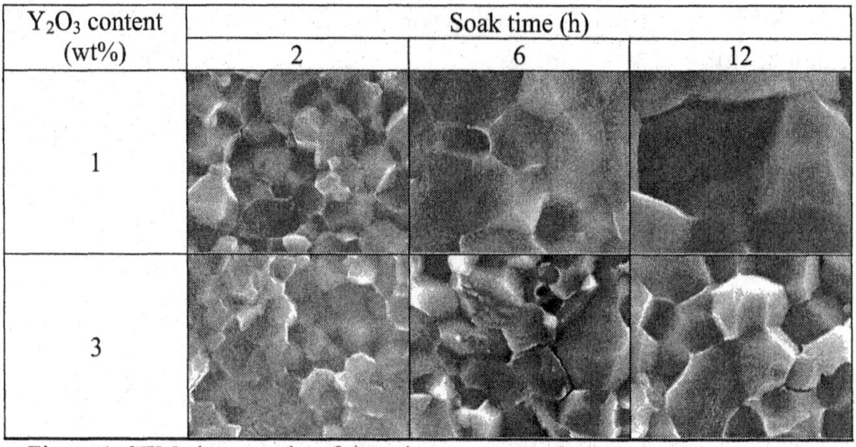

Y$_2$O$_3$ content (wt%)	Soak time (h)		
	2	6	12
1			
3			

Figure 1. SEM photographs of the microstructure of AlN ceramics with 1 to 3 wt% Y$_2$O$_3$ fired at 1850°C for 2 to 12 h..

5µm

40%, is dependent on the concentration of the sintering aid, and on the soak time during sintering.

Tatami et al.[8] analyzed the influence of the crack path on the fracture toughness of polycrystalline ceramics using a stochastic model based on the difference between the energies released during intergranular and transgranular crack propagation. They subsequently developed a formula to calculate the fraction of transgranular fracture on the fracture surface, $\hat{f}$, and the critical energy release rate, $\hat{G}_{IC}$, for a polycrystalline ceramic as functions of grain size, d, the fracture toughness of grain, $G_{IC}{}^{grain}$, and grain boundary ,$G_{IC}{}^{boundary}$ as follows :

$$\hat{f} = \frac{d - 2y_C}{d} \tag{1}$$

$$\hat{G}_{IC} = G_{IC}{}^{grain}\left(1 - \frac{2y_C}{d}\right) + G_{IC}{}^{boundary}\left\{1 + \frac{p}{6d}y_C{}^2 + \frac{q}{2d}y_C\right\}\frac{2y_C}{d} \tag{2}$$

where,

$$p = \frac{p'}{2a} \tag{3}$$

$$y_C = \begin{cases} \dfrac{-q + \sqrt{q^2 + 2\left(\dfrac{G_{IC}^{grain}}{G_{IC}^{boundary}} - 1\right)pd}}{p} & \left(d \geq d^*\right) \\[4ex] \dfrac{d}{2} & \left(0 < d < d^*\right) \end{cases} \tag{4}$$

$$d^* = \frac{8\left(\dfrac{G_{IC}^{grain}}{G_{IC}^{boundary}} - 1\right) - 4q}{p} \tag{5}$$

p', q : constants

This theory assumes that, when a crack propagates in the selective region of intergranular fracture with a length of y_C, the crack propagates intergranularly. d^* is the critical grain size for transgranular fracture, and it is calculated for the condition where $y_C \leq d/2$. This theory has been verified in polycrystalline alumina with equiaxed grains. The AlN sintered bodies in this paper also have equiaxed grains, so the above theory seems to be satisfactory. However, it should be noted that, in this study, the intergranular fracture includes not only the fracture at the interface between AlN grains and the secondary phase, but also the fracture within the secondary phase.

The fracture toughness of the grains, G_{IC}^{grain}, and of the grain boundary, $G_{IC}^{boundary}$, was calculated from the experimentally determined grain size, d, the fraction of transgranular fracture, f, and the critical energy release rate, G_{IC}. G_{IC} was determined from Irwin's equation assuming a value of 280 GPa for the Young's modulus, and a value of 0.25 for the Poisson ratio. The crack length, a, was set to 100 μm, which is almost equal to the crack length for the SCF method.

Tables I and II show the grain and grain boundary fracture toughness of the AlN specimens. The fracture toughness of the grains improves as the soak time increases. Considering the purification of the AlN grains as soak time increases, it is expected that the fracture toughness of the grain is related to the lattice constant, c, of the AlN. Figure 2 shows the relationship between the lattice constant, c, and the fracture toughness of the grains. The fracture toughness of the grains increases with increasing lattice constant, c. Therefore, it is concluded that the purification of the AlN grains leads to the strengthening of the AlN grains.

According to the calculations, the fracture toughness of the grain boundary is almost constant in the 1 wt% Y_2O_3 series. On the other hand, in the AlN-3 wt%Y_2O_3 system, the fracture toughness increases from 8 to 11 J/m^2 as soak time increases. From the XRD analysis, the secondary phases are $Al_5Y_3O_{12}$ in the 1 wt% Y_2O_3 series, and $Al_5Y_3O_{12}$ and $AlYO_3$ in the 3 wt% Y_2O_3 series. Considering the change in the secondary phases, the fracture toughness of the grain boundary depends on the secondary phase, especially the existence of $AlYO_3$. Additionally, the decrease in $Al_5Y_3O_{12}$ with increasing soak time in the 3 wt% Y_2O_3 series improves the fracture toughness of grain boundary.

CONCLUSION

From XRD analysis, it was found that the main phase in AlN sintered with Y_2O_3 additions is AlN, and the secondary phases depend on the amount of additive. The lattice constant, c, of AlN increases with increasing soak time during sintering with both additive concentrations. Soaking at high temperature results in the purification of the AlN grains. The grain size of the AlN ceramics increases with increasing soak time. AlN ceramics with 1 wt% Y_2O_3 have larger grains than those with 3 wt% Y_2O_3. The bend strength of the specimens with 1 wt% Y_2O_3 is almost independent of the soak time. With 3 wt% Y_2O_3, bend strength increases with increasing soak time. Fracture toughness increases with increasing soak time in both systems. From the fractography, the fracture mode is composed of transgranular and intergranular fracture. The fraction of transgranular fracture varies from 20 to 40% depending on the additive concentration, and on the soak time. The calculated fracture toughness of the grains improves as the soak time increases. It is concluded that the purification of the AlN grains leads to the strengthening of the AlN grains. In the AlN ceramics with 1 wt% Y_2O_3, the fracture toughness of the grain boundary is almost constant. In the AlN-3 wt% Y_2O_3 system, the fracture toughness increases from 8 to 11 Jm^{-2} as soak time increases. This can be explained by the differences in the secondary phase present.

REFERENCES

[1]Y. Baik and R.A.L. Drew, "Aluminum Nitride: Processing and Applications," *Key Engineering Materials*, **122-124** 553-70 (1996).

[2]K.A. Blakely, S.C. Martin and M.T. Spohn, "Material Supplier's Role In Market Development," *American Ceramic Society Bulletin*, **74** 55-58 (1995).

[3]L.M. Sheppard, "Aluminum Nitride," *American Ceramic Society Bulletin*, **69** 1801-12 (1990).

[4]N. Kuramoto, H. Taniguchi and I. Aso, "Development of Highly Thermal Conductive AlN Substrate by Green Sheet Technology," *Proceeding of the 36th Electronic Components Conference. IEEE Compon. Hybrids Manuf. Tech. Soc., ,Piscataway*, NJ, 424-29 (1986).

[5]K. Komeya, H. Inoue and A. Tsuge, "Effect of Heat Treatment on Sintering of Aluminum Nitride," *Journal of the Ceramic Society of Japan,* 58-64 (1981).

[6]T.B. Jackson, A.V. Virkar, K.L. More, R.B. Dinwiddie, Jr., and R.A. Cutler, "High-Thermal-Conductivity Aluminum Nitride," *Journal of the American Ceramic Society*, **80** 1421-35 (1997).

[7] K. Komeya, R. Terao, J. Tatami, and T. Meguro, "Search for New Additives to Improve Mechanical Strength of AlN Ceramics," *Ceramic Transaction*, **106** 501-07 (2000).

[8]J. Tatami, K, Yasuda , Y. Matsuo and S. Kimura, "Stochastic Analysis on Crack Path of Polycrystalline Ceramics Based on the Difference Between the Released Energies in Crack Propagation," *Journal of Materials Science*, **32** 2341-46 (1977).

[9]K. Shinozaki, N. Mizutani, and Y. Sawada, "Thermal Conductivity and Oxygen Impurity in AlN Sintered Body," *Bulletin of the Ceramic Society of Japan*, **25**, 1055-59 (1990).

MICROSTRUCTURE OF AlN FIBERS SYNTHESIZED FROM Al_2O_3 BY GAS-REDUCTION-NITRIDATION

Yusuke Matsumoto, Junichi Tatami,
Takeshi Meguro, and
Katsutoshi Komeya
Yokohama National University
Hodogaya-ku
Yokohama 240-8501, Japan

Toyohiko Yano
Tokyo Institute of Technology
Meguro-ku
Tokyo 152-8550,Japan

ABSTRACT

AlN fibers were synthesized from Al_2O_3 by gas-reduction-nitridation. An interesting advantage of this method is that the micrometer scale morphology of the alumina precursor can be maintained at the micrometer level during the gas-solid reaction. In this study, the microstructures of AlN fibers were characterized using a scanning electron microscope (SEM), a transmission electron microscope (TEM), and a scanning probe microscope (SPM). Additionally, the particle size and shape of the nano-size particles that make up the AlN fibers were characterized in detail using SPM. SEM analysis shows that the morphology of the nitrided fibers is the same as that of the alumina raw material. SPM images of initial Al_2O_3 fibers and AlN fibers synthesized at 1400°C for 2 h in NH_3-C_3H_8 gas were taken. TEM and SPM analysis determined that the particle size of AlN in the synthesized fibers is about twice as large as that in the starting Al_2O_3 material. Viscoelastic-atomic force microscopy (VE-AFM), which is one of the techniques in SPM, shows nitridation from the surface of each alumina particle.

INTRODUCTION

Aluminum nitride (AlN) has been used in applications where high thermal conductivity, good electrical insulation, and a thermal expansion coefficient close to Si are required.[1,2] Recent work[3] indicates that an AlN-filled polymer system can achieve an excellent property balance for advanced microelectronic packaging applications, i.e. the high thermal conductivity and low thermal expansion coefficient of the AlN, coupled with the moderately low dielectric constant and processability of the polymer. Additionally, it is expected that heat transfer in such a composite will be greatly improved by utilizing acicular AlN particles, mainly due to the increase in contact between AlN filler particles contributing to unimpeded heat flow through the matrix.[2]

The fabrication of AlN whiskers or fibers has received considerable attention in the literature.[4-9] Most research addresses combustion synthesis[7-9] or carbothermal reduction at high temperatures to produce whiskers.[4-6] Both techniques involve the vapor-liquid-solid (VLS) mechanism of growth, from which volume the fraction of whiskers in the reaction product is reported to be very low.[6,8]

Recently, Suehiro et al.[10] synthesized AlN fibers by reduction-nitridation of alumina fibers using NH_3 and C_3H_8 as the reactant gases. The overall reaction can be written as:

$$Al_2O_3 + 2 NH_3(g) + C_3H_8(g) \rightarrow 2AlN + 3CO(g) + 7H_2(g) \tag{1}$$

An interesting advantage of this method is that, on the micrometer level, the original morphology of the starting alumina can be maintained during the gas-solid reaction. Consequently, the morphology of the AlN product can be tailored by controlling the morphology of the precursor material. However, this has not been demonstrated for a nanometer scale microstructure.

In the present work, we characterized the nano-scale microstructure of AlN fibers synthesized by gas-reduction-nitridation, and compared it to the nano-scale microstructure of the raw alumina to investigate the gas-solid reaction in more detail.

EXPERIMENTAL PROCEDURE

The starting material used in this study was a wool-like alumina short fiber composed predominantly of γ-Al_2O_3. Table I lists the main characteristics of the raw material, and its morphology is shown in Figure 1. SEM shows that the raw alumina fiber has a smooth surface (Figure 1a). A dynamic force mode (DFM)

Table I. Main characteristics of the Al_2O_3 fiber raw material

Purity	> 99.9%
Crystal phase	mainly γ-Al_2O_3
Mean fiber diameter	5.8 μm
Mean crystal size	6 nm
Specific surface area	81 m^2/g
Transition to α-Al_2O_3	1140°C

Figure 1. The morphology of the Al_2O_3 precursor: a) SEM image; and b) DFM image.

image taken using scanning probe microscopy (SPM) shows that the raw alumina fiber is composed of the very fine particles, and that many holes exist (Figure 1b).

To synthesize AlN, a known mass of Al_2O_3 was placed in a high-purity alumina boat, and placed in an alumina tube furnace reactor (inner diameter of 42 mm). The reactor was first flushed with argon to eliminate oxygen from the system during the heat-up period. When the temperature reached 900°C a gas mixture of NH_3 (99.97% purity) and C_3H_8 (99.99%), with a molar ratio of $C_3H_8/NH_3 = 5 \times 10^{-3}$, was flowed through the reactor at a rate of 4 L/min (stp). The furnace was subsequently heated to the reaction temperature (1100-1400°C) at a rate of 8°C/min. After reacting for 0.5 h, the sample was cooled to room temperature at approximately 6°C/min in an ammonia atmosphere. The product was then reweighed, and the amount of Al_2O_3 converted to AlN was calculated from the mass loss.

The phases of the reaction products were characterized by X-ray diffraction (XRD, RAD-IIR, Rigaku) using CuKα radiation operated at 40 kV and 20 mA. XRD was completed on a powder sample obtained by grinding the products to a suitable size. The morphology of the products was characterized using a scanning electron microscope (SEM, JSM-5200, JEOL), a transmission electron microscope (TEM, H-9000, HITACHI), and a scanning probe microscope (SPM, SPI3800N, SII). Although there are various modes in SPM, DFM (dynamic force microscopy) and VE-AFM (viscoelastic-atomic force microscopy) were used in this work. DFM is the mode used to observe the sample surface by intermittently contacting the vibrated cantilever on the sample surface, and moving the scanner in the Z-axis so that the amplitude becomes fixed. VE-AFM is the mode used to image differences in elasticity using the bending amplitude of the cantilever, and

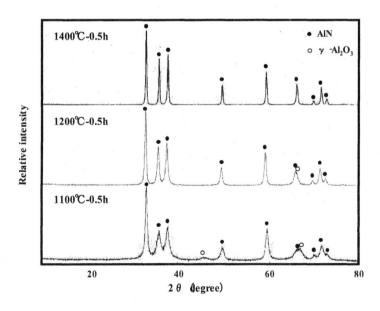

Figure 2. XRD pattern for the fibers synthesized at 1100, 1200, and 1400°C and 0.5 h.

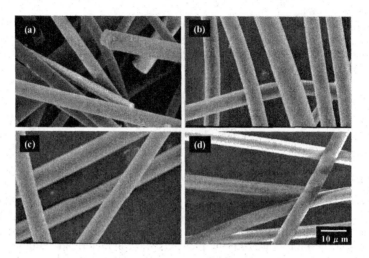

Figure 3. SEM photographs of the: a) raw material, and of the AlN fiber synthesized at: b) 1100; c) 1200; and d) 1400°C for 0.5 h.

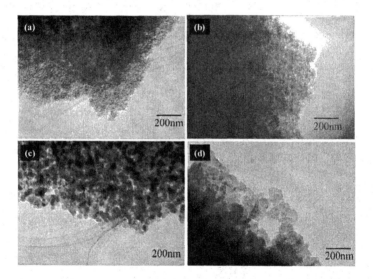

Figure 4. TEM photographs of the: a) raw material, and the AlN fiber synthesized at: b) 1100; c) 1200; and d) 1400°C for 0.5 h.

differences in viscosity determined by the phase shift between the sample and the cantilever when we let the sample vibrate in the Z-axis direction.

RESULTS AND DISCUSSION

Figure 2 shows the XRD pattern for the AlN fibers synthesized at 1100 to 1400°C for 0.5 h. The fibers synthesized at 1200 and 1400°C are almost entirely AlN; the percentage of Al_2O_3 converted to AlN is about 90%. The fiber synthesized at 1100°C consists of non-reacted γ- Al_2O_3 and AlN; the percentage of Al_2O_3 converted to AlN is about 60%.

Figure 3 shows a SEM photograph of the precursor Al_2O_3 material, and of the AlN fibers synthesized from it at 1100, 1200, and 1400°C for 0.5 h. SEM confirms that the AlN fibers synthesized at all firing temperatures maintain the fibrous morphology of the precursor material on the micrometer level. In this system, nitridation occurs at temperatures below the α-Al_2O_3 phase transition to form AlN directly from the highly reactive γ- Al_2O_3. Nitridation can thus be promoted, and grain growth can be suppressed to maintain the morphology of the precursor material.

Figure 4 shows TEM photographs of the precursor material, and of the AlN fiber synthesized from it at temperatures of 1100, 1200, and 1400°C for 0.5 h. TEM confirms that the particle size increases as the firing temperature increases. So the structure on the nanometer level was found to change.

Figure 5 shows DFM images of the surfaces of the precursor alumina fiber and of the AlN fiber synthesized from it at temperatures of 1100, 1200, and 1400°C for 0.5 h. The size of these pictures is 1×1 μm, a magnification that is

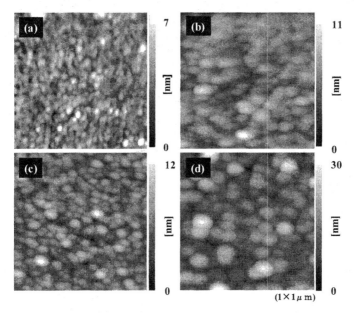

Figure 5. DFM images of the: a) raw material, and of the AlN
fiber synthesized at: b) 1100; c) 1200; and d) 1400°C for 0.5 h

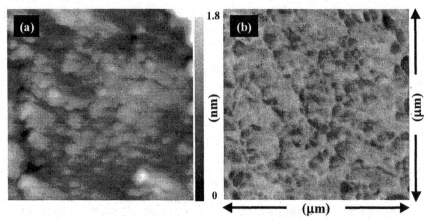

Figure 6. VE-AFM images of the cross section of AlN fibers synthesized at 1100°C for 0.5 h: a) topographic image; and b) viscoelastic image.

almost the same as in the TEM images. The contrast differences in the figure correspond to differences in height; the lower positions look dark, and the brightness increases with height. These images confirm that the grains grow as the firing temperature increases.

Figure 6 shows VE-AFM images of the cross section of the AlN fiber synthesized at 1100°C for 0.5 h, which contains about 60% AlN. To prepare the samples for analysis, a polymer resin and the AlN fibers were mixed and solidified. The solidified samples were ground and polished, and then observed by VE-AFM. The two images presented are from the same position on the same sample; on the left is a topographical image, and on the right is a viscoelastic image.

From the topographic image, it is found that the surface roughness is extremely small (Ra = about 1.8 nm). From the viscoelastic image, differences in brightness were identified between grains. The difference between the Young's modulus of AlN ($E_{AlN} \sim 280$ GPa) and γ-Al_2O_3 ($E_{AlN} \sim 250$ GPa) may be the cause of the differences in brightness. Higher elastic modulus leads to a brighter image. The bright grains in the viscoelastic image may be AlN, and the dark grains may be γ-Al_2O_3. The brightness increases from the periphery to the center of the grain, indicating that nitridation occurs from the grain surface.

SUMMARY

AlN fibers were synthesized from an Al_2O_3 precursor material by gas-reduction-nitridation. SEM analysis shows that the morphology of nitrided fibers is the same as that of the alumina raw material. SPM images of initial Al_2O_3 fibers and of the AlN fibers synthesized at 1400°C for 2 h under NH_3-C_3H_8 gas were taken. SPM analysis determined that the particle size of AlN in the synthesized fibers is about twice as large as that in the starting Al_2O_3 material.

CONCLUSIONS
1. Observations by TEM and DFM demonstrate that the AlN particle size increases with increasing in reaction temperature and time.
2. VE-AFM images of the cross section of the synthesized fibers indicate that the nitridation progress starts at the particle surface of the γ-Al$_2$O$_3$.

REFERENCES
[1]L.M. Sheppard, "Aluminum Nitride: A Versatile but Challenging Material," *Bull. Am. Ceram. Soc.*, **69** [11] 1801-12 (1990).
[2]"Annual Minerals Review: Aluminum Nitride," *Bull. Am. Ceram. Soc.*, **73** [6] 77 (1994).
[3]J.-W. Bae, W. Kim, S.-H. Cho, and S.-H. Lee, "The Properties of AlN-filled Epoxy Molding Compounds by The Effects of Filler Size Distribution," *J. Mater. Sci.*, **35**, 5907-13 (2000).
[4]P.G. Caceres and H.K. Schmid, "Morphology and Crystallography of Aluminum Nitride Whiskers," *J. Am. Ceram. Soc.*, **77** [4] 977-83 (1994).
[5]W.-G. Miao, Y. Wu, and H.-P. Zhou, "Morphologies and Growth Mechanisms of Aluminum Nitride Whiskers," *J. Mater. Sci.*, **32**, 1969-75 (1997).
[6]R.Fu, H. Zhou, L. Chen, and Y. Wu, "Morphologies and Growth Mechanisms of Aluminum Nitride Whiskers Synthesized by Carbothermal Reduction," *Mater. Sci. Eng.*, **A266**, 44-51 (1999).
[7]S.M. Bradshow and J.L. Spicer, "Combustion Synthesis of Aluminum Nitride Particles and Whiskers," *J. Am. Ceram. Soc.*, **82** [9] 2293-300 (1999).
[8]K.-J. Lee, D.-H. Ahn, and Y.-S. Kim, "Aluminum Nitride Whisker Formation during Combustion Synthesis," *J. Am. Ceram. Soc.*, **83** [5] 1117-21 (2000).
[9]G. Jiang, H. Zhuang, J. Zhang, N. Ruan, W. Li , F. Wu, and B. Zhang, "Morphologies and Growth Mechanisms of Aluminum Nitride Whiskers by SHS Method - Part II," *Journal of Materials Science*, **35**, 63-69 (2000).
[10]T. Shuehiro, J. Tatami, T. Meguro, K. Komeya, and S. Matsuo, "Aluminum Nitride Fibers Synthesized from Alumna Fibers Using Gas-Reduction Nitridation Method," *J. Am. Ceram. Soc.*, 85 [3] 715-17 (2002).

SINTERING BEHAVIOR OF AlN

Junichi Tatami, Katsutoshi Komeya,
Tomoaki Hoshina, Tomohiro Hirata,
and Takeshi Meguro
Yokohama National University
79-7 Tokiwadai, Hodogayaku,
Yokohama 240-8501, Japan

Akihiko Tsuge
Fine Ceramics Research Association
3-7-10 Toranomon, Minatoku
Tokyo
150-0001, Japan

ABSTRACT

In this study, we investigated the sintering behavior of AlN, focusing on the changes in the lattice constant of the AlN caused by the dissolution of oxygen, and the dihedral angle between the AlN grains and the liquid phase. The density of the sintered body increases with the firing temperature. The simultaneous addition of Y_2O_3 and CaO leads to densification at a lower temperature than with Y_2O_3 alone. Densification improves with the amount of oxygen dissolved in the AlN grains, and the wettability of the AlN grains by the liquid phase. Consequently, dissolved oxygen plays an important role in the sintering of AlN.

INTRODUCTION

Aluminum nitride (AlN) is a useful material for the production of electronic devices because it has high thermal conductivity, good electrical resistivity, and a thermal expansion coefficient close to that of silicon.[1-3] However, AlN is difficult to densify due to its strong covalent bonding. Densification of AlN has been achieved by adding sintering aids, such as Y_2O_3 and/or CaO that promote liquid phase sintering. Sintering aids are known to react with the Al_2O_3 on the AlN grains at high temperatures to form a liquid phase.[4-7] Kasori[6] reported that the simultaneous addition of Y_2O_3 and CaO (at a ratio of 1:3) leads to densification at lower temperatures. In a previous paper, we showed that the density of AlN containing Y_2O_3 and CaO increases with decreasing lattice constant, c, due to an increase in the oxygen content of the AlN grains. Although changes in the oxygen content of AlN probably improve the wettability of the AlN grains by the liquid phase, there is no information available on the solid-liquid interface. This study investigated the sintering behavior of AlN, focusing specifically on the change in the lattice constant of the AlN caused by the dissolution of oxygen in the AlN, and the dihedral angle between the AlN grains and the liquid phase. The sintering aids selected for this study were 5 wt% Y_2O_3 and 3.75 wt% Y_2O_3-1.25 wt% CaO.

EXPERIMENTAL PROCEDURE

High-purity, fine grain size powders of AlN (Grade F, 0.6 μm, Tokuyama Co., Ltd.), Y_2O_3 (RU, 0.6 μm Shin-Etsu Chemical Co., Ltd.) and $CaCO_3$ (High Purity

Reagent, 0.6 µm, Junsei Chemical Co., Ltd.) were used as the raw materials. Since $CaCO_3$ decomposes to CaO and CO_2 during heating, the amount of additive was calculated in terms of the CaO equivalent. Y_2O_3 (3.75 or 5 wt%) and CaO (1.25 wt%) were added to the AlN powder as sintering aids. The raw powders were mixed by wet ball milling in ethanol for 40 h. After mixing, the slurry was dried completely and passed through a 48-mesh sieve. The powder mixture was pressed into pellets (15 mm diameter x 5 mm tall) using a stainless-steel die, followed by cold isostatic pressing at 200 MPa for 60 s. The green bodies were fired in an induction-heated furnace in a graphite crucible under N_2 flowing at 1 l/min. The samples were fired at temperatures between 1600 and 1900°C for 0 to 6 h. To preserve the high temperature microstructure, the graphite crucible containing the fired specimen was removed from the furnace to allow rapid cooling. The dihedral angle is known to be influenced by the cooling rate.[8]

The density of the sintered body was measured by the Archimedes method, using butanol. The phases present after sintering and the lattice constant, c, of the AlN were determined using X-ray diffraction (XRD) analysis, with Si powder as a reference standard. The dihedral angle between the liquid and the solid interface was determined from polished and chemically etched surfaces at the same location. Chemical etching was performed by soaking in 10% HCl at 70°C for 330 s. About 200 measurements were made and then averaged to approximate the real dihedral angle in three dimensions.[9-10]

RESULTS AND DISCUSSION
Dihedral Angle in the AlN with 5.0 wt% Y_2O_3

The relationship between sintered density and firing temperature in the AlN-5.0 wt% Y_2O_3 is shown in Figure 1. The density increases with temperature, and a value of almost 100% is achieved at 1750°C. Many researchers have reported that AlN with Y_2O_3 is fully densified by firing at temperatures from 1750 to 1850°C.[4-7] Figure 2 illustrates the relationship between the lattice constant, c, and the firing temperature for the AlN-5.0 wt% Y_2O_3. The highest value of c is obtained for the raw powder. The lattice constant of the fired specimens rapidly decreases between 1600 and 1700°C, and reaches a minimum value at 1700°C. This decrease is attributed to an increase in the oxygen content in the AlN lattice. Dissolution of oxygen into AlN is known to form point defects based on the following process.[1,5,6]

$$Al_2O_3 \rightarrow 2Al_{Al} + 3O'_N + V'''_{Al} \tag{1}$$

These point defects generate a staking fault with a coordination number of 6, which has a shorter lattice constant than the AlN stacking with a coordination number of 4. As a result, the lattice constant of the AlN decreases with an increase in the amount of oxygen dissolved in the AlN. In this study, an increase in sintered density was observed with a decrease in the lattice constant, similar to our previous study.[7] Furthermore, the lattice constant increases with temperature above 1700°C, indicating that the oxygen dissolved in the AlN lattice is removed at higher sintering temperature by the oxygen trapping effect of the liquid phase.[5]

Figures 3 and 4 show SEM micrographs of polished and etched surfaces of the AlN specimens fired at various temperatures. In the sample fired at 1600°C, the yttrium aluminate grains (white grains) are interspersed in the AlN, although it is difficult to distinguish the individual AlN grains. From XRD analysis, the yttrium aluminate is identified as $Y_3Al_5O_{12}$ (YAG). By comparing polished and etched surfaces, we were able to distinguish between solid-vapor (i.e., pores) and solid-liquid interfaces. Figure 5 shows the relationship between the firing temperature and the average dihedral angle. The lowest average dihedral angle is measured between 1650 and 1700°C, and it gradually increases with increasing firing temperature up to 1800°C. This trend indicates that the wettability of the AlN grains by the liquid phase improves at ≥1700°C, which is where densification also improves (Figure 1). Consequently, it can be concluded that the densification of AlN with Y_2O_3 is enhanced by improving the wettability of the

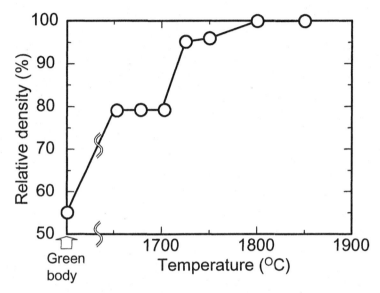

Figure 1. The density of AlN-5.0wt%Y_2O_3 as a function of sintering temperature. The soak time is 2 h for each sample.

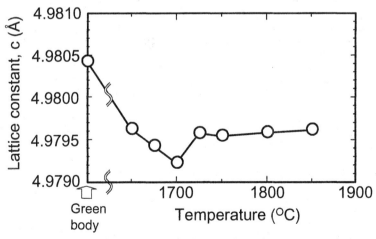

Figure 2. The lattice constant, c, of AlN for the AlN-5.0 wt% Y_2O_3 system as a function of sintering temperature. The soak time is 2 h for each sample.

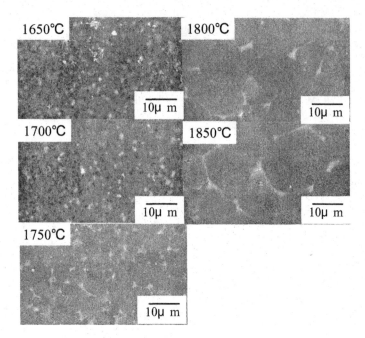

Figure 3. SEM photographs of polished surfaces of AlN-5.0 wt% Y₂O₃ fired at different temperature for 2 h.

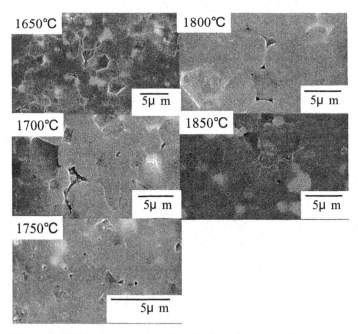

Figure 4. SEM photographs of etched surfaces of AlN-5.0 wt% Y₂O₃ fired at different temperatures for 2 h.

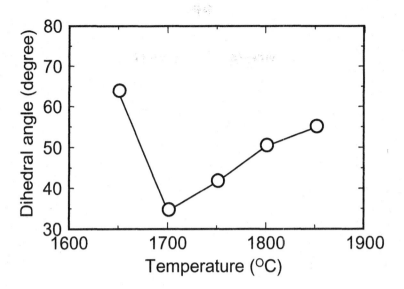

Figure 5. The Dihedral angle between the solid and liquid interfaces in the AlN-5.0 wt%Y$_2$O$_3$ system as a function of firing temperature. Soak time was 2 h for each sample.

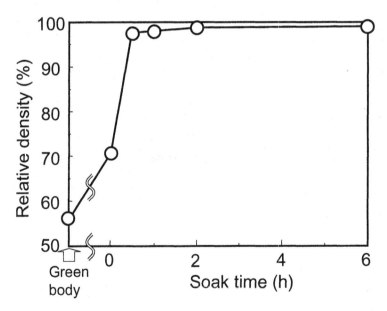

Figure 6. The density of the AlN-3.75 wt% Y$_2$O$_3$-1.25 wt% CaO as a function of sintering time at 1650°C.

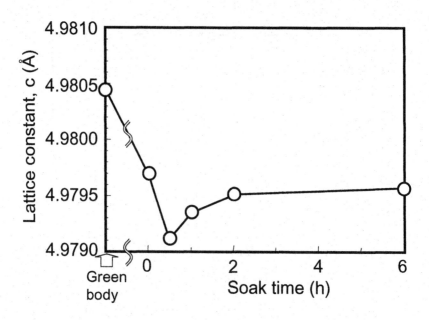

Figure 7. The lattice constant of AlN in sintered AlN-3.75 wt% Y_2O_3-1.25 wt% CaO as a function of sintering time at 1650°C.

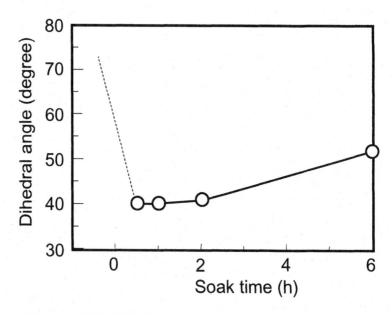

Figure 8. The dihedral angle between the solid and liquid interface in AlN-3.75 wt% Y_2O_3-1.25 wt% CaO as a function of sintering time at 1650°C.

AlN by the liquid phase. It is believed that the improved wettability and densification resulted from the dissolution of oxygen into the AlN grains.

Dihedral Angle in AlN-3.75 wt% Y_2O_3-1.25 wt% CaO

In the AlN-3.75 wt% Y_2O_3-1.25 wt% CaO system, the effects of sintering soak time on density, the lattice constant, and the dihedral angle at 1650°C were investigated. Figure 6 shows the density of the sintered AlN-3.75 wt% Y_2O_3-1.25 wt% CaO as a function of soak time. The density increases significantly by extending the soak time between 0 and 0.5 h. Close to 100% of theoretical density is achieved at 0.5 h.

Figure 7 shows the AlN lattice constant varies with soak time in AlN-3.75 wt% Y_2O_3-1.25 wt% CaO. The lattice constant decreases rapidly between 0 and 0.5 h to a minimum at 0.5 h, which correlates inversely with density (Figure 7). The lattice constant increases with soak times greater than 0.5 h, indicating that, as described above, the dissolved oxygen in the AlN is being removed. Consistent with this trend, the wettablity of the AlN grains by the liquid phase sets worse at longer soak times. This is evidenced in Figure 8, which shows the dihedral angle in the AlN-3.75 wt% Y_2O_3-1.25 wt% CaO as a function of soak time. The dihedral angle at 0.5 h is low, and increases with increasing soak time. The dihedral angle of the system appears to be lower than that of the AlN-5.0 wt%Y_2O_3 system (Figure 5). The improved wettablity evidenced by the lower dihedral angle is consistent with the improved sinterablity of AlN using Y_2O_3 and CaO versus Y_2O_3 alone.

CONCLUSIONS

In this study, we investigated the sintering behavior of AlN, with special consideration for the changes in the lattice constant and the dihedral angle of the solid-liquid interface as a function of the sintering aid, sintering temperature, and sintering time. The major findings from this study are as follows.

1) As determined from the change in the AlN lattice constant relative to pure AlN, the oxygen content is higher in the 5.0 wt% Y_2O_3 doped system at 1700°C, and in the 3.75 wt% Y_2O_3-1.25 wt% CaO system at 1650°C after soaking for 0.5 h.
2) As determined from dihedral angle measurements in comparison to the measured lattice constant, wettability correlates with oxygen content.
3) In AlN doped with 5.0 wt% Y_2O_3 or 3.75 wt% Y_2O_3-1.25 wt% CaO, sinterability improves at temperatures ≥1700°C, and at times longer than 0.5 h, respectively.
4) 3.75 wt% Y_2O_3-1.25 wt% CaO improves wettability, and is a better sintering aid for lower temperature sintering of AlN than 5.0 wt% Y_2O_3.

REFERENCES
[1]G. A. Slack, "Nonmetallic Crystals with High Thermal Conductivity," *J. Phys. Chem. Solids*, **34**, 321-335 (1973).
[2]G. Long, and L. M. Foster, "Aluminum Nitride, a Refractory for Aluminum to 2000°C," *J. Am. Ceram. Soc.*, **42**, 53-59 (1959).
[3]K. M. Taylor and C. Lenie, "Some Properties of Aluminum Nitride," *J. Electorochem. Soc.*, **107** 308-314 (1960).
[4]K. Komeya, H. Inoue and A. Tsuge, "Effect of Various Additives on Sintering of Aluminum Nitride," *J. Ceram. Soc. Japan,* **89**, 330-336, (1981).
[5]K. Shinozaki, N. Mizutani and Y. Sawada, "Thermal Conductivity and Oxygen Impurity in AlN Sintered Body" *Ceramics*, **26**, 25 (1990).
[6]M. Kasori, H. Sumino, A. Horiguchi and F. Ueno, "Mechanical and Thermal

Properties of Low Temperature Sintered AlN," *Ceramics Transaction*, **83**, 485-492 (1998) .

[7]T. Hirata, A. Tsuge, J. Tatami, K. Komeya and T. Meguro, "Liquid phase sintering behavior of AlN-Y_2O_3-CaO," *Proceedings of The 17th International Korea-Japan Seminar on Ceramics*, 52-55 (2000).

[8]W. J. Kim, D. K. Kim and C. H. Kim, "Morphological Effect of Second Phase on the Thermal Conductivity of AlN Ceramics" *J. Am. Ceram. Soc.*,**79** [4], 1066-72 (1996).

[9]S. Hara, M. Hanao and K. Ogino, "Interfacial Tension between Solid Cu and Liquid Cu--Bi and Cu--Pb Alloys," *J. Japan Inst. Metals*, **57** [2], 164-69 (1993).

[10]O.K. Riegger and L. H. Van Vlack, "Dihedral Angle Measurement", *AIME Trans.* **218** [10], 933-35 (1960).

CHARACTERIZATION OF INTERNAL INTERFACES IN TRANSLUCENT POLYCRYSTALLINE ALUMINA

G. C. Wei
Osram Sylvania
71 Cherry Hill Drive
Beverly, MA 01915

ABSTRACT

Internal interfaces or grain boundaries in translucent polycrystalline alumina play an important role for mass transport during sintering, and for transport of reactants and products for chemical reactions that occur in service as lamps. The effects of the MgO sintering aid, and other dopants such as Y_2O_3 and ZrO_2 on the sodium resistance of alumina in high-pressure sodium lamps were studied via microstructural analyses including electron microprobe and transmission electron microscopy. The relative changes in grain-boundary MgO levels can be calculated from the bulk MgO levels and relevant microstructural features. The enriched MgO level at grain boundaries is important for high-temperature properties. Segregation of various dopants to grain boundaries can affect the rate of boundary diffusion; the literature data on creep deformation, which is also controlled by grain boundary transport, is consistent with sodium resistance. This paper reports on microstructural and microchemical characterization of grain boundaries in alumina.

INTRODUCTION

Translucent polycrystalline alumina (PCA), used as envelopes in high-intensity discharge lamps, is typically made by hydrogen sintering of compacts of high purity, finely divided Al_2O_3 powder doped with MgO-based additives.[1] The MgO dopant (>200 ppm) prevents exaggerated grain growth, and allows annihilation of pores pinned at grain boundaries to attain nearly full density during sintering. It is widely recognized that the mass transport required for densification involves grain-boundary diffusion,[2] and oxygen vacancies play an important role. The retardation of grain growth has been attributed to grain-boundary drag[3] and homogenization of the microstructure[4] due to MgO solute segregation. A non-uniform distribution of the sintering aids, such as due to agglomeration[5] and impurities including SiO_2 and CaO,[6] which can be introduced by agglomerates, can cause local abnormal grain growth with entrapped pores.

The performance of high intensity discharge lamps is influenced by reactions of the lamp fills (gas, liquid, and solid) with the PCA arc tubes during service.[7] The material transport and reaction sequences involve multiple steps and paths.[8] In high-pressure sodium (HPS) lamps, Na reacts with the PCA and the magnesium aluminate spinel ($MgAl_2O_4$) second phase inside PCA to form sodium β-alumina ($Na_2O \cdot 11Al_2O_3$) and sodium-magnesium β'''-alumina ($Na_2O \cdot 4MgO \cdot 15Al_2O_3$). Eliminating the spinel second phase, and doping with $MgO+ZrO_2+Y_2O_3$ additives, in which the tetravalent Zr^{+4} charge-compensates the Mg^{+2}, improves the resistance of PCA to Na attack.[8,9] The rate of Na attack depends on the grain-boundary transport of the product, Al, of the following reactions:[9]

$$6Na + 34\ Al_2O_3 \rightarrow 3\ (Na_2O \bullet 11Al_2O_3) + 2\ Al \qquad (1)$$

$$6Na + 12\ (MgAl_2O_4) + 34\ Al_2O_3 \rightarrow 3\ (Na_2O \bullet 4MgO \bullet 15Al_2O_3) + 2Al \qquad (2)$$

Improved sodium resistance is achieved by retarding grain-boundary transport of the Al species brought about by: 1) Y and Zr segregation to grain boundaries that blocks boundary transport: and 2) having less Mg from the lattice available to segregate to grain boundaries, due to the attraction of Mg_{Al}' by $Zr_{Al}^{\bullet}$, which further slows down the grain-boundary diffusion of aluminum and oxygen species.[9]

Analytical microscopic instruments such as in-situ transmission electron microscopy (TEM)[10] and high-resolution secondary ion mass spectrometry (SIMS)[11-13] have contributed significantly to the understanding of grain-boundary segregation, second phases, and the related kinetics of mass transport in PCA, which are critical in the sintering process. Electron microscopy is useful in the study of the reactions of Na with PCA.[9] This paper reviews the characterization of internal interfaces and second phases, in as-sintered and post-test PCA. The changes in grain-boundary MgO levels, which are difficult to quantify even with state-of-the-art instrumentation, are computed from the bulk levels and measured microstructure parameters.

EXPERIMENTAL PROCEDURES

Alumina bodies consisting of high-purity, finely-divided powder doped with MgO-based additives, were used. Consolidation to full density involved hydrogen sintering at ~1850°C, with a cooling rate of ~30°C/min.[9] The average grain size of the sintered bodies is 10-30 μm. An accelerated HPS-PCA lamp test was run at a higher wall temperature than usual (~1350°C vs. 1200°C).[9] Microstructures of as-sintered and post-test PCA specimens were characterized with TEM equipped with energy dispersive x-ray analysis (EDXA),[14] and with electron microprobe analysis (EMPA) aided with wavelength dispersive x-ray analysis (WDXA). Microprobe analysis was performed to measure the spinel content. Analyses were conducted on cross sections at a magnification of 540X, which is high enough to show images of the Mg-rich spots ($MgAl_2O_4$ particles), and sufficiently low enough to give an overall average of the spinel number

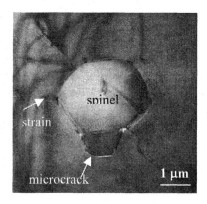

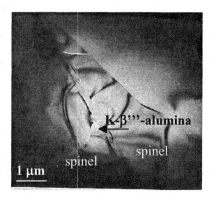

Figure 1. Transmission electron micrograph showing spinel second phase, residual strains, and microcracks in an alumina matrix

Figure 2. Transmission electron micrograph of K-β'''-alumina and spinel in an as-sintered PCA doped with MgO.

densities.[9] Induction-coupled-plasma (ICP) chemical analysis was used to determine the bulk dopant levels.

RESULTS AND DISCUSSION
Grain Boundaries and Second Phases in As-sintered PCA

The MgO dopant is distributed in four ways in PCA: residing as spinel second phase; segregating at grain boundaries; segregating to the internal surface of pores; and dissolving in the lattice. Spinel particles are usually located at grain boundaries, with no detectable residual stresses due to their relatively small size and their orientation with respect to the Al_2O_3 matrix grains.[14] However, some large spinel particles can cause strains[15] and microcracks at the interface of the spinel and alumina matrix grains (Figure 1). The spinel second phase content in sintered tubes generally varies in accordance with the dopant formulation. Due to the volatility of MgO in H_2, a large portion of the MgO dopant in the starting powder is lost during sintering. For a composition of ≤200 ppm MgO + 350 ppm Y_2O_3, there is no detectable spinel phase in the sintered tube.[9]

The solid solubility of MgO in a 10-30 μm-grain size alumina lattice is 95-100 ppm at 1800°C.[16,17] A recent study has shown that the solid solubility of magnesia in alumina is a function of the grain size; the solid solubility at 1250°C is >220 ppm in 0.5 μm grain size PCA,[18] which is much higher than the ~10 ppm value for large grain size (10-30 μm) PCA.[16,17] The segregation of MgO at grain boundaries, and the total volume of the boundary interface region in the 0.5 μm grain size alumina, accommodates >220 ppm MgO.[18] The grain size dependence of the MgO solid solubility in alumina indicates that, during the critical initial-stage of sintering, where the majority of the pores must be annihilated in order to achieve a fully dense body, MgO solute drag (instead of second-phase drag) is operating and retarding the boundary motion.

High resolution SIMS shows direct evidence of Mg segregation at the grain boundaries of alumina.[11,12] The grain boundary segregation is typically expressed by a term called the grain-boundary enrichment factor, which is defined as the ratio of the grain-boundary Mg concentration to the lattice Mg concentration inside the grains. The Mg grain-boundary enrichment factor is about 88 in 1800°C-sintered Al_2O_3 doped with 250 ppm MgO.[11] It is 400 in 1650°C-sintered Al_2O_3 doped with 400 ppm MgO.[12] The grain boundary width is still a topic of study; some studies calculated the grain boundaries as two monolayers, while others proposed a larger width.[19] A recent study showed that the MgO enrichment factor decreases significantly when a tetravalent dopant such as SiO_2 is present.[12] In principle, the triple points at three-grain junctions in less densely packed structures should have a higher MgO enrichment factor than the bi-planes between two grains. This is an area of continuing study.

In Na-resistant PCA compositions doped with MgO, Y_2O_3, and ZrO_2, such as 200 ppm MgO + 400 ppm ZrO_2 + 20 ppm Y_2O_3 (AMZY) and 200 ppm MgO + 350 ppm Y_2O_3 + 180 ppm ZrO_2 (AMYZ), Zr and Y are found by TEM/EDXA to segregate to the grain boundaries.[9] Similarly, Y and Zr segregation to the grain boundaries of alumina is reported in the literature.[19-22] Sintered PCA of the AMZY composition has an equiaxed grain microstructure with no detectable second phases such as spinel, ZrO_2, and yttrium aluminate garnet (YAG).[9] The absence of second phases in AMZY is due to a complete charge compensation of the Mg^{+2} with Zr^{+4} ions. PCA of the AMYZ composition sintered under the same condition as the AMZY has a bimodal grain size distribution, with interspersed YAG and zirconia (ZrO_2) second phases.[9] The presence of the second phases in AMYZ is because of an incomplete charge compensation between the Mg^{+2} and Zr^{+4} ions, and the relatively high level of the Y_2O_3 dopant. Since Y enrichment at grain boundaries[21] decreases with increasing grain size, the boundaries in the smaller grain size region of a bimodal microstructure can have Y enrichment factors significantly different from those in the large grain size region of the same microstructure. However, this is not detected in the sintered AMYZ composition, presumably because the differences between the average small grain size (4 μm) and average small grain size (40 μm) of the bimodal grain size alumina, is not sufficiently large.

A K-β'''-alumina ($K_2O \cdot 4MgO \cdot 15Al_2O_3$) in the form of an elongated, plate-like phase containing stacking faults (Figure 2), as well as K segregation to grain boundaries, similar to what has been reported in the literature,[23] are found in certain as-sintered PCA tubes. The potassium is believed to come from either an impurity in the starting powder, or from the K-doped W heating elements in the sintering furnace. Microcracks can form at the ends of the elongated K-β'''-alumina phase due to the thermal expansion mismatch during the heating-and-cooling lamp operation. HRTEM (high-resolution TEM) and EDXA show segregation of Na at grain boundaries in some as-sintered PCA (Figure 3). In the case of PCA doped with MgO, Y_2O_3, and CaO, Na and Y are found at the boundaries between adjacent alumina grains, and Ca and Y are detected at triple points (Figure 4).

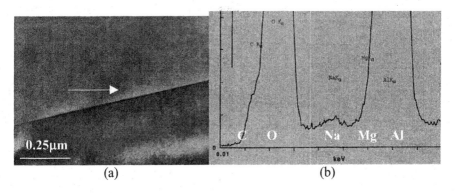

Figure 3. a) Transmission electron micrograph of a bi-plane grain-boundary in as-sintered MgO-doped PCA, and b) EDXA spectrum showing Na segregation at the boundary.

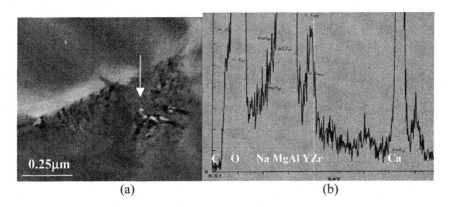

Figure 4. a) Transmission electron micrograph, and b) EDXA spectrum showing Ca and Y at triple points in an as-sintered PCA doped with MgO, Y_2O_3, and CaO.

Grain Boundaries and Second Phases in Post-test PCA

The presence of spinel second phase by itself does not affect the grain-boundary diffusivity of alumina. The effect of a spinel second phase on Na resistance is that Na preferentially reacts with spinel as compared to alumina. Thus, spinel second phase in areas away from the inner surface of the arc tube can react with Na transported along grain boundaries to form the Na-Mg-β'''-alumina ($Na_2O \cdot 4MgO \cdot 15Al_2O_3$) (Figure 5). Discrete spinel particles in the middle of the PCA wall react with Na to form the β'''-alumina.[9] Clearly, grain boundaries are the paths for the transport of the reactants and products. The above reactions, which involved a relatively large volume change, can cause cracks, exposing

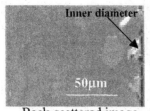
Back-scattered image

Na map

Mg map

Figure 5. EMPA maps of the near-inner-diameter cross section of a post-test PCA showing discrete Na-Mg-β'''-alumina formed inside the PCA wall. The dark areas in the Na and Mg maps (indicated by the arrows) contain Na and Mg, respectively.

more of the alumina internal surface to the Na. Microcracks are associated with the Na-Mg-β'''-alumina (Figure 6). The Na-Mg-β'''-alumina becomes a local source of Na for diffusion farther into the bulk, which thereby significantly decreases the diffusion distance. Thus, the effect of a higher level of spinel second phase is two-fold: a) more reaction products including Na-Mg-β'''-alumina and Al per Eq. (1) and (2); and b) a shorter diffusion distance due to the availability of the Na atoms (for further diffusion into the bulk of the PCA wall, and for reactions with the PCA) from the Na-Mg-β'''-alumina, as well as the microcracks around the Na-Mg-β'''-alumina.

TEM shows grain-boundary degradation characterized by a Na-containing, amorphous alumina phase[9] at the grain boundaries (Figure 7). Additionally, there are microcracks in the near surface region, which might represent an early stage of Na attack, prior to the formation of β-alumina ($Na_2O \cdot 11Al_2O_3$). The amorphous phase does not appear to be due to artifacts such as an electron beam induced phase transformation. Finding Na segregation at the grain boundaries

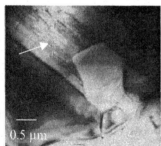

Figure 6. Na-Mg-β'''-alumina and microcracks in a post-test PCA sample.

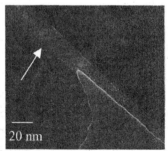

Figure 7. Amorphous Na-Al-O phase at a grain boundary in a post-test PCA.

(not amorphous phase) in some as-sintered PCA lends support to the presence of a real amorphous Na-containing phase (not an artifact) at grain boundaries in post-test PCA.

In the inner diameter region of post-test PCA samples, Na typically reacts with alumina to form β-alumina[9] (Figure 8). Several different types of β-alumina[9] are observed in the near-surface region, including Na-β-alumina, Na-Ba-β-alumina, Na-Ca-β-alumina, Na-Ca-Ba-β-alumina, and Na-Mg-β'''-alumina. The Ba and Ca come from the Ba-Ca-W-oxide emitter inside the lamp, and the Mg comes from the out-diffusion of MgO from the PCA or the frit seals. The β-alumina phase is typically large in size (~100 μm) and has associated microcracks caused by the volume change from the reactions.

Precipitation and growth of second phases such as ZrO_2 and YAG occur in PCA of AMZY composition during lamp tests.[9] In AMZY sintered at 1850°C, there are no detectable second phases. During a lamp test at 1350°C, the Zr and Y concentrations above their 1350°C solid solubility limits in alumina, start to precipitate as discrete second phases: ZrO_2 and YAG. Growth of YAG is observed[9] in post-test PCA doped with 200 ppm MgO and 350 ppm Y_2O_3. The observed precipitation and growth of second phases indicates that a great deal of atomic transport occurs during the lamp test. For example, Al species diffuse along grain boundaries in a direction, counter to the fluxes of Y, Zr, or Mg species moving in the opposite direction.

Effect of Heat Treatment on Sodium Resistance

Maekawa[8] first reported that a post-sintering heat treatment of PCA significantly improves Na resistance in lamps, and it is experimentally confirmed here. Sintered PCA doped with 200 ppm MgO + 350 ppm Y_2O_3 were heat treated in an H_2 atmosphere, and this treatment improved Na resistance (Figure 9). No significant grain growth occurs during the 1700°C/10h anneal. The MgO contents of the tubes were measured before and after the anneal. The as-sintered tubes are free of spinel second phase, and the bulk MgO level is ~50 ppm. After annealing, the bulk MgO level decreases to 40 ppm. The amount of decrease is small (~10 ppm); however, it demonstrates the trend of MgO depletion during annealing. Due to the absence of a spinel second phase, the depletion must be

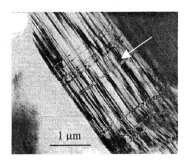

Figure 8. Na-β-alumina in a post-test PCA.

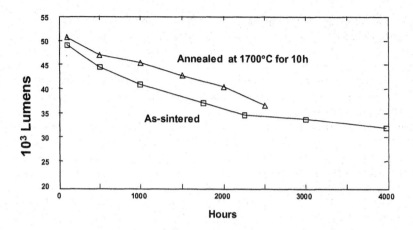

Figure 9. Lumens of PCA (doped with 200 ppm MgO and 350 ppm Y_2O_3) - Na lamps vs time showing the effect of a post-sinter anneal.[9]

occurring at grain boundaries, and maybe in lattice. For a depletion of 10 ppm of MgO at the bulk level, the reduction of the boundary MgO level is large, as shown in the following. The boundary MgO level (C_{gb}) can be related to the bulk level (C_b):[9]

$$C_{gb} = C_b \, (\varepsilon/(1+ \varepsilon \, d/GS)) \qquad (3)$$

where d is the grain boundary width (2 nm), GS is the grain size (10 μm), and ε is the boundary enrichment factor, which is ~88 at 1800°C[11] and ~400 at 1650°C.[12] Taking derivatives of Eq. (3) gives:

$$\Delta C_{gb}/ C_{gb} \sim (1/\varepsilon)(GS/d) \, \Delta C_b /C_b \qquad (4)$$

Using the measured $\Delta C_b /C_b$ values, Eq. (4) shows that the grain-boundary MgO level can decrease by a factor of ~20-100 by annealing at high temperatures. The decreased C_{gb} of MgO slows down grain-boundary diffusion, which in turn, improves Na resistance.

The grain-boundary MgO level is a major limiting factor for Na resistance. In PCA triply doped with MgO, Y_2O_3, and ZrO_2, Na resistance is outstanding. This is explained as follows. Zr^{+4}doping ties up Mg^{+2}, as the positively charged Zr at Al sites compensates and pairs with the negatively charged Mg at Al sites. When this pairing occurs, less Mg becomes available to reside at grain boundaries. Energetically, the charge compensation adds another term (the electrostatic energy for the interaction of the solutes: ZrAl$^{\bullet}$ and MgAl') to the segregation energy to decrease the total free energy for segregation, lowering the grain-boundary enrichment factor (by a factor of 6)[12] for MgO in PCA. Vacuum or hydrogen annealing can reduce the MgO level at grain boundaries by a factor of 20-100.

SUMMARY

Grain boundaries in PCA doped with MgO-based additives play an important role for mass transport during sintering, as well as for transport of reactants and products for chemical reactions that occur during service in lamps. A high resistance to Na attack requires the absence of spinel second phase, and a low grain-boundary MgO level. Reducing the grain-boundary MgO content by co-doping with tetravalent cations (e.g., Zr^{+4}) to charge-compensate the Mg^{+2} at Al^{+3} lattice sites, or post-sintering annealing, can slow down grain-boundary transport, and thereby, increase the resistance to Na attack. The enriched MgO level at grain boundaries is an important parameter for high-temperature properties. The solid solubility of MgO in alumina is a function of grain size in the submicron to 1 μm range. A solute drag mechanism that retards boundary motion appears to be operative during initial-stage sintering, where the majority of the pores are required to be annihilated in order to achieve a fully-dense, translucent body.

ACKNOWLEDGEMENTS

The author thanks C. Sung of Univ. Mass. Lowell (TEM); J. Gagne, A. Hecker, J. Neil, V. Perez of OSI; D. Goodman, S. Nordahl, W. Rhodes, and R. Thibodeau, formerly of OSI; OSI-technical analytical lab, and OSI-PCA plant at Exeter, NH, for contributions in various aspects of PCA research.

REFERENCES

[1]R.L. Coble, "Transparent Al_2O_3 and Method of Preparation," U.S. Pat. No. 3026210 (1962).

[2]R.L. Coble, "Sintering of Crystalline Solids: II. Experimental Test of a Diffusion Models in Porous Compacts," *J. Appl. Phys.* **32** [5] 793-99 (1961).

[3]S.J. Bennison and M.P. Harmer, "A History of the Role of MgO in the Sintering of α-Al_2O_3," *Ceram. Trans.* **7** 13-49 (1990).

[4]C.A. Handwerker, J.M. Dynys, R.M. Cannon, and R.L. Coble, "Dihedral Angles in MgO and Al_2O_3," *J. Am. Ceram. Soc.* **73** [5] 1371-7 (1990).

[5]S. Cho, Y. Lee, H. Lee, S. Sim, and M. Yanagisawa, "Chemical Inhomogeneity in Commercial Al_2O_3 Powders and its Effect on Abnormal Grain Growth during Sintering," *J. Euro. Ceram. Soc.* **23** 2281-88 (2003).

[6]I.J. Bae and S. Baik, "Abnormal Grain Growth of Alumina," *J. Am. Ceram. Soc.* **80** 1149-56 (1993).

[7]J.J. de Groot and J. A. J. M. van Vliet, *The High-Pressure Sodium Lamp*, p. 290, MacMillan, London (1986).

[8]K. Maekawa, "Recent Progress in Ceramic Materials for Lamp Application," 293-302, *Proc. of 7th Light Source Conf.* Kyoto, Japan (1995).

[9]G.C. Wei, A. Hecker, and D. Goodman, "Translucent Polycrystalline Alumina with Improved Resistance to Na Attack, " *J. Am. Ceram. Soc.* **84** [12] 2853-62 (2001).

[10]M. Komatsu, H. Fujita, and H. Matsui, "A New Type of Very-high Temperature Heating Stage and In-situ Experiments on α-Al$_2$O$_3$ Crystals," *Japan J. Appl. Phys.* **21** [8] 1233-37 (1982).

[11]K K. Soni, A. M. Thompson, M. P. Harmer, D. B. Williams, J. M. Chabala, and R. Levi-Setti, "Solute Segregation to Grain Boundaries in MgO-doped Al$_2$O$_3$," *Appl. Phys. Lett.* **66** [21] 2795-2797 (1995).

[12]K.L. Gavrilov, S.J. Bennison, K.R. Mikeska, and R. Levi-setti, "Grain Boundary Chemistry of Al$_2$O$_3$ by High-resolution Imaging SIMS, " *Acta Mater.* **47** [15] 4031-39 (1999).

[13]I. Sakaguchi, V. Srikanth, T. Ikegami, and H. Haneda, "Grain Boundary Diffusion of Oxygen in Al$_2$O$_3$ Ceramics, " *J. Am Ceram. Soc.* **78** [8] 2557-59 (1995).

[14]G.C. Wei, S. Jean, C. Sung and W. H. Rhodes, "Characterization of Second Phases in Translucent Al$_2$O$_3$ by Analytical Transmission Electron Microscopy, " pp. 311-22, *Ceramic Microstructure: Control at the Atomic Level*, Plenum Press, NY (1998).

[15]G.C. Wei, "Characterization of Translucent Polycrystalline Alumina Ceramics," *Ceram. Trans.* **133** 135-144 ed. M. Matsui et al, Am.Cer.Soc. Westerville, OH (2002).

[16]C. Greskovich and J.A. Brewer, "Solubility of Mg in Polycrystalline Al$_2$O$_3$ at High Temperatures, " *J. Am. Ceram. Soc.* **84** [2] 420-25 (2001).

[17]K. Ando and M. Momoda, "Solubility of MgO in Single-crystal Al$_2$O$_3$," *Yogyo-Kyokaishi* **95** [4] 381-86 (1987).

[18]G.C. Wei, "Sintering and Grain Growth in Translucent Alumina," *Proc. Ceram. Pac-Rim-5 Conf., J. Ceram. Soc. Japan*, Nagoya, Japan (in press).

[19]C.M. Wang, G.S. Cargill III, H.M. Chan, and M.P. Harmer, "Structural Features of Y-saturated and Supersaturated Grain Boundaries in Alumina," *Acta. Mater.* **48** 2579-91 (2000).

[20]S. Lartrigue, C. Carry, and L. Priester, "Grain Boundaries in High Temperature Deformation of Y$_2$O$_3$ and MgO Co-Doped Al$_2$O$_3$," *Coll. de Phys.* C1-985 (1990).

[21]J. Cho, C.M. Wang, H.M. Chan, J.M. Rickman, and M.P. Harmer, "Role of Segregating Dopants on the Improved Creep Resistance of Al$_2$O$_3$," *Acta. Met.* **47** [15] 4197-4207 (1999).

[22]M. A. Gülgün, V. Putlayev, and M. Rühle, "Effects of Y Doping a-Al$_2$O$_3$: I, Microstructure and Microchemistry," *J. Am. Ceram. Soc.* **82** [7] 1849-56 (1999).

[23]M. Blanc, A. Mocellin, and J.L. Strudel, "Observation of K-β'''-Al$_2$O$_3$ in sintered Al$_2$O$_3$," *J. Am. Ceram. Soc.* **60** [9-10] 403-409 (1977).

JOINING OF Si_3N_4 TO A NICKEL-BASED SUPERALLOY USING ACTIVE FILLERS

S.P. Lu and Y. Guo
Institute of Metal Research, Chinese Academy of Sciences
Shenyang, Liaoning, P.R.China, 110016

ABSTRACT
Pressureless-sintered Si_3N_4 and a nickel-based superalloy were joined at $920^\circ C$ using AgCuTi as a single interlayer filler, or at $1160^\circ C$ using AgCuTi/NiCrCoWMoSiB as a double interlayer filler. The Si_3N_4/filler bonding interface was characterized using electron-probe microanalysis (EPMA) and X-ray diffraction (XRD). The bend strength reaches 74 MPa and decreases with increasing titanium content in the filler for the AgCuTi single interlayer joint. The bend strength for the double interlayer joint is in the range of 30-35 MPa. EPMA shows that the active element (Ti) in the filler segregates to and reacts with the Si_3N_4 ceramic at high temperature. TiN, Ti_5Si_3, and Cu_2Ti form at the interface between the Si_3N_4 ceramic and the filler. Dissolution and diffusion occur between the AgCuTi single interlayer filler, and between the NiCrCoWMoSiB double interlayer filler during high temperature bonding. The interface between the filler disappears after bonding to become a homogeneous Ni-based brazing seam between the Si_3N_4 and the nickel base superalloy.

INTRODUCTION
Silicon nitride is a perspective structural ceramic with superior properties, such as refractoriness, oxidation resistance, and wear resistance. However, its poor machinability, due to its hardness and brittleness, is a barrier to manufacturing complex structures directly. Consequently, ceramic to ceramic or ceramic to metal bonding is often necessary to form a complex shape structure. Active brazing is one technique that has received extensive attention for bonding ceramics to metals.[1]

AgCuTi was developed as an active filler to join ceramics to metal. Chen and Nogi[2] succeeded in joining Si_3N_4 to Inconel 600 using Ag71Cu27Ti2, with emphasis on the metallurgical behavior at the interface between the filler metal and the base metal. Their results showed that nickel from the base metal combines the titanium to decrease its activity and reactivity with Si_3N_4. Titanium is an active element that plays an important role in ceramic to metal joining. Three AgCuTi fillers with different titanium contents are designed in this experiment to investigate the effect of titanium content on bond strength. However, a ceramic/metal joint made with the AgCuTi filler cannot be used in a high temperature surrounding, which limits the application of the high-temperature ceramic. Consequently, there is interest in searching for a new filler alloy or a new technology to join ceramics and metals for high-temperature applications. Naka and co-workers[3,4] studied the joining of SiC/SiC and SiC/Nb/W using a Ni-Ti active filler. Their work increased the test temperature for the joint shear strength to 973 K. Mcdermid and Co-workers[5] investigated directly joining SiC to Inconel 600 using a high temperature nickel-based brazing alloy, and they analyzed the interface reaction using thermodynamics. Their conclusion showed that the direct joining of SiC to Inconel 600 using a nickel-based brazing alloy is not feasible.

In the present study, the joining of Si_3N_4 to a nickel-based superalloy using a single AgCuTi interlayer filler or a double AgCuTi/NiCrCoWMoSiB interlayer filler is investigated.

EXPERIMENTAL

Pressureless-sintered Si_3N_4 ceramic, and nickel-based superalloy (LDZ125) were used as the substrates. The LDZ125 superalloy is an important material for aircraft engines. The substrate size was 3 mm × 4 mm × 20 mm. The Si_3N_4 and LDZ125 substrates were joined with a single or double interlayer filler. The AgCuTi or NiCrCoWMoSiB filler is first melted in vacuum to form a thin belt. The thickness of the AgCuTi and NiCrCoWWMoSiB belt is 60 μm and 100 μm, respectively. The solid-liquid temperature range for NiCrCoWMoSiB is 1050 ~ 1100°C. The compositions of the LDZ125 superalloy, the 3 different fillers, and the NiCrCoWMoSiB filler are shown in Table I.

Before the experiments, the Si_3N_4 specimens were polished with SiC powder paste on a glass plate, and the LDZ125 specimens were polished with sandpaper. Figure 1 shows a schematic of the assembled specimens before bonding. The bonding process was completed in a resistance-heated cylindrical vacuum furnace heated with a molybdenum element. The effective chamber size was φ100 × 1200 mm. The assembled specimen was first placed in the chamber with an 80 g weight on the top of the ceramic bar. Then the chamber was evaluated to 1.0×10^{-3} Pa, followed by heating at the rate of 20°C/min. The single interlayer specimen was heated to 920°C and held for 8 min. The double interlayer specimen, was first heated to 920°C for 5 min, and then heated to 1160°C and held for 5 min. The specimens were cooled to room temperature in the chamber at 20°C/min. The bond strength of the butt joint was determined from a four-point bend test at room temperature[6,7]. The interface of the ceramic/metal bond was characterized using electron-probe microanalysis (EPMA) and X-ray diffraction (XRD).

Table I. Compositions (wt%) of LDZ125 superalloy and fillers

	C	Cr	Co	W	Mo	Ta	Al	B	Si	Ni	Ti	Ag	Cu
LDZ125	0.1	9	10	7	2	3.8	4.8	0.01		bal.	3.5		
AgCuTi filler											5	68	27
											7	67	26
											10	65	25
Nickel-based filler	0.9	16	8	5	4			2	5.3	bal.			

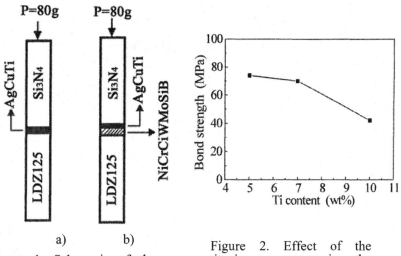

a)   b)

Figure 1. Schematic of the:
a) single interlayer: and
b) double interlayer assembly
used to join Si₃N₄ and LDZ125

Figure 2. Effect of the
titanium content in the
AgCuTi interlayer filler on
the bond strength of a
Si₃N₄/LDZ125 joint.

RESULTS AND DISCUSSION
Si₃N₄/LDZ125 Bonding with a single AgCuTi Interlayer Filler

Three AgCuTi fillers with different titanium contents were examined (Table I). Figure 2 shows the effect of the titanium content on the bonding strength of the Si₃N₄/LDZ125 joint. The four-point bend strength reaches 74 MPa when the titanium content in the filler is 5 wt%. The bond strength decreases with increasing titanium content in the filler. Residual stress in the ceramic/metal joint is generated by differences in the thermal expansion coefficients between the ceramic and the metal, and this becomes a negative factor that decreases the bond strength.[8] Relaxation of the residual stress depends, to a large extent, on the plasticity of the interlayer in the cooling process.[9] The AgCu eutectic alloy has good plasticity. However, the addition of titanium decreases its plasticity. The higher the titanium content, the lower the plasticity. Therefore, the ability to relax residual stress in the insert filler decreases as the titanium content increases in the AgCu alloy.

The distribution of Ti and Ni at the Si₃N₄/AgCuTi/LDZ125 joint was mapped by EPMA, and is shown in Figure 3. The active element, titanium, segregates to and reacts with the ceramic to produce a reactive bond. The titanium also diffuses to the LDZ125 superalloy. It is clear from the nickel map shown in Figure 3b that the nickel from the LDZ125 superalloy also dissolves in the filler during the bonding process. Since the titanium has a strong affinity for the nickel, the nickel diffuses to the ceramic together with the titanium.

Figure 3. Distribution of: a) Ti; and b) Ni at the Si₃N₄ / AgCuTi / LDZ125 joint

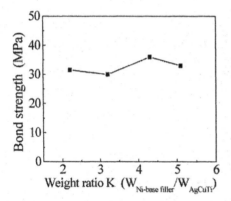

Figure 4. Effect of the weight ratio of the two fillers on bond strength.

Figure 5. The distribution of titanium at the Si₃N₄ / bond seam

Si₃N₄/LDZ125 Bonding with a Double AgCuTi/NiCrCoWMoSiB Interlayer Filler

As shown in Figure 1b, the AgCu7Ti filler is adjacent to the Si₃N₄ ceramic, and NiCrCoWMoSiB is adjacent to the LDZ125 superalloy. During the bonding process, the AgCuTi filler first melts around 920°C, and reacts with the Si₃N₄ ceramic. When the temperature increases to 1160°C, the NiCrCoWMoSiB filler melts. At high temperature, diffusion occurs between the AgCuTi and NiCrCoWMoSiB fillers. During cooling, a homogeneous nickel-based bond seam forms between the Si₃N₄ ceramic and the LDZ125 superalloy.

The effect of the weight ratio, K, which has been defined as the weight ratio of NiCrCoWMoSiB filler ($W_{\text{Ni-base filler}}$) to AgCu7Ti filler ($W_{\text{AgCuTi}}$), on the bond strength of the joint is shown in Figure 4. K has been defined as the weight ratio of NiCrCoWMoSiB filler ($W_{\text{Ni-base filler}}$) to AgCu7Ti filler ($W_{\text{AgCuTi}}$). The four-point bend strength is consistently around 35 MPa. Figure 5 shows the titanium distribution at the Si₃N₄/bond seam interface. Similar to the results for the AgCuTi single interlayer joint shown in Figure 3, titanium segregates to the Si₃N₄.

The majority of the failures that occur during the bend tests occur at the interface of the Si_3N_4/bond seam. The reactive layer at the Si_3N_4/bond seam interface is a weak area. The X-ray diffraction analysis of the fracture surface on the Si_3N_4 side of the bond is shown in Figure 6. TiN, Ti_5Si_3, and $CuTi_2$ are found on the fracture surface, which indicates that titanium in the filler reacts with the Si_3N_4 during bonding to form reaction products that create a weak joint.

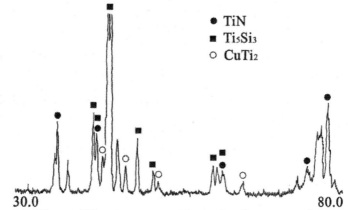

Figure 6. X-ray diffraction pattern from the fracture surface at the interface of the Si_3N_4 / bond seam in Si_3N_4/LDZ125 joined with a double interlayer filler

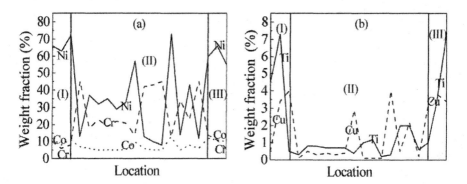

Figure 7. Distribution of: a) Ni, Cr, and Co; and b) Cu and Ti in the LDZ125 /bond seam / Si_3N_4 joint, where I = the LDZ125/bond seam interface, II = the Bond seam, and III = the Si_3N_4/bond seam interface.

Quantitative elemental analysis of the LDZ125/braze seam/Si_3N_4 joint by EPMA is shown in Figure 7. It is clear that the Ni, Cr, and Co in the Ni-based filler diffuse to the Si_3N_4/AgCuTi interface during joining, and the Ti and Cu elements in the AgCuTi filler diffuse to the LDZ125 superalloy/Ni-base filler interface. The interface between the AgCuTi and the NiCrCoWMoSiB interlayer filler disappears. Therefore, the bond seam becomes a homogeneous nickel-base bond seam composed of Ni, Cr, Co, Ag, Cu, and some other minor elements. Titanium and nickel enrichment occurs at the Si_3N_4/bond seam interface, and at the LDZ125/bond seam interface, as shown in Figure 7. This phenomenon is similar to the results for the AgCuTi single interlayer bond shown in Figure 3, and is attributed to the strong affinity between nickel and titanium.

Comparison of the Bond Strength

The bond strength by the AgCuTi single interlayer filler is higher than that by the AgCuTi/NiCrCoWMoSiB double interlayer filler. There are two main factors affecting the bond strength. One is the plasticity of the bond seam, and the other is the bond temperature.

The bond seam between the Si_3N_4 and the LDZ125 produced with the double interlayer fillers becomes a homogeneous nickel-based alloy, because there is considerable diffusion between the AgCuTi filler and NiCrCoWMoSiB filler during the high temperature bonding process. The plasticity of the nickel-base bond seam is lower than that of AgCuTi bond seam. As such, the ability to relax residual stress (σ_R) by plastic deformation is poorer in the nickel-based bond seam than in the AgCuTi seam. Additionally, the bonding temperature for the double interlayer filler (1160°C) is higher than for the single interlayer filler (920°C). The coefficient of thermal expansion of the LDZ125 superalloy increases with increasing temperature, while the coefficient of thermal expansion of Si_3N_4 does not change a lot with temperature. Therefore, the difference in the thermal expansion ($\Delta\alpha$) between the LDZ125 superalloy and the Si_3N_4 increases with the increasing bonding temperature. The residual stress (σ_R) in the ceramic to metal joint can be approximated from the following formula:[10]

$$\sigma_R = \frac{E_1 E_2}{E_1 + E_2} \Delta\alpha\Delta T \tag{1}$$

$$\Delta T = T_B - T_0 \tag{2}$$

where, T_0 = room temperature;
T_B = bonding temperature;
E = elastic modulus of substrate;
$\Delta\alpha$ = the difference in thermal expansion coefficients of the substrates.

The higher the bonding temperature (T_B), the larger the ΔT and $\Delta\alpha$ values, and hence, a larger residual stress is generated in the joint. Because of the higher bonding temperature and the lower plasticity of the nickel-based bond seam, the bond strength of the AgCuTi/NiCrCoWMoSiB double interlayer filler is lower than that by the AgCuTi single interlayer filler.

CONCLUSIONS

(1) The four-point bend strength of Si_3N_4/LDZ125 joined with a AgCuTi single interlayer filler with 5 wt% Ti is 74 MPa, and strength decreases with the increasing titanium content in the filler;

(2) The bend strength of Si_3N_4/LDZ125 joined with an AgCuTi/NiCrCoWMoSiB double interlayer filler is around 35 MPa, which is lower than that of the AgCuTi single interlayer filler joint; the lower strength is attributed to the higher bonding temperature, and the lower plasticity of the nickel-base bond seam;

(3) During the bonding process with the AgCuTi/NiCrCoWMoSiB double interlayer filler, the AgCuTi filler melts first and the active element, Ti, segregates to and reacts with the Si_3N_4 ceramic. TiN, Ti_5Si_3 and $CuTi_2$ compounds form a reaction layer between the Si_3N_4 and the AgCuTi filler. As the temperature increases to 1160° C, the NiCrCoWMoSiB filler melts and diffuses into the AgCuTi filler. The AgCuTi filler also diffuses into the NiCrCoWMoSiB filler. Upon cooling, a homogeneous nickel-based bond seam forms between the Si_3N_4 and the LDZ125 superalloy.

REFERENCES

[1]A.M. Hadian and R.A.L. Drew, "Thermodynamic Modeling of Wetting at Silicon Nitride / Ni-Cr-Si Alloy Interfaces," *Materials Science and Engineering*, **A189** 209-17 (1994).

[2]J.H. Chen, G.Z. Wang, K. Nogi, M. Kamai, N. Sato, and N. Iwamoto, "The Metallurgical Behavior During Brazing of Ni-base Alloy Inconel 600 to Si_3N_4 with Ag71Cu27Ti2 Filler Metal," *Journal of Materials Science*, **28**[11] 2933-42 (1993).

[3]M. Naka, H. Taniguchi and I. Okamoto, "Heat-Resistant Brazing of Ceramics (Report I)," *Transactions of JWRI*, **19** [1] 25-31 (1990).

[4]M. Naka, H. Taniguchi and I. Okanoto, "Heat-Resistant Brazing of Ceramics (Report II)," *Transactions of JWRI*, **19** [2] 29-34 (1990).

[5]J.R. Mcdermid, M.D. Pugh, and R.A.L. Drew, "The Interaction of Reaction-Bonded Silicon Carbide and Inconel 600 with a Nickel-based Brazing Alloy," *Metallurgical Transactions*, **20A** [9] 1803-10 (1989).

[6]A.P. Xian and Z.Y. Si, "Joining of Si_3N_4 Using $Ag_{57}Cu_{38}Ti_5$ Brazing Filler Metal," *Journal of Materials Science*, **25** [10] 4483-87 (1990).

[7]A.P. Xian, "Joining of Sialon Ceramics by Sn-5 at % Ti Based Ternary Active Solders," *Journal of Materials Sciences,* **32** [23] 6387-93.

[8]D.Y. Chen, S.P. Lu and Y. Guo, "The Calculation of the Residual Stress in C/M Joint," *Chinese Journal of Materials Research*, **14** [Suppl.] 49-52 (2000).

[9]K.H. Thiemann, H.J. Weinert, and W. Rauchle, "Soldering Foil Joining of Ceramic Bodies to Metal," GB Pat. No. 2151173A, 1985.

[10]O.M. Akselsen, "Review Advances in Brazing of Ceramics," *Journal of Materials Science*, **27** [8] 1989-00 (1992).

FABRICATION OF GRAIN REFINED 7475 Al ALLOY FOILS UTILIZING RF MAGNETRON SPUTTERING

T. Shibayanagi, M. Maeda, M. Naka,
and D. Watanabe
JWRI, Osaka University
11-1 Mihogaoka
Ibaraki
Osaka 567-0047, Japan

Y. Takayama
Department of Mechanical Engineering
Utsunomiya University
7-1-2 Yoto
Utsunomiya
Tochigi 321-8585, Japan

ABSTRACT

The present work deals with the fabrication and characterization of the grain refined 7475 Al alloy. RF magnetron sputtering, utilizing a 7475Al alloy target, successfully yielded grain-refined foils having a grain size less than $1\,\mu m$, which has not been achieved using conventional production methods. The foils are a super-saturated solid solution with the FCC structure, and there is a newly discovered amorphous phase in the grain boundary region. The non-equilibrium amorphous phase disappears when the foil is annealed at elevated temperature. Vickers microhardness testing reveals that the deposited foils are softer than the starting alloy, and become softer upon annealing at 643 K. No cracks are observed around the indentations, evidencing the ductility of the grain-refined foil fabricated by RF magnetron sputtering.

INTRODUCTION

Grain refinement is attracting much attention for "nano-materials" and the development of "nano-technology". Heavy deformation processes such as equal channel angular extrusion (ECAE)[1], mechanical alloying (MA)[2], and accumulated role bonding (ARB)[3] are well known techniques for grain refinement that divide the solid into very small arrays of dislocation walls or groups of tangled dislocations, and that alter the grain orientation distribution in a localized region. Grain refinement is also possible with vapor deposition methods such as plasma vapor deposition (PVD) and chemical vapor deposition (CVD). These methods normally produce foils, and the character of the grain boundary is different from that in heavily deformed materials. In addition, deposition methods are useful because the foil making process can provide the parts of micro-machines.

7475 Al alloy is a commercial alloy that was successfully designed to have superplasticity at elevated temperatures.[4] Superplastic deformation is based on grain boundary sliding at elevated temperature. Grain refinement is essentially to realize such a unique deformation mode, and a finer grain size is required for materials to have better deformation characteristics, including lower temperature superplasticity or a higher strain rate for the superplasticity. Commercial 7475 Al alloy has a grain size about 30 μm. Grain refinement of this alloy is expected to show better ductility, although there is no technical data for this alloy with finer grain sizes, such as sub-micron or nano-size.

The present work focuses on fabricating grain-refined 7475 Al alloy foils with a grain size <1 μm as an example of an engineered nano-structures and to make parts for micro-machines.

EXPERIMENTAL PROCEDURE

RF magnetron sputtering was utilized to fabricate foils. The starting material was 1.5 mm thick 7475 Al alloy, which was used as a target. The average grain size of the alloy is 30 μm, which was determined by a linear intercept method. The alloy plate was machined into a 100 mm diameter disk shape, and placed in the vacuum chamber. The distance between the sputtering target and the 2 mm thick copper substrate was about 40 mm, and the background pressure in the chamber was less than 2×10^{-5} Pa. Sputtering was performed in an 8.0 Pa Ar gas atmosphere. The sputtering frequency and power were 13.56 MHz and 400 W, respectively. The sputtering time ranged from 12.6 ks to 36 ks. The average sputtering rate was 2.8×10^{-3} μm/s. The present system cannot monitor the temperature of the water-cooled substrate during the sputtering.

The microstructure of each foil was characterized using scanning electron microscopy (SEM), transmission electron microscopy (TEM), and X-ray diffraction (XRD). The XRD measurement was performed using the theta-2theta method. Differential scanning calorimetry (DSC) was utilized to evaluate the thermal stability of the foils. The DSC measurements were made in flowing Ar gas using a heating rate of 10 K/s.

Vickers microhardness testing at room temperature was used to characterize the mechanical properties of the foils. The load of indentation was 490 mN, and the loading time was 15 s.

RESULTS AND DISCUSSION

Microstructure of Deposited Foils

Figure 1 shows a SEM image of the cross section of a deposited foil. The thickness of the foil is 30 μm, and no voids or other defects are detected in the image. Longer sputtering time yields thicker foils without significant grain growth. Electron probe micro-analysis (EPMA) revealed that the elemental distribution of each constituent is almost uniform throughout the film thickness as (Figure 2). Based on the EPMA data, the film composition is shown in Table I along with the composition of the sputter target. A comparison of the two compositions shows they are similar, indicating that sputtering does not significantly affect the composition. Each specimen made has a similar chemical composition.

X-ray diffraction was performed to identify the phases in the sputtered foil, and the result is shown in Figure 3. The diffraction pattern of the sputter target clearly shows the FCC structure accompanied by weak peaks corresponding to precipitates or other inclusions in the alloy. The small peaks between the 111 and 200 FCC peaks correspond to the other crystallographic structures that

Table I. Chemical composition of specimens (mass%)

	Zn	Mg	Cu	Cr	Fe	Si	Ti	Mn	O	Al
Target (7475)	5.74	2.57	1.50	0.20	0.07	0.04	0.02	0.00	--	bal.
Deposited foils	5.81	2.97	1.41	0.20	0.05	0.04	0.01	0.00	3.07	bal.

Figure 1. SEM image of a cross section of a foil. produced by RF sputtering.

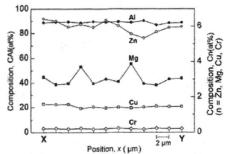

Figure 2. Elemental distribution of each constituent in the deposited foil. The analysis was performed along the line X-Y line indicated in Figure 1.

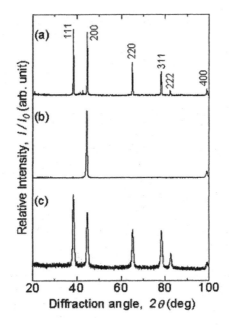

Figure 3. XRD (Cu-Kα) pattern for the: a) 7475Al alloy; b) deposited foil: and (c) filed powder from the deposited foil.

originate from precipitates in the alloy. These peaks are not observed in the deposited foil.   In addition, the deposited foils show a strong texture component in the (200) plane, i.e., the RF sputtering produces {100} texture.

To elucidate the crystallographic structure, the foil was filed to form a powder, and the powder was characterized using XRD. The results (Figure 3c) clearly show the existence of the FCC structure without the precipitates observed in Figure 3a. Therefore the deposited foil is a super-saturated solid solution having the FCC structure. The lattice parameter of the foil was calculated to be 0.4046 nm, utilizing the peak data shown in Figure 3c. The sputter target material has a lattice constant of 0.4055 nm.

TEM characterization provided important and new information on the film microstructure. The selected area diffraction pattern (SADP) shown in Figure 4a confirms the FCC structure determined by XRD, and also shows the existence of a weak hallo inside the first diffraction ring of the 111 reflection. A dark field image of the foil obtained utilizing the 111 and 200 reflections clearly shows many small grains of bright contrast, indicating they have almost the same orientation. This is consistent with the results obtained by XRD shown in Figure 3. The average grain size on a planar view section of the foil is 495 nm. Columnar grains are observed in the cross section of the 101 μm thick foil obtained after 36 ks of sputtering.

The existence of another bright contrast region is observed in the grain boundary region in Figure 4b. This grain boundary phase is unique to the deposited foil, and has not been observed in the 7475 Al alloy. The rapid quenching from the vapor state to solid state apparently changes the grain boundary nature in comparison to material fabricated by a heavy deformation process. This should have some affect on the mechanical properties.

Thermal Stability of the Deposited Foils

Figure 5 shows the DSC curves obtained for the target material and the deposited foils. There are three thermal events on the DSC curve for the sputter target material, including events at 753, 764, and 803 K. These correspond to the phase change from $Al + MgZn_2$ to liquid[4], from $Al + Mg_3Zn_3Al_2$ to liquid,[4] and the solidus temperature, respectively. The deposited foil shows only two distinct

Figure 4. TEM plan-view, dark field image of a RF sputtered foil with a SADP in the upper left corner.

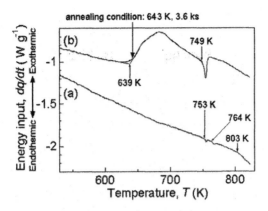

Figure 5. DSC curves for; (a)7475 Al alloy and (b) the deposited foil.

events, including ones at 639 and 749 K. The latter signal corresponds to the Al + MgZn$_2$ to liquid phase change observed in the target material. However, the former signal does not appear in the starting material, suggesting that a different phase change occurs in the deposited foils. Taking into account the existence of the non-equilibrium phase observed in Figure 4, the exothermic peak at 639 K indicates the disappearance or the crystallization of the amorphous phase in the grain boundary region. TEM characterization of the foil after annealing at 643 K for 3.6 ks (Figure 6), clearly shows coarsened grains with an average grain size of 850 nm and no grain boundary amorphous phase. This result strongly supports a crystallization process such as indicated by the exothermic peak at 639 K. Taking the 10 K/s DSC heating into account, the DSC measurement shows that the deposited foil is stable up to 639 K, which is the crystallization temperature.

Hardness of the Deposited Foil

Vickers micro-indentation tests on a deposited foil having a thickness of 101 μm produces no cracks around the indentation, suggesting the foils are ductile. The hardness of the foil is 133 Hv (1.30 GPa), which is slightly harder than the starting material (153 Hv, 1.50 GPa). Annealing the foil at 643 K for 3.6 ks reduces the hardness to 125 Hv (1.23 GPa). Pictures of the indents are shown in Figure 7. No cracks are observed around the indentation in any specimen tested, suggesting that the deposited foils are ductile.

It is well known that mechanical properties such as yield stress and flow stress of polycrystalline materials obey the Hall-Petch relationship. Many nano-crystalline materials have been reported to show an inverse Hall-Petch relationship with decreasing grain size below a critical value that depends on the material.[5] This peculiar characteristic of nano-materials has been reported in Cr-B alloy foils with a grain size less than 20 nm produced by RF sputtering.[6] In this alloy foil, a grain boundary amorphous phase was also observed, and this phase contributes significantly to the softening of the foil. Since the volume fraction of grain boundary region is larger than 20% when the grain size becomes smaller than 20 nm, the relatively softer amorphous and/or grain boundary region can behave like a second phase that can increase the ductility. Although the grain size of the present alloy foil is much larger than that of the Cr-B alloy foil, and the inverse Hall-Petch rule is not applicable, the amorphous phase in the grain boundary region should still affect the mechanical properties.

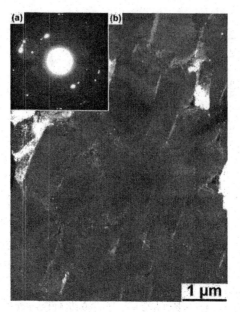

Figure 6. TEM dark field image of a foil annealed at 643 K for 3.6 ks, with the SADP in the upper left corner.

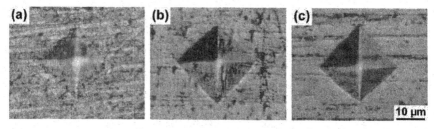

Figure 7. A Vickers diamond indentation in the: a) 7475 Al alloy; b) deposited foil; and c) foil annealed at 643 K for 3.6 ks.

The grain boundary amorphous phase possibly originates from some oxides,[7] or from a reaction between moisture and the vaporizing atoms from the sputtering target.[8]

The softening in the film is possibly due to the texture of the foil since the number of active slip systems and the required resolved shear stress on each slip plane depend on the orientation of each grain.

SUMMARY

Fine-grained 7475 Al alloy foils were successfully fabricated utilizing RF magnetron sputtering. The average grain size of the deposited foil is 495 nm, and the foil is a super-saturated solid solution with a highly textured FCC structure. An amorphous phase is present in the grain boundary region. The non-equilibrium grain boundary phase is stable up to 639 K. The Vickers microhardness of the target alloy and the deposited foil are 153 and 133 Hv, respectively. The deposited foils are softer than the starting alloy, suggesting an effect of the grain refinement and the existence of non-equilibrium grain boundary phase. The foil seemed to be ductile, since no cracks were observed around the Vickers indentation.

ACKNOWLEDGEMENT
The present work was supported by grants-in-aid of Ministry of Education, Culture, Sports, Science and Technology. (Project No. 13450295 and 15656185)

REFERENCES
[1]M. Furusawa, Z. Horita, M. Nemoto, R.Z. Valiev, and T.G. Langdon, "Microhardness Measurements and the Hall-Petch Relationship in an Al-Mg Alloy with Submicrometer Grain Size," *Acta Mater.*, **44** [11] 4619-4629 (1996).

[2]Y. Kimura, H. Hidaka and S. Takaki, "Work-Hardening Mechanism During Super-Heavy Plastic Deformation in Mechanically Milled Iron Powder," *Mater. Trans. JIM*, **40** [10] 1149-57 (1999).

[3]Y. Saito, H. Utsunomiya, N. Tsujiand, and T. Sakai, *Acta Mater*, "Novel Ultra-High Straining Process for Bulk Materials-Development of the Accumlative Roll-Bonding (ARB) Process," **47** [2] 579-83 (1999).

[4]Y. Takayama, T. Tozawa, and H. Kato, "Superplasticity and Thickness of Liquid Phase in the Vicinity of Solidus Temperature in a 7475 Aluminum Alloy," *Acta. Mater.*, **47** [4] 1263-70 (1999).

[5] A.H. Chokshi, A. Rosen, J. Karch, and H. Gleiter, "On the Validity of the Hall-Petch Relationship in Nanocrystalline Materials," *Scripta Metall.*, **23** [11] 1679-84 (1989).

[6]M. Mori, T. Shibayanagi, M. Maeda, and M. Naka, "Characteristics of Nanostructured Cr-B Films Produced by RF Magnetron Sputtering," *Scripta Mater.*, **44** [8] 2035-38 (2001).

[7]P.B. Barna and M. Adamik, "Fundamental Structure Forming Phenomena of Polycrystalline Films and the Structure Zone Models," *Thin Solid Films*, **317** [1] 27-33 (1998).

[8]J.M. Schneider, A. Anders, I.G. Brown, B. Hjorvarsson, and L. Hultman, "Temporal Development of the Plasma Composition of a Pulsed Aluminum Plasma Stream in the Presence of Oxygen," *Applied Phyics Letters*, **75** [5] 612-14 (1999).

Hot Gas Cleaning Technology

ANALYSIS OF ASH ADHESION BEHAVIOR AT HIGH TEMPERATURE CONDITIONS - RELATIONSHIP BETWEEN POWDER BED STRENGTH AND MICROSCOPIC BEHAVIOR

Mayumi Tsukada, Hajime Yamada, and Hidehiro Kamiya
Tokyo University of Agriculture and Technology, BASE
Koganei, Tokyo 184-8588, Japan

ABSTRACT

The change in tensile strength of ash powder beds was measured using a split-cell type tester at temperatures up to 900°C. Changes in tensile strength are discussed in comparison to microstructure changes observed using a FE-SEM and EDS.

The tensile strength of PFBC coal ash increases markedly above 800°C. EDS analysis shows higher sodium and aluminum contents in regions of PFBC coal ash that melt at ≤870°C. For RDF combustion ash, the liquid phase forms down around 400°C. Evidence of vaporization is observed which decreases tensile strength at ~700°C.

INTRODUCTION

The cohesion of ash particles at high temperature creates problems in power generation plants such as blocking high temperature filters, choking plant lines, and corroding heat transfer surfaces and/or gas turbines. These problems affect the stable operation of various high-efficiency power-generation plants fueled by burning coal, biomass, and/or waste. The cohesion of ash particles at high temperature is caused by the formation of low melting compounds. In previous work,[1] model ash particles composed of amorphous silica with a small fraction of alkaline metal showed a marked increase in cohesiveness at temperatures higher than 900°C, similar to what occurs in combusted coal ash. This work showed that low temperature alkaline metal eutectics form liquid bridges between the ash particles to increase the cohesion of ash powder beds. Field emission scanning electron microscopy (FE-SEM) with a computer control stage (CCSEM) proved to be a good tool to characterize the submicron ash particles.[2] To verify the formation of a localized liquid phase at high temperatures, a new system consisting of a computer controlled FE-SEM with a heated chamber was developed by Kamiya et al.[3,4] This equipment enables one to observe the exact same position on a sample before and after heating under specific gas composition and pressure conditions. The formation of microscopically localized, inhomogeneous, traces of liquid phase between coal ash particles has been observed directly with this system. When biomass is combusted, due to its higher alkali metal and chlorine content, the powder cohesion starts to increase around 400°C, which is much lower than for coal ash.[5] This limits the operating temperature of a biomass energy plant to a rather low temperature to avoid the corrosion of heat transfer surfaces and gas turbines, and to avoid the plugging of the reactor and transport lines.

In this work, pressurized fluidized bed combustion (PFBC) fly ash particles were characterized using FE-SEM before and after heating, and the eutectics formed between particles were characterized using energy dispersive spectroscopy (EDS). The results are compared with those of biomass combustion ash, and are discussed in terms of the changes in cohesion with temperature as measured using a split-type powder tensile strength tester.[6] The behavior of refuse derived fuel (RDF) combustion ash is also presented for comparison.

EXPERIMENTAL
Ash Samples

The samples examined include fly ash from a commercial scale pressurized fluidized bed combustion (PFBC) power plant, and RDF (refuse derived fuel) fly ash from a circulating fluidized bed combustion (CFBC) pilot plant in Japan. The PFBC ash has a higher calcium content than raw coal. The RDF was made from municipal waste, and calcium oxide was added. Data from bark fly ash[7] from a stoker type biomass combustion plant in Germany are referred to for comparison.[5] Thermal gravimetric and thermomechanical analyses were performed on the fly ash samples using a TGA-TMA system. All the data were collected in heating at 5 K/min in air. TMA dilatometry data were taken under a static load of 5.00 kPa.

Powder Bed Tensile Strength

The tensile strength of ash beds at temperatures up to 900°C was determined using a powder bed tester developed by Kamiya et al.[6] A powder bed sample in a cylindrical 50ϕ mm $\times$ 10 mm quartz glass cell was preliminarily pressed at 2.66 kPa for 10 min at room temperature, and then heated to a specified temperature. The relationship between the tensile load and the displacement of the movable half of the bed was measured, and the tensile strength of each powder bed was determined from the maximum tensile load and the cross section of the powder bed.

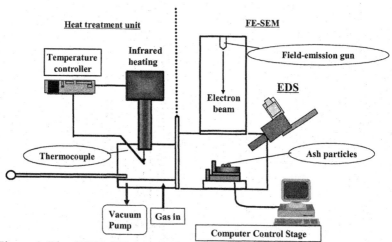

Figure 1. The CCFE-SEM analysis system with EDS and an atmosphere controlled heating chamber.[5]

Microstructure Characterization

The system used in this study was developed by Kamiya et al.[3,4] and was also used by Tsukada et al.[5]   It consists of a computer controlled FE-SEM with EDS, and it has an atmosphere controlled heating chamber. A schematic of the system is shown in Figure 1. Ash samples that were not coated with platinum or carbon were characterized before and after heat-treating for 3 min at a specified temperature in a chamber. The computer-controlled stage was used to ensure that the exact same region was characterized each time.

EDS was used to determine the local composition of the ash particles, and to locally map elemental distributions in samples after heating to a temperature that resulted in marked changes in particle shape. For the EDS analysis, samples were coated with sputtered platinum after the heat treatment.

RESULTS AND DISCUSSION

Coal PFBC Fly Ash

Figure 2 shows the powder bed tensile strength of coal fly ash samples from PFBC plant A, pulverized coal (PC) combustion ash[6], and pure silica,[6] and ash from PFBC plants B, C and D.[4] At temperatures higher than 800°C a marked increase in tensile strength is observed in conjunction with shrinkage of the powder bed, as shown in TMA data in Figure 3.

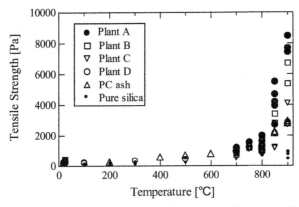

Figure 2. Effect of temperature on powder bed tensile strength.

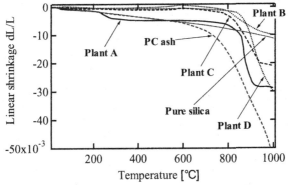

Figure 3. Powder bed (TMA) shrinkage as a function of temperature.

Characterization & Control of Interfaces for High Quality Advanced Materials   337

Preliminary FE-SEM results[3,4] show some changes in particle shape occur between 800 and 900°C, which suggest localized melting of ash particles in PFBC processes. SEM observations were made before and after heat-treating at 870°C (Figure 4), and EDS analysis was performed after the 870°C heat treatment (Figure 5). At point A in Figure 4b, which melts at ≤870°C, a higher sodium, potassium, and aluminum content is detected in comparison to Point B, that does not melt. This proves the existence of localized low melting eutectic compositions in coal combustion ash. Elemental maps in Figure 6 also show aluminum and sodium rich regions in melted ash particles, and the same trend is observed throughout the sample.

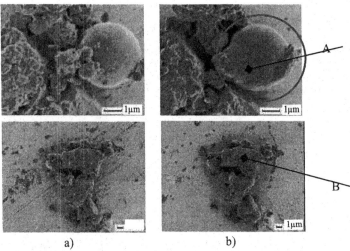

a)   b)

Figure 4. FE-SEM photomicrograph of PFBC fly ash: a) before, and b) after heat-treating at 870°C. After heating, the top image shows melting, while the bottom image does not.

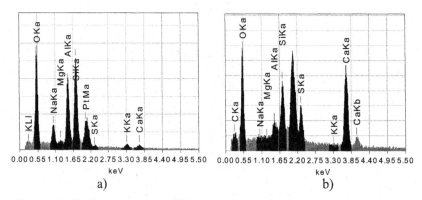

a)   b)

Figure 5. EDS spectrum of PFBC fly ash after heat-treating at 870°C: a) position A in Figure 4b that shows melting, and b) position B in Figure 4b that does not show melting.

The Na_2O-SiO_2-Al_2O_3 phase diagram constructed using FactSage 5.0 is shown in Figure 7. It shows how the melting temperature of sodium aluminosilicate decreases with increasing alumina content in low alumina regions.

RDF Ash

Figure 8 shows the tensile strength of RDF ash beds. The scatter in the data is considerable. Data for biomass ash, which also has more chlorine than coal ash, are also shown in Figure 8 for comparison. The tensile strength of the biomass increases at around 400°C and then at above 800°C.

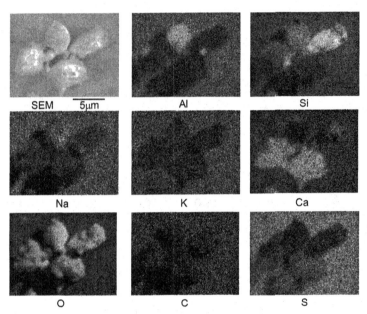

Figure 6. Elemental maps of PFBC ash after heat-treating at 870°C.

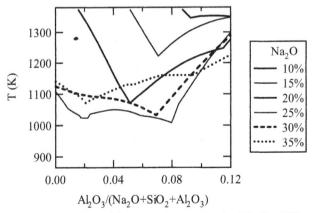

Figure 7. The decrease in melting temperature in the Na_2O - SiO_2 - Al_2O_3 system with increasing Na_2O content.

From the TGA and TMA data shown in Figure 9, sample weight decreases in two steps, first from 500 to 650°C, and then above 850°C. Sample shrinkage occurs over the entire temperature range above 400°C. The microstructure of RDF combustion ash before and after heat-treating at 500, 600, and 700°C is shown in Figure 10. Evidence of volatilization is observed on heating at 600°C (Figure 9), which is consistent with the measured TGA weight loss data. Vaporized gases may be evolving from low melting eutectics with chlorine and alkaline metals. Such evaporation has also been observed in biomass combustion ash.[5] The vaporization makes the sample more porous, and decreases its tensile strength around 700°C (Figure 8).

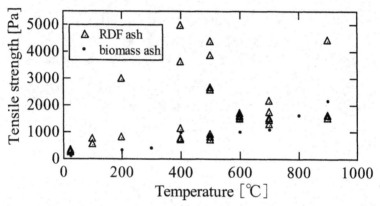

Figure 8. Tensile strength of an RDF ash powder bed as a function of temperature in comparison to biomass data.[5]

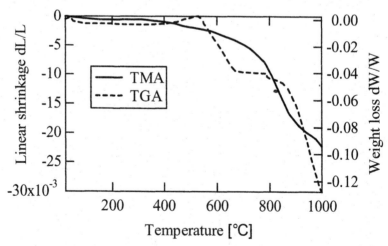

Figure 9. TGA and TMA analysis of RDF ash powder bed.

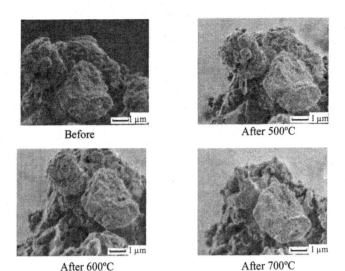

Before	After 500°C

After 600°C	After 700°C

Figure 10. FE-SEM micrographs of RDF combustion ash particles before and after heat-treating.

CONCLUSIONS

The temperature dependence of the tensile strength of PFBC coal fly ash and RDF combustion fly ash beds was determined and compared to the microstructure changes the ash particles undergo after heating at different temperatures. For PFBC coal ash, the powder bed tensile strength increases markedly above 800°C. This is accompanied by localized melting in regions composed of higher amounts of sodium and aluminum in the microstructure. The low temperature melting regions seem to have eutectics such as sodium aluminosilicate. For the RDF combustion ash, which has higher chlorine content, the liquid phase forms down around 400°C. Evidence of vaporization is observed in RDF combustion ash, which decreases the tensile strength at ~700°C.

ACKNOWLEDGMENT

This work was supported by the NEDO GRANT international collaboration project "High temperature gas cleanup", and by the Tanigawa Fund.

REFERENCES

[1] H. Kamiya, A. Kimura, M. Tsukada, and M. Naito, *Energy and Fuel*, "Analysis of the High-Temperature Cohesion Behavior of Ash Particles Using Pure Silica Powders Coated with Alkali Metals," **16** 457-61 (2002).

[2] C.A. O'Keefe, T.M. Watne, and J.P. Hurley, "Development of Advanced Scanning Electron Microscopy Techniques for Characterization of Submicron Ash," *Powder Technol.*, **108** 95-02 (2000).

[3] H. Kamiya, M. Tsukada, H. Yamada, and M. Naito, "Analysis of Ash Adhesion Behavior at High Temperature Condition by Using Computer Controlled FE-SEM with Heat Treatment Unit," Proc. of 5th International Symposium on Gas Cleaning at High Temperature, Paper No.3.11 (2002) (in CD-ROM).

[4] H. Kamiya, H. Yamada, M. Tsukadam, and M. Naito, "Analysis of the Mechanism of Coal Ash Cohesiveness at High Temperature Using Computer

Controlled FE-SEM with a Heat Treatment Unit," *Energy and Fuel*, under preparation (2004).

[5]M. Tsukada, H. Yamada, and H. Kamiya, "Analysis of Biomass Combustion Ash Behavior at Elevated Temperatures," *Advanced Powder Technol.*, **14** 707-17 (2003).

[6]H. Kamiya, A. Kimura, T. Yokoyama, M. Naito, and G. Jimbo, "Development of a Split-Type Tensile-Strength Tester of Inorganic Fine Powder Bed for High-Temperature Conditions," *Powder Technol.*, **127** 239-45 (2002).

[7]G. Hemmer, D. Hoff and G. Kasper, "Thermo-Analysis of Fly ash and Other Particulate Materials for Predicting Stable Filtration of Hot Gases," *Advanced Powder Technol.*, **14** 631-55 (2004).

A REPORT ON COAL GASIFICATION TECHNOLOGY THAT INCLUDES A SYNTHETIC CLEAN-UP SYSTEM

Sadao Wasaka
New Energy and Industrial Technology
Development Organization
Energy and Environment Technology
Development Dept.

Junichi Suhara
Electric Power Development Co., Ltd
Wakamatsu Research Center
1,Yanagasaki-machi,Wakamatsu-ku,
Kitakyusyu-shi, 808-0111 Japan

ABSTRACT
EPDC (Electric Power Development Co., Ltd.) has conducted the EAGLE (Coal Energy Application for Gas, Liquid & Electricity) project, funded by the Ministry of Economy, Trade and Industry (METI), to establish coal gasification technology that includes a synthetic gas clean-up system. One of the final objectives of this project is to obtain operational data on the Integrated Coal Gasification Fuel Cell Combined Cycle (IGFC) system to assess its performance.

INTRODUCTION
History
    With a goal of efficient coal gasification power generation, EPDC built EAGLE. Construction started in 1998, and test operations, including unit tests started in July 2001. Gasification operations have been conducted since March 2002. A project schedule is shown in Figure 1.

EAGLE Pilot Plant
    The EAGLE plant consists of a coal gasification unit, a gas clean-up unit, an air separation unit, and a gas turbine. The system flow diagram for the EAGLE

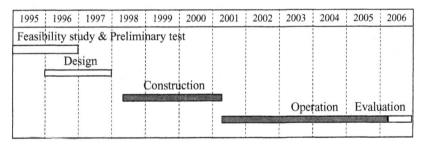

Figure1. Project Schedule for EAGLE

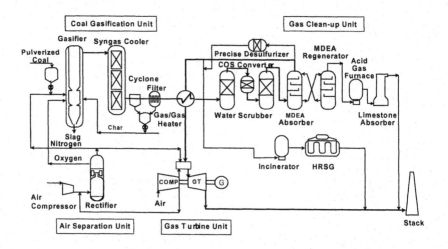

Figure 2. System flow diagram for the EAGLE pilot plant

pilot plant is shown in Figure 2.

Pulverized coal is fed into the gasifier, reacted with 95% pure oxygen (the gasifying agent) at high temperature and high pressure, and converted into CO-and H_2-rich coal gas. The raw syngas generated in the gasifier is about 1100°C, and the heat is recovered down to 400°C with a SGC (syngas cooler). Char in the raw syngas from the SGC is recovered with the char recovery unit composed of a cyclone and a char filter. Impurities such as sulfide compounds are also removed in the clean-up facility.

An air separation facility supplies oxygen and nitrogen to the system. Oxygen is supplied to the gasifier. Nitrogen is used as the coal and char-feed carrier gas, air ration gas for several hoppers, and purge gas for the gasifier. Excess nitrogen is used to reduce NOx in the gas turbine combustor.

Clean coal syngas goes to a gas turbine where it is combusted to produce electricity.

CHAR FILTRATION FOR EAGLE
Porous Filter Elements

In the EAGLE pilot plant, the syngas filtering system is comprised of a cyclone and a porous filter. The raw syngas goes through the cyclone first. The relatively coarse char is separated from the gas in the cyclone. Then the fine char is removed by the porous filter elements. The char filter flow diagram is shown in Figure 3.

The char recovery system composed of the cyclone and the filter elements has the important role to:
1. Recycle the char that remains in the raw syngas from the gasifier to improve gasification efficiency; and
2. Remove char to prevent operation problems; such as clogging of the slipstream.

The sintered metallic filter at the exit of the syngas cooler is an important piece of equipment that removes particulates (i.e., such as char in coal gasified raw syngas) so that they will not be carried to the wet gas cleaning process. The

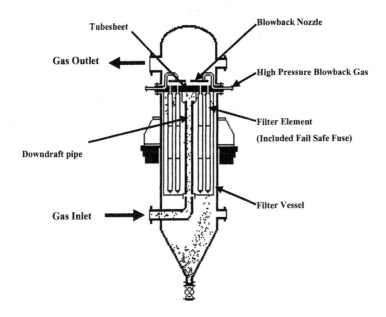

Figure 3. A schematic of the char filtration system in EAGLE

filter must be reliable under the high temperature and pressure corrosive gas environment.

The char filter is a sintered metallic filter element that was adopted as the porous filter as a result of various studies. The sintered metallic filter has high strength compared to a ceramic filter, and it also has excellent heat shock characteristics.

Porous Filter Element Performance

The EAGLE coal gasification pilot plant has operated approximately 900 h, with 19 start-stop times up to August 2003. One kind of coal from the U.S. was gasified during the 18-month operation. Although failure of one of the blowback valves occurred during the gasification operation, there haven't been any setbacks that have lead to the shut-down of the pilot plant. We checked the surface of the char filter elements with a fiber optic scope in November 2002, and in August 2003. No abnormalities in the filter elements or in the other parts of the char filtration system were observed. We expected to see less than 5 mg/m^3 at the char filter outlet. In actual operation, we observed less than 1 mg/m^3.

The differential pressure of the sintered metallic char filter as a function of gasifier load and cumulative operation hours is shown in Figure 4. The differential pressure is the average pressure drop from the gas inlet to the gas outlet of the char filter after blowback. The differential pressure rises linearly with increasing gasifier load, regardless of the cumulative operation hours. The differential pressure after 480, 700, and 800 of cumulative hours of operation is higher compared to after 5 and 140 h. The similarity in the differential pressure after 480, 700, and 800 h is indicates that a stable adhesion layer forms on the surface of the porous filter element after 480 h through the plant shakedown.

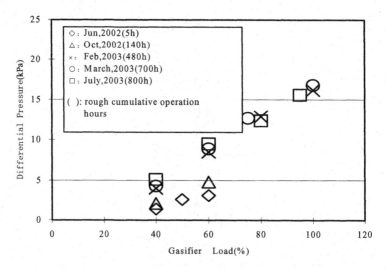

Figure 4. The differential pressure across the sintered metallic char filter in the EAGLE plant

SUMMARY

The EAGLE coal gasification started operation in March 2002. The EAGLE pilot plant has operated approximately 900 h as of August 2003. The char filtration system satisfied the design performance, and has been operating satisfactory.

ACKNOWLEDGEMENT

The EAGLE project is being conducted as a national project. We would like to express our deep appreciation for the support and guidance we have received from all concerned parties, including the Agency of Natural Resources and Energy of METI.

REFERENCES

[1]H. Sasatsu, N. Misawa, K. Kobori, and J. Iritani, "Hot Gas Particulate Cleaning Technology Applied for PFBC/IGCC – The Ceramic Tube Filter(CTF) and Metal Filter," 5th International Symposium on Gas Cleaning at High Temperature, Sept.17-20, 2002.

ASH PARTICLE FORMATION AND METAL BEHAVIOUR DURING BIOMASS COMBUSTION IN FLUIDIZED BED BOILER

Esko I. Kauppinen
VTT Processes
P.O. Box 1602
FIN-02044 VTT
Finland
and
Center for New Materials
Helsinki University of Technology
P.O. Box 1602
FIN-02044 VTT
Finland

Jorma K. Jokiniemi and Terttaliisa Lind
VTT Processes
P.O. Box 1602
FIN-02044 VTT
Finland

Jouko Latva-Somppi
ÅkerKvaerner
P.O. Box 10
00048 FORTUM
Finland

ABSTRACT

Ash particle formation and the behaviour of Cd, Pb, Cu, and Zn were studied in a 35 MW circulating fluidized bed boiler. The fuels used were a forest residue, and a Swedish willow (Salix). Quartz sand was used as an additional bed material. The size distributions of the fly ash particles examined are bimodal. The fine fly ash particles are non-agglomerated. The coarse fly ash particles are large agglomerates that consist of up to thousands of submicron primary particles. These agglomerates are very effective at capturing volatile metals. Almost no Cd is found in the bottom ash, whereas Zn is clearly enriched in the bottom ash. 15 - 27% of the Pb (for both fuels) and ~25 % of the Cu (for forest residue only) are found in the gas phase downstream of the cyclone at T = 810°C. Less than 3% of Cd or Zn are found in the gas phase. All of the metals are found entirely in the coarse fly ash particles, downstream of the convective pass at T = 150°C. Mechanisms for ash formation and metal behaviour during circulating fluidized bed combustion (CFBC) of biomass fuels are presented.

INTRODUCTION

The behaviour of ash forming constituents from solid fuels is a matter of interest when fluidized bed combustion is used for heat and power production. Ash from the combustion process affects plant operation and emissions. Ash deposition on various parts of the boiler can reduce heat transfer, and can cause corrosion and blockage, which can lead to plant shutdowns.[1] Also, a fraction of the ash accumulates on the bed material particles, so periodic replacement of the bed material is required to avoid agglomeration.[1] The particle emissions from the combustion process depend on the composition and size distribution of the fly ash particles, the gas composition, and the particle removal devices used.[2-4] Due to recent results concerning the adverse health effects of fine ambient aerosol particles, the control of fine particle emissions is of increasing importance.[2,4]

During combustion, ash-forming compounds are volatilized and released into the gas phase. The fraction volatilized depends on the fuel characteristics, the combustion temperature, the gas atmosphere, and the combustion technology in use.[5-7] A high combustion temperature and a reducing atmosphere enhance volatilization. Even elements present as refractory oxides, such as SiO_2, CaO, and MgO, with very high boiling temperatures, can volatilize in the reducing conditions inside the burning char. The volatilized fraction typically ranges from one to a few percent of the total amount of ash produced for pulverised combustion of high-rank coals, and it is up to 10% for the combustion of low-rank coals. During char particle combustion, metal oxide nanoparticles are formed by nucleation. Nucleation is induced when the volatile suboxides and metals are transported through the char boundary layer into the oxidizing gas conditions of the furnace.[4] The nucleated particles then grow by coagulation and condensation. The size of the resulting ash particles, D_p, is usually <1 µm. These particles form the so-called fine particle mode in the bimodal distribution of ash particles.

The compounds remaining in the char form residual ash particles inside the char particle, and on the char surface.[8-7] Mineral particles, which are typically abundant in the high-rank coals, usually melt and coalesce inside and on the surface of the char particles. This results in the ash particles with a wide range of compositions, depending on the composition of the parent mineral particles.[1] Low-rank coals and biomass contain a large amount of ash-forming compounds that are organically bound in the carbon structure, and as salts.[1] They are more mobile than mineral-bound ash compounds, and therefore, are more readily available for reaction inside the char particle. During low-rank coal and biomass combustion, the organically bound ash-forming compounds coalesce to form beads on the char surface. These beads can vary in diameter from submicron size to several micrometers.[9,10] The beads coalesce on the char surface, and are released from the char as residual ash particles during char burn-out.

During fluidized bed combustion, ash-forming compounds may either attach to the bed material particles, or be released from the furnace as fly ash particles or inorganic vapours. The ash compounds can be deposited on the bed material by two mechanisms: i) by physical attachment onto the bed material surface; or ii) by chemical reaction (of the volatile ash compounds) with the bed material. The bed particles grow by ash attachment. When large enough, the particles are removed from the furnace as bottom ash.[1,2]

The aim of this paper is to discuss the mechanisms of ash formation, and the subsequent transformations of the ash-forming compounds during fluidized bed combustion of solid fuels. Special emphasis is placed on the formation of fine ash particles (PM 2.5) as well as the volatilization of the heavy metals (Cd, Pb, Cu and Zn), and their subsequent transformations.

CASE STUDY ON FINE ASH FORMATION AND HEAVY METAL BEHAVIOUR DURING CFBC OF BIOMASSES

Experiments

Ash formation and heavy metal behaviour during circulating fluidized bed combustion (CFBC) of forest residue and willow were studied. The experiments were carried out at a combined heat and power plant with a capacity of 26 MW heat, and 9 MW electricity. The boiler conditions were kept as stable as possible during the measurements. For the experiments with forest residue, the boiler load was about 80%. With willow, it was about 70% of the full load. Willow was only used as a fuel during the day, and the fuel was switched to forest residue (for the night) between the measurement periods.

The fly ash formed in the boiler was characterized via aerosol measurements carried out at two locations in the flue gas channel. Measurement location #1 was downstream of the cyclone, and upstream of the convective heat exchanger section. The flue gas temperature was about 810 - 830°C. Location #2 was downstream of the convective heat exchanger section, and upstream of the electrostatic precipitator. The flue gas temperature was about 150°C. Soot-blowing was not carried out in the boiler during the measurements.

Results and Discussion
Ash Particle Size: The fly ash particle size distributions (by mass), as determined by with a low-pressure impactor method[11,12] are clearly bimodal. Submicron average particle diameters (D_p) of approximately D_p = 100 nm at T = 830°C and of approximately D_p = 200 nm at T = 150°C are measured for the forest residue. For the willow, the submicron average particle diameter was found to be 200 – 300 nm at both locations (Figure 1). The main elements in the submicron mode are K and Cl for the forest residue, and K and S for the willow. The particle size distribution upstream of the heat exchanger for the forest residue and for the willow have a major size mode of approximately 15 µm and around 7 - 8 µm, respectively (Figure 1). Particles larger than about 10 µm are almost non-existent in the particle size distribution measured downstream of the heat exchanger section. This is due to deposition of the fly ash particles in the convective pass in the absence of soot-blowing.[3,4] Deposition in the convective pass is approximately 60-80%. The total particle mass concentration is higher,

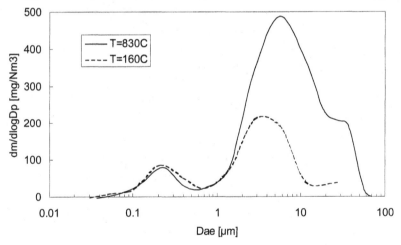

Figure 1. Fly ash mass size distributions measured during CFB combustion of willow as a function of particle aerodynamic diameter, D_{ae}. The aerodynamic diameter of a particle is the diameter of a unit-density sphere that has the same terminal settling velocity as the particle. Particle mass size distributions are usually presented as differential mass size distributions. Thus, an integrated area between the mass size distribution curve and the axis of abscissas represents the particle mass concentration (mg/m³; in NTP) within that particle size range. The notation dlogDp on the axis of ordinates means that the "length" of the integrated area is the dimensionless ratio $\log(D_{p,2}/D_{p,1})$, where $D_{p,2}$ and $D_{p,1}$ are the end point and the starting point of the integration, respectively.

and the submicron particle concentration is lower with forest residue than with willow.

Ash Particle Morphology: The coarse mode ($D_p > 1$ µm) fly ash particles formed during the CFB combustion of both forest residue and willow are irregular shape agglomerates consisting of up to thousands of primary particles of varying sizes (Figure 2a). The primary particles are partially melted and sintered together. The primary particles in a single agglomerate have varying composition. The coarse fly ash particles are presumably formed inside and on the surface of the burning char particles from the ash-forming species that do not volatilize during combustion. This ash fraction is called the 'residual ash' in the pulverised coal combustion (PCC) literature.[13] In pulverised coal combustion, the residual ash particles are mainly spherical, and are formed from molten ash species.[13] Clearly, the morphology of the residual ash particles from CFBC of biomass is different from the morphology of the residual ash particles from pulverised coal combustion. The fine particles ($D_p < 1$ µm) are mainly individual particles (Figure 2b). This fine particle morphology is different from the morphology of the fine particles formed in coal combustion,[14] where the fine ash particles are typically chain-like agglomerates formed from a few to tens of primary particles.

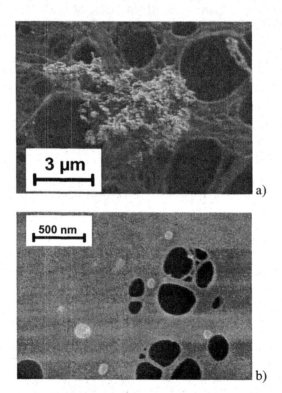

Figure 2. SEM micrograph of the; a) coarse; b) fine fly ash particles formed during CFB combustion of forest residue. The coarse fly ash particles are agglomerates. The fine fly ash particles mostly single particles.

Ash Particle Composition: The elemental particle size distributions were determined for the matrix elements.[4,2] Also, ash transformation and heavy metal behaviour was studied for FB boilers burning mixtures of wood-based biomass and waste sludge from pulp and paper mill operations.[1] The size distributions measured show that the main elements in the small particle size mode (< 0.3 μm) are K and Cl from the combustion of forest residue, and K, S and Cl from the combustion of willow. Consequently, the main compounds are K_2SO_4 and KCl. The chlorine is almost entirely in the gas phase at 830°C and condenses as KCl during collection. K_2SO_4 has already condensed in the flue gas channel prior to the measurements.

All the studied heavy metals (Cd, Pb, Zn and Cu) are effectively captured by the coarse fly ash particles. Figure 3 shows an example the elemental size distributions of Cd and Pb, both upstream and downstream of the heat exchanger. The concentrations of the heavy metals downstream of the convective pass at 150°C do not show any enrichment in the fine particles. In fact, the concentrations are relatively size independent in the size range $D_p > 0.5$ μm, and even decrease with decreasing particle size in the smaller particle size range. The behaviour is similar with both fuels. Cd is volatile in the bed, and it is found mainly in the coarse fly ash particles at measurement location #1 (Figure 3).

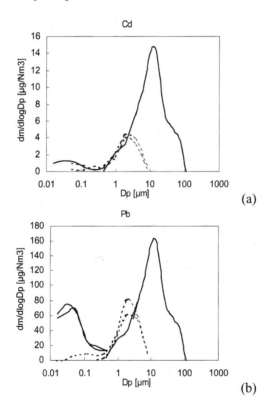

(a)

(b)

Figure 3. a) Cd, and b) Pb elemental size distributions from CFB combustion of forest residue. The measurements were carried out downstream of the process cyclone (T = 830°C; solid line) and downstream of the heat exchanger section (T = 150°C; dashed line).

This is due to an unknown chemical reaction between volatile Cd species and the coarse fly ash particles. The reaction occurs preferentially with the coarse fly ash particles, because the porous agglomerates of coarse fly ash have a relatively large surface area. Detailed chemistry is difficult to determine due to the low concentration of Cd.[2]

Pb is at least partially volatile during combustion. 15 - 27 % of the lead is found in the fine size mode particles, and in the gas phase at T = 810 - 830°C. At

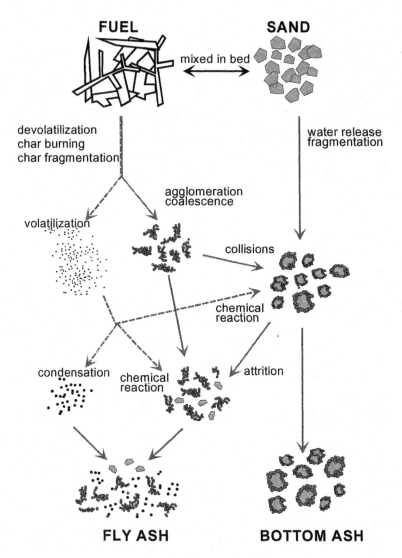

Figure 4. A diagram of the proposed ash formation mechanisms during CFB combustion of solid wood-based biomass.[2]

least this fraction volatilizes during combustion. In the convective pass, the volatilized fraction reacts with coarse fly ash particles, and only minor amounts of Pb are found in the fine mode particles downstream of the convective pass.

Zn is clearly enriched in the bottom ash from both fuels. The enrichment is presumably due to a reaction of volatile Zn with the quartz sand particles or the ash layer on the surface of the sand particles. Less than 3% of Zn is found in the fine fly ash particles at measurement location #1.

Only the behaviour of Cu is different between the two fuels. During combustion of forest residue, 24 - 27% of the copper is found in the gas phase at 830°C. During combustion of the willow, there is only approximately 2%. Based on these results it is not possible to determine whether the Cu in the coarse fly ash particles is due to non-volatile behaviour, or due to a chemical reaction of volatile Cu with the surface of coarse fly ash particles.

ASH FORMATION MECHANISMS DURING FB COMBUSTION OF BIOMASS FUELS

Based on the results presented above as well the results presented in the literature, we propose the ash formation mechanisms as shown in Figure 4.

ACKNOWLEDGEMENTS

This work has been funded by TEKES, VTT and the CEC through a contract JOR3-CT95-0001, which is gratefully acknowledged.

REFERENCES

[1]J. Latva-Somppi, "Experimental Studies on Pulp and Paper Mill Sludge Ash Behaviour in Fluidized Bed Combustors," PhD Thesis, Helsinki University of Technology, VTT Publications 336, Espoo, Finland, 1998.

[2]T. Lind, "Fly Ash Formation During Circulating Fluidized Bed Combustion of Coal and Biomasses," PhD Thesis, Helsinki University of Technology, VTT Publications 378, Espoo, Finland, 1999.

[3] T. Lind, E.I. Kauppinen, G. Sfiris, K. Nilsson, and W. Maenhaut, "Fly Ash Deposition onto the Convective Heat Exchangers during Combustion of Willow in a Circulating Fluidized Bed Boiler," in *The Impact of Mineral Impurities in Solid Fuel Combustion.* Proceedings of the Engineering Foundation Conference, Kona, HI, November 2-7, 1997.

[4]T. Valmari, "Potassium Behaviour During Combustion of Wood in Circulating Fluidised Bed Power Plants," PhD Thesis, Helsinki University of Technology. VTT Publications 41, Espoo, Finland, 2000.

[5]D.D. Taylor and R.C. Flagan, "The Influence of Combustor Operation on Fine Particles from Coal Combustion," *Aerosol Sci. Technol.*, **1** 103-117, (1982).

[6]C.L. Senior and R.C. Flagan, "Synthetic Chars for the Study of Ash Vaporization"; pp. 921-929 in *Twentieth Symposium (Int'l) on Combustion*. The Combustion Institute, Pittsburgh, PA, 1984.

[7]R.J. Quann, M. Neville, and A.F. Sarofim, "A Laboratory Study of the Effect of Coal Selection on the Amount and Composition of Combustion Generated Submicron Particles," *Combust. Sci. and Tech.*, **74** 245-265, (1990).

[8]W.P. Linak and T.W. Peterson, "Mechanisms Governing the Composition and Size Distribution of Ash Aerosol in a Laboratory Pulverized Coal Combustor"; pp. 399-410 in *Twenty-first Symposium (International) on Combustion*, The Combustion Institute, Pittsburgh, PA, 1986.

[9]R.J. Quann and A.F. Sarofim, "A Scanning Electron Microscopy Study of the Transformations of Organically Bound Metals during Lignite Combustion," *Fuel*, **65** 40-46, (1986).

[10]M.J. Wornat, R.H. Hurt, N.Y.C. Yang, and T.J. Headley, "Structural and Compositional Transformations of Biomass Chars during Combustion," *Combust. & Flame*, **100** 131-13, (1995).

[11]E.I. Kauppinen, "On the Determination of Continuous Submicron Liquid Aerosol Size Distributions with Low Pressure Impactors," *Aerosol Sci. Technol.*, **16** 171-197 (1992).

[12]E.I. Kauppinen and T.A. Pakkanen, "Coal Combustion Aerosols: A Field Study," *Environ. Sci. Technol.*, **24** 1811-1818 (1990).

[13]C.L. Senior and R.C. Flagan, "Synthetic Chars for the Study of Ash Vaporization"; pp. 921-929 in *Twentieth Symposium (Int'l) on Combustion*. The Combustion Institute, Pittsburgh, PA, 1984.

[14]E.I. Kauppinen, T.M. Lind, T. Valmari, S. Ylätalo, J.K. Jokiniemi, Q. Powell, A.S. Gurav, T.T. Kodas, and M. Mohr, "The Structure of Submicron Ash from Combustion of Pulverized South African and Colombian Coals"; pp. 471-484 in *Applications of Advanced Technology to Ash-Related Problems in Boilers*. Edited by L. Baxter and R. DeSollar. Plenum Press, New York, 1996.

AN OVEVIEW OF SILICA DUST POLUTION AND CONTROLS

Haiying Qi, Changfu You, and Yingjie Bao
Institute of Thermal Engineering
Tsinghua University
Beijing 100084, P.R. China

ABSTRACT

Silica dust emissions from a silicon smelting furnace has been investigated. Preliminary site tests were completed to control silica dust pollution using a cyclone system. It was verified that silica powder has a strong tendency to agglomerate, which can be used in the dust removal process. Particle behavior in a gas flow was studied in the laboratory to better understand the thermophoresis effect on fine particles. The results confirm that fine particles tend to move towards a cooled wall when a temperature difference between the wall and the flow exits. The results indicate that it may be possible to use thermophoresis to concentrate the silica particles in flue gas near the cool wall to promote agglomeration and particle removal.

SILICA DUST EMISSIONS AND CONTROLS

Silica Dust Emissions from Smelting

Ultra-fine silica (SiO_2) dust is a byproduct of manufacturing metallurgical grade silicon by smelting a mixture of silica and charcoal in an arc furnace. During cooling, SiO_2 that is reduced by the carbon, and silicon vapor in the flue gas can oxidize and condense to form silica fume particles with a diameter of <1 μm. Under normal operation conditions, 120-200 kg of silica dust is produced from 20000-30000 m^3 of flue gas by smelting 1000 kg of metallurgical grade silicon. The average concentration of silica (powder) in the flue gas reaches 7-9 g/m^3 and can be up to 17-20 g/m^3.[1,2]

The silica dust generated becomes environmental pollution. Research has shown that this kind of silica powder can have serious health effects if inhaled (e.g. "silicosis").[3]

Silica Dust Pollution in China

Based on the distribution of mineral resources in China, most of the smelting industry for metallurgical grade silicon is located in the western area of China, like in Yunnan, Guizhou, Sichaun, Qinghai, and Inner Mongolia. Facilities are also located in the southeast provinces such as Hunan and Fujian. Silicon smelting has become one of the key industries for local economic growth that is encouraged and aided by local government. It plays an important role in the local economy, such as in resource utilization, production value, revenue, transportation, employment, and standard of living for local residents.

In a typical silica smelting plant, especially in the underdeveloped mountainous area of China, the furnace works continuously for 24 h, and there is no dust removal equipment installed. Silica dust emissions are especially bad when the operating conditions are not well controlled. Silica fume smoke not only exhausts through the chimney, but also disperses in the atmosphere directly from the working floor at the opening of the arc furnace.

According to the statistical data issued by Bureau of Environmental Protection of Dehong County in the Yunnan Province, there are 7 smelting plants with 16 arc furnaces in the county. Their capacity totaled 61300 kilovolt-amperes. Every day about 11.4×10^6 m^3 of flue gas are discharged into atmosphere, which produces about 28.5×10^3 kg of silica dust. In a year (an 8 month production period) 2736×10^6 m^3 of flue gas and 6840×10^3 kg of SiO_2 dust are produced. Baoshan County also has 7 smelting plants with 10 furnaces. In other areas such as Diqing, Lijiang, and Wenshan many different scales of smelting plants have been installed. One can imagine the enormous amount of SiO_2 dust that is discharged from all of those plants is. Additionally, plants that produce alloys with 75% and 50% silicon also generate a large amount of flue gas with silica dust.[4]

To control dust pollution from the smelting industry, and to recover silica powder, the government has issued more stringent policies and emission standards. For example, the concentration of silica powder in flue gas after dust removal must be controlled to be below 200 mg/m^3, i.e., the dust removal rate must reach 95% or much higher. However, the following special conditions of the smelting industry, e.g., in Yunnan Province, has to be considered before implementing pollution control technology:

- Smelting facilities are small or middle scale, and are privately owned and lack funds to install large-scale dust removal equipment.
- Smelters run discontinuously: because they consume a large amount of hydroelectricity, they have to stop production in low water periods, to maintain equipment, and to prepare raw materials.
- A conventional simple smelting process is generally used, i.e., manual operation in an open system, which makes it difficult to control operation conditions and emissions effectively.
- Smelters are located on hills or in hollows, and they often operate in humid local climates.
- As an additive with a high additional value for other industries such as cement production, silica powder should be collected from the flue gas in a dry process.

Considering the above conditions, the pollution control technology should feature a dry, cheap, and available process.

Pollution Controls for Silica Dust

Bag Filters: In most cases, a sack precipitator with a membrane is regarded as the best choice to remove ultra-fine SiO_2 powder.[5,6] However, this solution is still too expensive for the Chinese smelting enterprise. The humid climate requires stringent maintenance of the MSP during production stops to avoid damage to the sacks due to solidification of silica on the sack surface due to the absorption of water from the humid atmosphere. This also increases operating costs.

In industrialized countries, the silicon smelting industry is quite different from that in China. The issue of silica dust emissions has been solved by using an advanced, closed smelting process, together with membrane bag filters and computer controls. As a result, there are hardly available technologies and fundamental researches for special emission situation in China.[7]

Agglomeration: According to site observations in a smelting plant, silica powder tends to agglomerate and deposit easily on the surface of the flue gas tube to form thick dust cakes from low flow velocity flue gas. This characteristic has been verified theoretically and experimentally.[8] The Van der Waals surface force dominates when the particle size is smaller than 30 μm, which causes agglomeration when particles collide. It is no doubt that this agglomerating behavior of silica powder avails itself to conventional dust-removal technologies.

Cyclone Separation: It is easy to separate particles with a diameter of >30 μm from the flue gas using a cyclone.

As mentioned above, the agglomeration effect for silica powder is one of possibilities that can be considered for dust removal. As a precondition for agglomeration, however, a high particle concentration that favors frequent collisions between particles is needed. Additionally, particles must be in contact long enough to adhere to one another.

Thermophoresis: A so-called "thermophoresis" force is caused by the temperature gradient around particles. When hot flue gas flows over a cooled solid surface, a force acts on the suspended particles towards the cool surface (Figure 1). Due to this force, particles will tend to move to the surface and assemble there. Compared with other forces like drag, the Saffman force, and the Magnus force, the thermophoresis effect dominates.[8,9]

CHARACTERIZATION OF SILICA DUST EMISSIONS IN THE LONG-CHUAN PLANT

Considering the special situation of the silicon smelting industry in the aforementioned areas of China, we have to face the challenge and try to develop a new dry technology with higher efficiency and lower costs. It should enable the target enterprises to bare the investment and operating costs, and bring them profits. Thus, it is possible to improve both the local pollution status, and economic conditions.

In this paper an investigation of silica dust emission from a facility in the Longchuan area of Yunnan Province has been performed. The results of preliminary tests are reported.

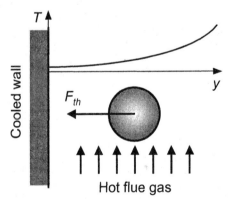

Figure 1. An illustration of the thermophoresis effect that attracts a particle in a hot gas to a cool wall.

Silica Deposits on Chimney Walls
In the Longchuan plant, silica powder agglomerates on the inner wall of the chimney tube. Even with a smoke velocity of about 12-14 m/s in the chimney with a diameter of 0.8 m, silica powder still agglomerates and deposits on the wall to form a cake with thickness of 80-100 mm. The average particle size of the cake sample is more than 30 μm, and the density is about 160 kg/m³.

Particle size and morphology: Figure 2 shows the size distribution of the silica powder measured using a Mastersizer 2000. The average diameter is 30-40 μm, and the diameter range is rather wide, i.e., from 0.25 to 600 μm. In fact, this is the size distribution of silica agglomerates, because samples were taken from the silica cakes on the pipe wall, and not directly from flowing smoke. The results in Figure 2 are somewhat misleading.
From the scanning electron microscope (SEM) image in Figure 3 it is clear that the silica dust is really ultra-fine particles and regular spheres. The agglomerates have a fractal geometry that can be characterized by a fractal dimension.[10,11] It may be possible to use the agglomerating potential and fractal character of the silica to develop the dust removal technology.

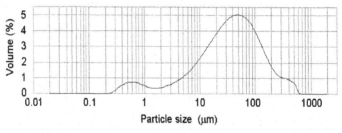

Figure 2. The size distribution of the silica powder from the cake on the chimney wall as determined using a Mastersizer 2000.

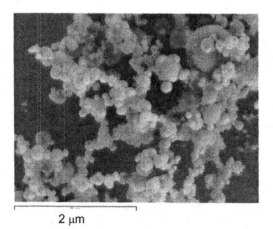

2 μm

Figure 3. SEM Image of the silica powder in the cake on the chimney wall

X-Ray Diffraction (XRD) shows the cake is a high-grade silica fume (Figure 4) with a SiO_2 content of 96.37% (Table I).

Pollution Control Using a Cyclone

A high-efficiency cyclone system was designed and connected to the chimney of the Longchuang plant through a horizontal bypass to determine if conventional technology would work well to remove silica powder from flue gas. Flue gas flows through the bypass into the pretreatment unit. Part of silica powder is collected under gravity by decelerating the gas flow. The rest of silica particles are removed in a cyclone. The clean flue gas returns to the chimney (Figure 5).

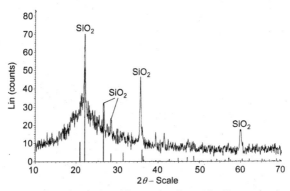

Figure 4. XRD pattern from the silica cake from the Longchuan plant

Table I. Chemical analysis of the silica cake from the chimney of the Longchuan plant (% by Weight).

SiO_2	Al_2O_3	Fe_2O_3	CaO	MgO	C	K_2O	Na_2O	Loss
96.37	0.38	0.40	0.17	0.40	0.46	1.01	0.22	2.18

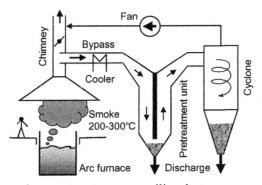

Figure 5. The cyclone system to remove silica dust

It was found that the cyclone system can effectively collect the powder that adheres to the chimney wall in the form of agglomerates like those in Figure 6. However the cyclone system does not work with fresh flue gas continuously online. The main reason for this is that the condensed fine silica particles are homogeneously dispersed in flue gas, and they are not agglomerated, i.e., the particles are too small for the cyclone to work.

According to calculation, the volume fraction of silica particles in flue gas is about 1.25×10^{-4}, with a dust content of 20 g/m^3 and a density of 160 kg/m^3.

Consequently, the silica powder in the flue gas is too dilute to undergo appreciable agglomeration.

Attempts were made to enhance agglomeration to make larger particles by increasing the turbulent disturbance of the flow, by changing the flue gas temperature, by spraying water in the flue gas, or by using a water cooler or hanging wire nettings (see Figure 7). However, it was concluded that alternative pollution control technology should be considered.

Figure 6. Silica agglomerates and cakes in the horizontal pipe after operating for 48 h with a very low flow rate of the flue gas.

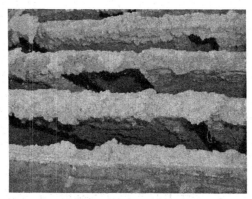

Figure 7. Silica dust collected on a screen filter at the Longchuan plant.

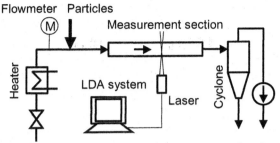

Figure 8. Experimental apparatus used to measure the thermophoresis effect.[9,12]

Evaluation of the Thermophoresis Effect for Pollution Control

An experimental apparatus was constructed to better understand the flow behavior and separation performance of the fine particles under the thermophoresis effect (Figure 8). The velocity and concentration of particles in a heated gas flow near a water-cooled side wall (size $920 \times 86 \times 86$ mm) was measured using Laser Doppler Velocimetry. Instead of silica dust, glass beads that do not agglomerate were used as the particles in the measurements. Their size distribution ranges from 0 to 100 μm, and the average volume fraction is 9.17×10^{-5}. The temperature difference, ΔT, between the heated gas flow and the cooled wall varied from 0 to 130.45°C, and the flow velocity varied from 12.56-16.03 m/s.

It was determined that the relative component, W_s/U_r, of particle velocity that is normal to the cooled wall, and that points towards the wall increases with the temperature difference, ΔT, especially when $\Delta T > 60$°C (Figure 9). U_r is the friction velocity of turbulence on the wall, and is calculated from Equation 1. The results demonstrate that the glass particles tend to move towards the cool wall under the thermophoresis effect.

$$U_r = \sqrt{\tau_w/\rho} \tag{1}$$

Where τ_w is the wall shear stress (Equation 2) and ρ is air density:

$$\tau_w = 0.03955 \rho U_0^2 \, Re^{-0.25} \tag{2}$$

Where U_0 is the local average air velocity, and the Reynolds number is defined as:

$$Re = \frac{U_0 D}{\nu} \tag{3}$$

Here D is the tube diameter of the measurement section (see Figure 8)

Enhancing the ΔT also results in a notable increase in the relative particle concentration in the near wall region (Figure 10). C_{av} is the average particle concentration in the local cross section. It should be noted that at $\Delta T = 60$°C, both the velocity and the concentration of particles near the cool wall ($y = 0.4$ mm) reach the minimum. This phenomenon may be caused by turbulence (Figure 11), because under the same conditions, the root mean square values (RMS) of the particle velocity fluctuation reach the maximum. However, the specific reason has got to be determined.

The above experimental results suggest that it may be possible to concentrate the silica in hot flue gas near a cooled wall, which could intensify particle agglomeration, and enable the removal of the silica dust from the gas.

SUMMARY

In this paper silica dust emissions from a silicon smelting plant in the west area of China has been investigated. Preliminary tests to characterize the silica dust, and to control pollution were performed. The pollution control technology needs to be adaptable to the local economy, and to production. High efficiency and low cost pollution control technology is needed to control emissions, but also to maintain an economic source of ultra-fine silica. Based on laboratory research, it has been found that silica powder tends to agglomerate. The thermophoresis effect can be used as one of the choices to promote agglomeration and control silica dust emissions.

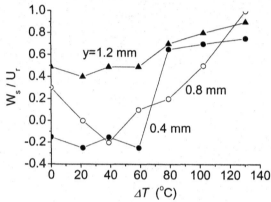

Figure 9. Relative increase in the relative particle velocity Ws/Ur with increasing temperature gradient ΔT between the gas flow and cool wall.[12]

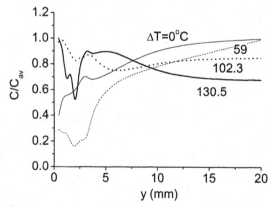

Figure 10. Influence of ΔT on the relative particle concentration C/C_{av} as a function of location in the gas flow.[12]

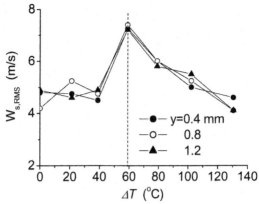

Figure 11. The change in particle velocity fluctuation with ΔT.[12]

REFERENCES

[1]Y.P. He and E.H. Wang, "Production of Industrial Silicon," *Metallurgy Industry Press,* Beijing, 1996 (in Chinese).

[2]Y.P. He, "Technical Reports on Industrial Silicon," *Metallurgy Industry Press,* Beijing, 1991 (in Chinese).

[3]R.L. Que, "Science and Technology of Silicon Material," *Zhejiang University Press,* Hangzhou, 2000 (in Chinese).

[4]M. Xiao, "New Recoverable Technology of Removal Ultra-fine SiO_2 Powder", Internal *Report of Kunmin Auxie Science and Technology Ltd.*, March 10, 2002.

[5]Z.Q. Tong, "Cleaning and Utilization of Industrial Exhaust Gases," *Chemical Engineering Press,* Beijing, 2001.

[6]N. Ono, "Theory and Practice of Dust Removal and Collection," Translated by W.C. Shan, *Science and Technology Literature Press*, Beijing, 1982, (in Chinese).

[7]J.T. Sun, "Thermal Process and Equipments of Silicate Industry," *Architecture Industry Press of China*, Beijing, 1985 (in Chinese).

[8]W.B. Zhang, "Study on Mechanism of Collision and Agglomeration of Fluidized Particles," Ph.D. Thesis, Tsinghua University, 2002.

[9]Y. Xue, "The Behavior of Fine Particles in the Flow and Temperature Boundary Layer," Master thesis, Tsinghua University, May 2002.

[10]X.D. Xiang, B.Z. Chen and C.Q. Zhang, "Characteristic of Dust Fractal Geometry and Discussion of Its Application in Dust Control", *Journal of Thermal Engineering in Architecture and Air Conditioning*, [2] 14-17 (1999).

[11]P. Wiltzius, "Hydrodynamic Behavior of Fractal Aggregates", *Phys. Rev. Lett.*, [58] 2371-2374 (1987).

[12]J.C. Zhang, "The Experimental Study of Super Fine Particles Movement in Turbulent Thermal Boundary Layer," Master thesis, Tsinghua University, 2001.

CHARACTERIZING AND MODELING FILTRATION IN A GRANULAR COKE BED

M. Furuuchi, C. Kanaoka,
M. Hata, and Y. Kawaminami
Department of Civil Engineering
Faculty of Engineering
Kanazawa University
2-40-20 Kodatsuno
Kanazawa 920-8667, Japan

R. Takahashi, J. Yagi, and R. Takehama
Graduate School of International
Cultural Studies
Tohoku University
Kawauchi, Aoba-ku
Sendai 980-8576, Japan

ABSTRACT

The dust collection performance of a coke granular bed filter that has been developed as a material classifier for a new process to recover material from the dust in the exhaust gas of an electric arc furnace (EAF) for the steel production was investigated theoretically and experimentally. The influence of granule size and dust load on dust collection efficiency was modeled on the basis of the depth filtration theory, in which the increase in the single granule collection efficiency with dust load was taken into account. Results are compared with experiments and data from the literature. The model predicts results that are consistent with experiments, when channeling of the dust is negligible. Granular beds were found to be more sensitive to dust loading than fibrous filters.

INTRODUCTION

Dust in exhaust gas from electric arc furnaces (EAF) for steel production generate ca. 5×10^8 kg/year in Japan (1999), and contain a large amount of metallic components such as Fe, FeO, and ZnO.[1,2] About 60% of the dust is re-melted under a reducing atmosphere to recover Zn, while the remaining 30% goes to the landfill.[2] Two problems with EAF dust treatment are: 1) high cost and energy consumption; and 2) heavy environmental loads.[1]

A new process to recover material from EAF dust, and to save energy has been developed by the Japan Research and Development Center for Metals (JRCM). This process has been tested on a pilot scale plant.[2] Figure 1 shows the concept of this process.[1] It uses "a coke granular bed filter" as a hot gas cleaning device. The exhaust gas from the EAF passes through the coke granular bed filter operated under a reducing atmosphere around 1000°C; solid components such as FeO dust and condensed Fe vapor are collected on the coke granules; ZnO is reduced to Zn vapor, which passes through the coke bed, and is collected in a metal condenser downstream.

Since the concentration of dust in the exhaust gas of an EAF is extremely high,[1,2] cake filtration (collecting dust on dust particles) can take place near the

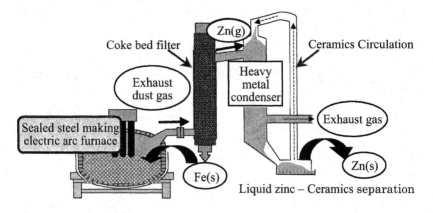

Figure 1. The concept of direct recovery of Zn and Fe from EAF exhaust gas.

filter inlet. Depth filtration (i.e., dust collection on coke granules) may be important downstream of the filter. The filter must be very effective in collecting the dust to ensure the purity of the Zn recovered in the vapor condenser. Theoretical descriptions of the dust collection performance of granular bed filters have mostly focused on the initial period of filtration, or on low dust concentration conditions.[3-6] There have been some analytical descriptions and numerical simulations on the influence of dust load on the collection performance of a fibrous filter.[7-11] Although application of the equations for a fibrous filter to coke bed filtration may make it simpler to characterize the filtration of the coke bed filter, difficulties may arise due to the smaller void volumes between granules, and less uniform structure due to the larger size of the collecting medium, e.g., there is higher possibility of re-suspension of collected particles[6] and the wall effect.

In this study, the dependency of collection performance in a fixed coke bed filter is discussed, taking into account the influence of the amount of collected dust. The dust load calculated is compared with cold experiments on a coke bed filter using different size test dust and coke granules. The limitations of the model, and the different influences of dust load for a granular bed and fibrous filters is also discussed.

EQUATIONS DESCRIBING DUST LOAD DISTRIBUTION INSIDE A COKE BED FILTER
Single Granule Collection Efficiency as a Function of Dust Load and Dust Load Distribution

Figure 2 shows the radius, Y_m, of the limiting trajectory of a particle with an original diameter, D_c, moving toward the collecting medium with an increased diameter, D_{cm}, due to the dust collected. The particle deposition pattern depends on the filtration condition.[7] However, we assume uniform dust deposition for simplicity.

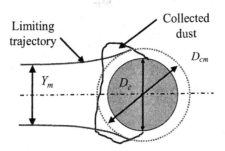

Figure 2 A schematic of the limiting trajectory of a particle with an increased diameter, D_{cm}, due to collected dust.

The collection efficiency, η_m', of a spherical coke granule of the increased diameter, D_{cm}, is defined as the ratio of the cross section of the area inside the limiting trajectory to the projected area of the dust loaded coke granule:

$$\eta_m' = \frac{\pi Y_m^2}{\pi D_{cm}^2} = \left(\frac{Y_m}{D_{cm}}\right)^2 \tag{1}$$

Adapting the depth filtration theory for a fibrous filter to a granular bed filter,[7] the dust concentration distribution inside a coke bed filter can be expressed by:

$$\frac{\partial C}{\partial x} = -\frac{\alpha_0}{1-\alpha_m} \cdot \frac{3}{2} \cdot \frac{1}{D_{cm}} \cdot \eta_m' \cdot C \tag{2}$$

where C is the dust concentration, x is the distance from the bed inlet, α_0 is the initial volume fraction of granules, and α_m is the volume fraction of granules with dust load. In the equation, α_m, D_{cm} and η_m' change with dust load, and thus, are simplified to reduce the variables following Yoshioka[10] and Kanaoka[11] as:

$$\alpha_m = \alpha_0 \left(\frac{D_{cm}}{D_c}\right)^3 \tag{3}$$

$$\eta_m' = \left(\frac{Y_m}{D_{cm}}\right)^2 = \left(\frac{Y_m}{D_c}\right)^2 \cdot \left(\frac{D_c}{D_{cm}}\right)^2 = \eta_m \left(\frac{D_c}{D_{cm}}\right)^2, \quad \eta_m \equiv \left(\frac{Y_m}{D_c}\right)^2 \tag{4}$$

$$\frac{\partial C}{\partial x} = -\frac{\alpha_0}{1-\alpha_m} \cdot \frac{3}{2} \cdot \frac{1}{D_{cm}} \cdot \eta_m' \cdot C = -\frac{\alpha_0}{1-\alpha_0} \cdot \left(\frac{1-\alpha_0}{1-\alpha_m}\right) \cdot \frac{3}{2} \cdot \frac{1}{D_{cm}} \cdot \left(\frac{Y_m}{D_{cm}}\right)^2 \cdot C$$

$$= -\frac{\alpha_0}{1-\alpha_0} \cdot \frac{3}{2} \cdot \eta_m \cdot \frac{1}{D_c} \cdot \left(\frac{1-\alpha_0}{1-\alpha_m}\right) \frac{D_c}{D_{cm}} \cdot \left(\frac{D_c}{D_{cm}}\right)^2 \cdot C = -\frac{\alpha_0}{1-\alpha_0} \cdot \frac{3}{2} \cdot \eta_m \cdot \left(\frac{1-\alpha_0}{1-\alpha_m}\right) \frac{\alpha_0}{\alpha_m} \cdot C$$

$$\cong -\frac{\alpha_0}{1-\alpha_0} \cdot \frac{3}{2} \cdot \eta_m \cdot \frac{1}{D_c} \cdot C \quad (\alpha_0 \cong \alpha_m) \tag{5}$$

Using a mass balance, the increasing rate dust load can be written as:

$$\frac{\partial C}{\partial x} = -\frac{1}{v}\frac{\partial m}{\partial t} \tag{6}$$

where m is the mass of dust collected per unit volume of filter, and v is the filtration (superficial) velocity. Similar to that for a fibrous filter[7], the collection efficiency, η_m, of a single coke granule is assumed to be:

$$\eta_m = (1 + \lambda m)\eta_0 \tag{7}$$

where η_0 is the initial collection efficiency of single coke granule, and λ is the increasing collection efficiency factor. If η_m is defined by Eq. (7), one can analytically solve Eq. (5) to obtain m as a function of filtration period, t, and distance from the bed inlet, x:

$$m = -\frac{1}{\lambda}\frac{\exp(-\lambda A C_i v t) - 1}{\exp(-\lambda A C_i v t) + \exp(Ax) - 1} \tag{8}$$

where,

$$A = \frac{3\alpha_0 \eta_0}{2D_c(1 - \alpha_0)} \tag{9}$$

and C_i is the inlet dust concentration.

Pressure Loss Through a Dust Loaded Filter

The pressure loss through the coke bed filter was approximated using Ergun's equation:[6]

$$\Delta P = \int_0^x \left[150\frac{\alpha_0^2}{(1-\alpha_m)^3}\frac{\mu}{D_c^2}v + 1.75\frac{\alpha_0}{(1-\alpha_m)^3}\frac{\rho}{D_c}v^2 \right] dx \tag{10}$$

where,

$$\alpha_m = \alpha_0 + \frac{m}{\alpha_{cm}\rho_p} \tag{11}$$

$\alpha_{cm}\rho_p$ is the apparent density of the deposited dust structure, ρ is the air density, and μ is the air viscosity. Eqs. (10) and (11) were used to relate m to the pressure drop through the dust loaded filter.

EXPERIMENTS

Figure 3 shows a diagram of the experimental apparatus used to characterize coke bed filtration. Coke granules having a mean diameter listed in Table I were packed inside a transparent acrylic resin column to make the coke bed filter. The volume faction of coke in the column is listed in Table I. The column has a 10 cm inner diameter, and has an effective height of 35 cm, which could be separated into two vertical half-cylinders to measure the dust hold-up. To monitor the change in pressure distribution inside the packed bed with time, pressure sensors were installed along the bed axis. Fine glass beads and Aerosil[®] particles (anhydride SiO_2) having a mean diameter listed in Table I were used as the test dusts. The mean size of a primary Aerosil[®] particle is 0.03 μm, but the mean size measured by laser diffraction is 2.50 μm. This may suggest the existence of agglomerates.

Table I. Experimental parameters and physical properties of the test dusts

	Test dust	
	Glass beads	Aerosil particles
50 % mean diameter (μm)	44.0	2.50(0.03)
50% mean coke size (mm)	2.9	9.1
mean coke volume fraction (-)	0.56	0.49
dust concentration (kg/m³)	0.280	7.33×10^{-4}

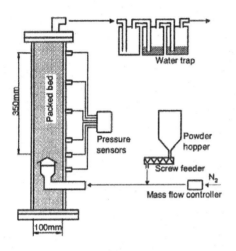

Figure 3 Schematic diagram of the experimental apparatus

The test dust was fed into flowing nitrogen gas using a screw feeder. Nitrogen gas was used to eliminate effects of humidity on the dust collection. The dust/gas mixture was introduced into the bottom of the coke bed at room temperature and 0.65 m/s. The dust that passed through the bed top was collected in a water trap. During the test run, the pressure was monitored on line.

After a test, the coke bed column was tilted so that the coke granules and accumulated dust in each bed section could be recovered into a sampling box that was divided into 19 sections corresponding to different bed heights. The concentration of dust collected is listed in Table I.

RESULTS AND DISCUSSION

Figure 4 shows the pressure drop with time in the model coke bed filter. The pressure drop was measured for the glass beads at four different heights in the coke bed under the conditions listed in Table I. The pressure drop initially increases, and then levels off. The pressure drop increases almost proportionally with distance from the bed inlet. The glass beads tested may bounce when colliding with the coke granules because their size exceeds the critical size for bouncing, which is determined from, e.g., the balance between the kinetic energy of an impacting particle and the wall-particle adhesion energy.[12] The critical size is roughly estimated as 20 μm for the impact velocity of 0.65 m/s.

In Figure 5, the pressure drop with time for Aerosil® is shown. Since Aerosil® particles are agglomerates of ultra-fine particles, as can be seen from Figure 6, and their mean size (2.5 μm) is much smaller than the critical size for bouncing, bouncing is not likely to take place. This leads to a rapid decrease in free void space between the coke granules, and a higher pressure drop.

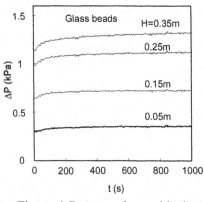

Figure 4 Pressure drop with time measured for glass beads through the coke bed. $D_p = 44$ μm, $D_c = 2.9$ mm, $\alpha_0 = 0.56$, $v = 0.65$ m/s, $C_i = 0.280$ kg/m³

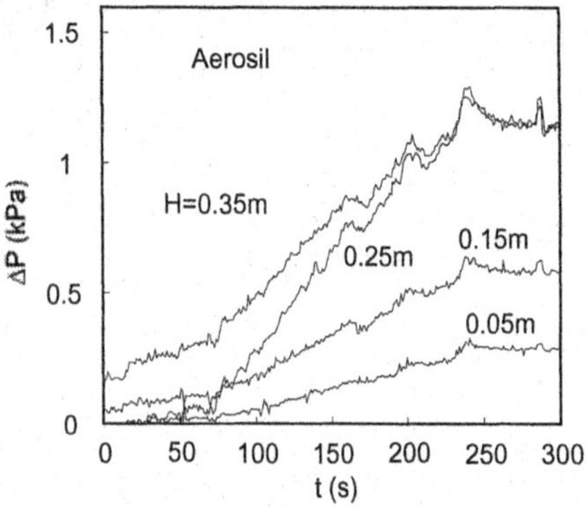

Figure 5 Pressure drop with time measured for Aerosil® through the coke bed. $d_p = 2.5$ μm, $D_c = 9.1$ mm, $\alpha_0 = 0.49$, $v = 0.65$ m/s, $C_i = 7.33 \times 10^{-4}$ kg/m³

Figure 6. SEM photograph of Aerosil® particles captured on coke granule

Glass dust load distributions in the coke bed filter as a function of distance, x, from the bed inlet are plotted in Figure 7 for different amounts of dust filtrated through a unit filter area, $C_i vt$, where t is the filtration period. The dust load, m, per unit volume of filter was evaluated from the pressure drop using Eq. (10) and (11) with $\alpha_{cm} = 0.44$. α_{cm} was determined by adjusting Eq. (10) to data for the dust load and pressure drop distributions measured at steady state ($t = 7700$ s). The dust load increases with the filtrated dust amount, and m decreases as the distance, x, from the filter inlet increases. The solid lines in Figure 7 are m values calculated from Eq. (8) using fitted values of η_0 and λ. The fitted η_0 is much lower than a predicted value (~1) for impact and intercept collection.[3] The change in dust load for the glass beads with time and position is described reasonably well using a constant λ and η_0. This is not the case for the Aerosil® dust, as shown in Figure 8, where α_{cm} was similarly obtained as 0.035. The

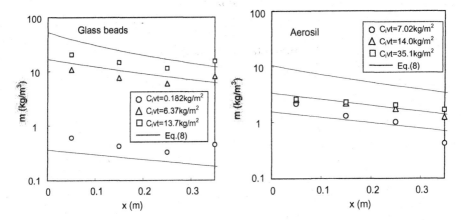

Figure 7. Dust load distribution in the coke bed filter from the measured pressure drop for the glass beads as compared with calculated results using Eq. (8) and constants $\lambda = 0.045$ m^3/kg and $\eta_0 = 0.0030$

Figure 8. Dust load distribution in the coke bed filter determined from the measured pressure drop for the Aerosil® particles as compared with results calculated using Eq. (8) and constants $\lambda = 0.010$ m^3/kg and $\eta_0 = 0.0130$

difference between the experiment and the model may be due to the channeling flow observed along the inner wall of the coke bed column, which may attributed to a non-uniform structure due to the wall effect. The values of α_{cm} are almost same with those from the apparent density of a packed bed, and are slightly larger than those from an experimental equation.[13] Hence, at least under the present condition, Eq. (10) describes the influence of dust load on the pressure drop.

Eq. (8), with constant λ and η_0, also was adapted to other published data on granular bed filters.[4,5] Figure 9 shows the dust load distribution in a fixed coke bed filter, and the agreement with the calculation from Eq. (8). Relationships between η_0 and λ were obtained by fitting Eq. (8) to data from the literature[5] (Figure 10). Data for fibrous filters[8,9] are also plotted. The solid lines show the results of numerical simulations of particle deposition on a single fiber[9], where $R = d_p/D_f$, and D_f is the fiber diameter. Data for granular beds seems to differ from those for fibrous filters: λ decreases at smaller η_0 than fibrous filters, indicating a higher sensitivity to the dust load. This may be due to the size difference in the pores, or the distance between the collecting media. 50% pore size is ca. D_c in the granular bed[14] for $\alpha_0 = 0.4 \sim 0.6$, while the ratio of a mean hydraulic radius of pores in fibrous filter to a fiber diameter is $3 \sim 4.5$[15]. The ratio of the particle diameter to the mean pore diameter, d_p/D, may be a measure of the sensitivity to the dust load. Using equations from a simulation,[12] the d_p/D is calculated to be ca. 1.6×10^{-2}, 2.7×10^{-4}, and 2.2×10^{-3} respectively for glass beads, Aerosil®, and the reported data[4]. From these, more sensitive behavior of the glass beads should appear, although we should have more data.

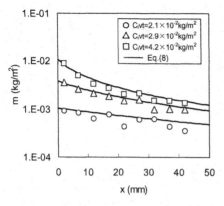

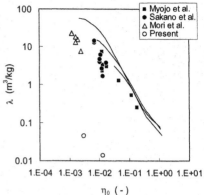

Figure 9. The dust load distribution in the granular bed filter (literature data ($D_c = 759$ μm, $d_p = 1.8$ μm, $\alpha_0 = 0.41$, $v = 0.029$ m/s) in comparison to the prediction from Eq. (8) using $\lambda = 7.5$ m^3/kg and $\eta_0 = 0.00246$

Figure 10. Relation between λ and η_0. for a granular bed (open symbols) and a fibrous filter (filled-in symbols) with different fiber R values.

CONCLUSIONS
1) We have adapted a filtration model to coke bed filters that takes into account the increase in coke collection efficiency with dust loading when the influence of flow channeling is small.
2) The behavior of granular beds differs from fibrous filters in that the granule bed filters appear to be more sensitive to dust loading.

ACKNOWLEDGEMENT

This research has been re-sponsored by JRCM, which has been supported by the New Energy and Industrial Technology Development Organization (NEDO).

REFERENCES
[1]T. Furukawa, H. Sasamoto, S. Isozaki and F. Tanno, "Process Technology of the Direct Separation and Recovery of Iron and Zinc Metals Contained in High Temperature EAF Exhaust Gas," Proc. The 6th International Symp. on East Asian Resources Recycling, p.1-5, Gyeongju, Korea (2001).
[2]H. Sasamoto, T. Furukawa, "EAF Dust Recycling Technology in Japan", Proc. The 6th International Symp. on East Asian Resources Recycling, p.9-17, Gyeongju, Korea (2001).
[3]Y. Otani, et. al., "Experimental Study of Aerosol Filtration by the Granular Bed Over a Wide Range of Reynolds Numbers," Aerosol Science and Technology, 10, 463-474 (1989).
[4]H. Mori, N. Kimura and S. Toyama, "The Performance of Granular Bed Filters with Dust Loading," J. Soc. Powder Technol. Japan, 29, 18-25 (1992).
[5] N. Kimura, M. Kanamori, H. Mori and M. Shirato, "Dust Load Distributions in Fibrous Mat and Granular Bed Filters," J. Chem. Eng. Japan, 15, 119-125 (1989).

[6]S. Toyama, H. Mori and Y. Mizutani, "Performance of Horizontal Gas Flow Granular Bed Filters," *J. Chem. Eng. Japan*, **13**, 621-627 (1987).

[7]C. Kanaoka, "Performance of an Air Filter at Dust-loaded Condition," pp.323-335 in *Advances in Aerosol Filtration*, K. R. Spurny ed., Lewis Pub., 1998.

[8]T. Sakano, Y. Otani, N. Namiki and H. Emi, "Prediction of Transient Collection Performance of Medium Performance Fibrous Filter for Mono-disperse Particles", *J. Aerosol Research, Japan*, **17**, 185-190 (2002).

[9]T. Myojo, C. Kanaoka and H. Emi, "Experimental Observation of Collection Efficiency of a Dust-loaded Fiber," *J. Aerosol Sci.*, **15**[4], 483-489 (1984).

[10]N. Yoshioka, H. Emi, M. Yasunami, H. Sato, "Filtration of Aerosol through Packed Column with Dust Load," *Kagaku Kogaku*, **33**, 1013-1018 (1969).

[11]C. Kanaoka, H. Emi, T. Myojo and M. Ohta, "Estimation of Collection Efficiency of an Air Filter with Dust Load," Int. Chem. Sym. Series No. 59, Solid Separation Processes, p.3:4/1-3:4/15, 1980.

[12]W. C. Hinds, "Particle Bounce," pp.146-147 in *Aerosol Technology*, 2nd ed., Wiley-Interscience, New York, 1999.

[13]J.C. Laborde, L. Del Fabro, V. M. Mocho, L. Ricciardi, "Contribution to the Modeling of Industrial Pleated Filters," Proc. Int. Workshop on Particle Loading & Kinerics of Filtration in Fibrous Filters, Karlsruhe, p.69-78, 2002

[14]K. Gotoh, M. Nakagawa, M. Furuuchi and A. Yoshigi, "Pore Size Distribution in Random Assemblies of Equal Spheres," *J. Chem. Phys.*, **85**, 3078-3080 (1986).

[15]H. W. Piekaar and L. A. Clarenburg, "Aerosol Filters – Pore Size Distribution in Fibrous Filters," *Chem. Eng. Sci.*, **22**, 1399-1408 (1965).

DIRECT MEASUREMENT OF THE ADHESION FORCE FOR SINGLE ASH PARTICLE AT HIGH TEMPERATURE

Hidehiro Kamiya, Takashi Aozasa, and Mayumi Tsukada
Graduate School of Bio-Applications and Systems Engineering, BASE
Tokyo University of Agriculture and Technology, Koganei
Tokyo 184-8588, Japan

Hiromitsu Matsuda and Hisao Makino
Central Research Institute of Electric Power Industry, CRIEPI
Yokosuka
Kanagawa 240-0196, Japan

ABSTRACT

A new measurement system was developed that can determine the adhesion force between a fine (~20 μm) spherical particle and a polished flat substrate at temperatures up to 1000°C. The equipment consists of a spot heating system with an infrared furnace, a quartz glass leaf spring, a high-resolution non-contact displacement sensor, and a high-precision displacement control stage. The resolution of the displacement and the force measurement between the particle and the plate is 0.1 nm and 60 nN, respectively. A spherical particle was mounted to the edge of a pointed quartz glass rod connected to the control stage of the system with the aide of a micromanipulator. Coal combustion ash spheres of about 20 μm in diameter were prepared by burning coal in a drop tube furnace. The adhesion force between an ash sample particle and the polished stainless steel (SUS 310S) plate was measured at temperatures up to 950°C. At relatively low temperatures below 800°C, the adhesion of an ash particle is very weak. Adhesion increases above 800°C.

INTRODUCTION

Adhesion between ash particles at high temperature prevents the long and stable operation of advanced power generation systems that fire and/or gasify solid fuels such as coal, waste, and biomass. For coal, there is a remarkable increase in the adhesion force above 800°C[1], which is caused by the formation of low temperature eutectics between alkaline metals and silica.[2] There is potential to reduce the adhesion of ash particles by modifying the ash properties; however, new testing equipment is needed to complete representative tests. Laboratory scale filtration tests are the most representative tests to evaluate ash adhesion to ceramic filters, and these tests have been performed with ash samples from a variety of fuels.[3,4] Additionally, convenient conventional analysis methods are available to characterize ash, including dilatometry tests on ash powder beds,[3] ash powder pellet fracture tests,[5] and ash powder bed tensile strength tests,[1] and all can be performed at high temperature. To investigate the interaction force between individual particles, a more fundamental adhesion force measurement between a single particle and a target plate is effective. In our work, a system with an extremely precise displacement control stage has been developed to measure adhesion forces up to 1000°C, however the size of the test particle is limited to about 1 mm.[6] To reduce the size of the test particle down to about 10 µm, which is on the order of the size of the ash particles emitted from an actual combustion furnace, a new test method is needed. Colloid probe AFM[7] can be used to determine particle interaction forces in this particle size range; however, the temperature is usually limited under 200°C.

The purpose of the present work is to develop a new, high resolution measurement system to determine the adhesion force between a single particle around 10 µm in diameter at elevated temperatures up to 1000°C.

EXPERIMENTAL
Ash Sample Preparation
The ash sample particles tested were prepared by coal combustion in a drop tube furnace. Pulverized coal under 150 µm in diameter and having 13.0 wt% ash content was used to prepare the ash samples. The ash composition is shown in Table I.

The drop tube furnace shown in Figure 1 was used to prepare the ash sample. The tube is 50 mm in diameter, with 800 mm in the reaction zone. Coal is fed into

Table I. Composition of coal ash.

Substance	Content (wt%)	Substance	Content (wt%)	Substance	Content (wt%)
SiO_2	66	MgO	0.50	P_2O_5	0.54
Al_2O_3	26	Na_2O	0.15	TiO_2	1.3
Fe_2O_3	3.1	K_2O	1.1	V_2O_5	0.06
CaO	0.87	SO_3	0.26	MnO	0.02

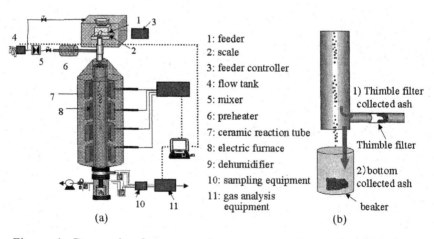

1: feeder
2: scale
3: feeder controller
4: flow tank
5: mixer
6: preheater
7: ceramic reaction tube
8: electric furnace
9: dehumidifier
10: sampling equipment
11: gas analysis equipment

1) Thimble filter collected ash

Thimble filter

2) bottom collected ash

beaker

(a)   (b)

Figure 1. Drop tube furnace used to prepare ash samples, showing: a) Drop tube furnace; and b) Ash sampling port.

the top of a drop tube furnace at a rate of 0.42 g/min and burned at 1400°C in ambient air flowing at 0.64 Nl/min. The gas velocity in the tube is 0.183 m/s, and the residence time of the coal in the reaction zone is 4.4 s. The stoichiometric air ratio is 1.24. Ash was collected with a thimble filter by air suction from the downstream end of the furnace. Residual ash drops and accumulates in an ash beaker at the bottom (Figure 1b).

Adhesion Force Measurement

Figure 2 shows the newly developed adhesion force measurement system. It is comprised of an extremely precise displacement control stage (Nano Stage, Nippon Laser & Electronic Co. Ltd) with a test particle mounted at the end of a pointed rod, a leaf spring with a flat target plate, a displacement sensor (Nanometric Sensor 613A, Mitz Co., Ltd.), and an infrared heating system (GA 198, Thermo Riko Co., Ltd.). The adhesion force between the test particle and the flat plate is determined from the deflection of the leaf spring. The resolution of a displacement control stage is 0.005 nm per pulse, and the controllable minimum linear speed is currently 0.5 nm/s. The resolution of the displacement sensor is 20 ~ 100 nm depending on the surrounding vibration. The leaf spring is made of quartz glass having the dimensions 1.0 mm^w × 100 mm^l × 0.3 mm^t. The rod connected to the precise displacement controller also is made of quartz glass. The relationship between the force acting on the leaf spring and the displacement detected by the sensor is determined by using precise weights to construct a calibration curve (Figure 4). The calibration curve is then used to determine the sensitively of the measurement, and the actual adhesion.

The precision of the measurement system can be assessed by how small of a particle's van der Waals force can be measured. Van der Waals force, F_{vdw}, for two facing spheres of the same diameter D_p is calculated by

$$F_{vdw} = -H_A D_p/(12h^2) \qquad (1)$$

Where H_A is Hamaker constant and h is the distance between surfaces. Substituting $H_A = 16.4 \times 10^{-20}$ J (the value for silica[8]), h = 4×10^{-10} m and $F_{vdw} = 60$ nN (the maximum resolution of the measured force for the present leaf spring material, dimensions, and displacement sensor resolution = 100 nm), D_p, is calculated to be 700 nm. This means that van der Waals force between 7 μm particles can be measured within an accuracy of one digit.

To attach a 10 – 30 μm ash particle to the 100 μm diameter pointed quartz glass rod, a micromanipulator (Shimadzu, MMS-77) is used with a video microscope. The process used is fundamentally the same as that used in our previous work to prepare colloid probes for the atomic force microscope.[9] The quartz glass rod is attached to one arm of the manipulator, and two thin metal wires with several ash particles attached with a small amount of adhesive cement (Aron Ceramic, E, Toagosei Co., Ltd.) are on the other arm. After memorizing the position of one of the ash particles, some adhesive cement is attached to the wire. Then the glass rod is moved into contact with the cement, and before the cement dries, the quartz glass rod is moved quickly to the memorized ash particle's position to attach the ash particle. To decrease the speed of evaporation of the solvent from the cement circumference during this operation, the manipulator is covered with a flexible transparent film.

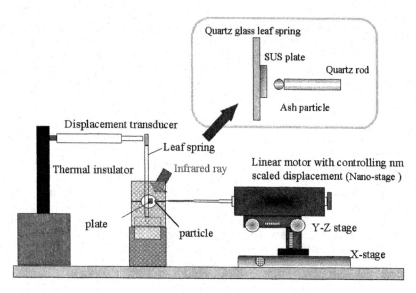

Figure 2. The testing system used to measure adhesion force.

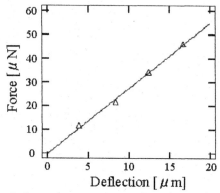

Figure 3. Force-deflection calibration curve for the quartz glass leaf spring (i.e. for a distance of 80 mm between the contact point and the point where the displacement is measured).

A polished, 2 mm × 2 mm × 0.2 mm stainless steel plate (SUS 310S, arithmetic mean deviation of the assessed profile, Ra = 1.1 μm) is attached to the leaf spring with adhesive cement. For both the quartz glass rod and the SUS plate, the adhesive cement is dried at 100°C for 1 h, and at 160°C for 1 h.

To complete an adhesion force measurement, the sample is heated to the specified temperature, and the sample is positioned such that it is almost touching the SUS plate. Then, the particle moved toward the SUS plate at a velocity of 10 nm/s using Nanostage while sampling displacement data at 200 Hz. An example of the signal detected by the displacement sensor, along a schematic of the particle-plate relative location, is shown in Figure 4. After the particle contacts the plate, the compression force increases to a maximum value determined by

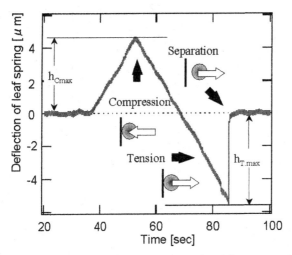

Figure 4. An Example of a raw signal obtained from an adhesive force measurement (850°C).

h_{Cmax} and the spring constant. Then the ash particle is moved away from the plate at the same velocity. The adhesion force is determined from h_{Tmax}, which is the force necessary to detach the particle from the plate.

RESULTS AND DISCUSSION

Figure 5 shows an SEM image of the ash particles produced in the drop tube furnace. The median particle size of the ash particle produced is 6.2 μm. We collected 10-30 μm particles from this sample to attach to the quartz glass probe (Figure 6). An example of the raw data used to determine the adhesion force, F_{Cmax}, and the maximum compression force F_{Tmax} is shown in Figure 4 for a measurement at 850°C. F_{Tmax} may affect the adhesion force when the deformation of an ash particle due to compression does not recover in tensile process quickly. Figure 7 shows F_{Tmax} divided by particle radius as a function of F_{Cmax}. An increase

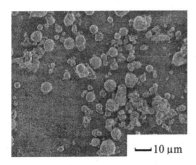

Figure 5. SEM image of the ash sample produced in the drop tube furnace.

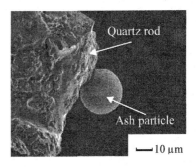

Figure 6. SEM image of an ash particle attached to a quartz glass rod.

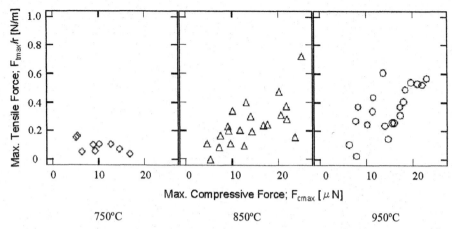

Figure 7. Effect of temperature and maximum compression force on adhesion force.

in adhesion force with increasing temperature from 750 to 950°C is observed at higher F_{Cmax}. This trend is consistent with powder bed tensile strength test data[1] and direct adhesion force measurements for granulated and melted ash particles on a substrate.[10]

CONCLUSION

New equipment was developed to characterize the adhesion force between a single ash particle and a substrate at temperatures up to 1000°C. With the aid of a micromanipulator system, it was possible to attach a single 20 μm ash particle onto the end of a pointed quartz glass rod. Using a non-contact displacement sensor with 100 nm resolution and a nanometer scaled displacement control stage, the minimum resolution of the adhesion force measurement is about 60 nN. The adhesion force of pulverized coal ash produced in a drop tube furnace was successfully determined using the newly developed measurement system. An increase in the adhesion force between ash particles above 800°C is observed with the present method.

ACKNOWLEGMENT

This work is supported by NEDO GRANT international collaboration project "High temperature gas cleanup", Proposal-Based New Industry-Type Technology R&D Promotion Program from NEDO in Japan, and Grant-in-Aid for Scientific Research (B), Japan.

REFERENCES

[1]H. Kamiya, A. Kimura, T. Yokoyama, M. Naito, and G. Jimbo, "Development of a Split-type Tensile-strength Tester of Inorganic Fine Powder Bed for High-Temperature Conditions," *Powder Technol.*, 127 239-45 (2002).

[2]H. Kamiya, A. Kimura, M. Tsukada, and M. Naito, "Analysis of the High-Temperature Cohesion Behavior of Ash Particles Using Pure Silica Powders Coated with Alkali Metals," *Energy and Fuel*, 16 457-61 (2002).

[3]G. Hemmer and G. Kasper, "Predicting of Operating Behavior of Ceramic Filters from Thermo-Mechanical Ash Properties," Proc. 5th Int. Symp. Gas Cleaning at High Temperature, Paper No. 3.5 (2002).

[4]H. Kamiya, K. Deguchi, J. Gotou, and M. Horio, "Increasing Phenomena of

Pressure Drop during Dust Removal Using Rigid Ceramics Filter at High Temperature," *Powder Technol.* **118** 160-65 (2001).

[5]H. Kamiya, A. Kimura, M. Horio, J. Seville, and E. Kauppinen, "Diametral Compression Characteristics of Cohesive Ash Powder Pellets at High Temperature," *J. Chem. Eng. Japan*, **33** 654-56 (2000).

[6]H. Kamiya, N. Yamaguchi, A. Koga, and M. Horio, "Development of a Direct Measurement System for a Single Particle Force at High Temperature," Preprints of Fall Meeting of Soc. Powder Technol. Japan, p.207-10 (1999)

[7]W.A. Ducker and T.J. Senden, "Measurement of Forces in Liquids Using a Force Microscope," *Langmuir*, **8**,1831-36 (1992).

[8]K. Okuyama, H. Masuda, K. Higasitani, M. Chikazawa, and T. Kanazawa, "Interaction between Two Particles," *J. Soc. Powder Technol. Japan*, **22** 451-75 (1985)

[9]H. Kamiya, J. Yoneyama, Y. Fukuda, H. Abe, and M. Naito, "Analysis of Anionic Polymer Dispersant in Dense Alumina Suspension with Various Additive Content by Using Colloidal Probe AFM," *Ceramic Trans.*, **133** 65-70 (2002).

[10]N. Yamaguchi, M. Tsukada and H. Kamiya, "Analysis for the Ash Adhesion Prevention at High Temperatures Using Surface Force Measurements," Preprints Fall Meeting Soc. Chem. Eng. Japan, B106 (2001)

DEVELOPMENT OF A REGENERABLE DESULFURIZATION SORBENT FOR HOT COAL-DERIVED GAS

Makoto Nunokawa, Makoto Kobayashi, and Hiromi Shirai
Central Research Institute of Electric Power Industry
Nagasaka 2-6-1, Yokosuka, Kanagawa, 240-0196 Japan

ABSTRACT

We developed a desulfurization sorbent containing zinc ferrite ($ZnFe_2O_4$) as the reactive material. Zinc ferrite is a double oxide of iron and zinc that has relatively high sulfur absorbing capacity and sulfur reducing ability. The reactivity was evaluated with simulated coal-derived gas at 723 K and 0.9 MPa in a laboratory scale fixed bed reactor. The sorbent reduces the concentration of sulfur compounds below 1 ppm. Spent sorbent is regenerated by reacting with oxygen gas at 723 K and 0.9 MPa. The sulfur capacity of the iron and zinc in the sorbent decrease in early desulfurization/regeneration cycles. However, the sulfur capacity of the iron and zinc become stable after the 40th desulfurization cycle. The breakthrough time is shortened with repeated cycles. The sorbent still can decrease the concentration of the sulfur compounds to below 1 ppm after the desulfurization cycle. An extruded honeycomb sorbent, which is necessary for large scale reactors, maintains the desulfurization performance of the zinc ferrite.

INTRODUCTION

Coal gas derived in a gasification process contains impurities such as dusts, halide compounds, and sulfur compounds.[1,2] The development of high performance gas cleanup technology is essential for the use of coal-derived gas in an advanced power generation system. To use coal-derived gas in a molten carbonate fuel cell (MCFC), the concentration of the sulfur compounds in the form of hydrogen sulfide (H_2S) and carbonyl sulfide (COS) must be reduced below 1 ppm to preserve fuel cell performance.[1,2] Hot gas cleanup technologies have advantages compared to conventional wet cleanup technologies, because the energy loss during gas cooling and reheating can be avoided.[3] The development of high performance sorbents, which absorb contaminants in coal-derived gas, is important for hot gas cleanup technologies. For economics, it must be possible to regenerate the sorbent for reuse. Additionally steps must be taken to prevent carbon deposition that can cause a significant drop in sorbent performance in a pressurized coal-derived gas environment.[4]

Many attempts have been made to develop high temperature desulfurization sorbents using iron oxide, zinc ferrite, zinc titanate, and other metal oxides.[5-7] We propose the desulfurization sorbent containing zinc ferrite ($ZnFe_2O_4$) as the reactive material.[8] Zinc ferrite is a double oxide of iron and zinc, that has a relatively high sulfur absorbing capacity and sulfur reducing ability. Spent sorbent can be regenerated by reacting with oxygen gas.[9] In situ X-ray diffraction (XRD) analyses has revealed that ZnS and FeS are produced when the H_2S concentration is 1 vol%; however only ZnS is produced when the H_2S concentration is 80 ppm. This indicates that the zinc is necessary for sulfur removal at lower H_2S concentrations.[10]

This paper describes the characterization of a $ZnFe_2O_4$ desulfurization sorbent in an advanced power generation system. The changes in sulfur absorbing capacity and sulfur reducing ability after multiple desulfurization cycles are evaluated. The desulfurization performance of a honeycomb $ZnFe_2O_4$ sorbent for large-scale reactors, is also evaluated with pressurized simulated coal-derived gas.

EXPERIMENTAL SECTION
Preparation of Desulfurization Sorbent
    The desulfurization sorbent used is composed of $ZnFe_2O_4$-SiO_2 powder and TiO_2 powder. The $ZnFe_2O_4$-SiO_2 powder was prepared by precipitation from aqueous salt solutions. Table I summarizes the preparation conditions and resulting properties of the desulfurization sorbent. The molar ratio of the zinc to iron was adjusted to the stoichiometric value to form $ZnFe_2O_4$. The precursor raw materials used were zinc nitrate, iron nitrate, and silica sol. Solutions of these salts were used for homogeneous precipitation with urea. The precipitate was calcined in an electric furnace. The concentration of zinc and iron in the calcined powder is 0.83 and 1.7 mmol/g-sorbent, respectively.

Apparatus and Procedures to Evaluate of Sorbents
    The performance of the desulfurization sorbent was evaluated in a laboratory scale fixed bed reactor. Figure 1 shows a schematic flow diagram of the fixed bed reactor. The test conditions are summarized in Table II. Sulfidation was performed by introducing a gas mixture that simulates the coal-derived gas produced by dry coal fed, air blown, coal gasification. H_2S concentrations of 1.1% and 1000 ppm were used to evaluate sulfur absorbing capacity and sulfur reducing ability, respectively. The breakthrough curve, which indicates the sulfur concentration changes in the outlet gas stream, was determined using a gas analyzer with a flame photometric detector. The data was used to calculate the amount of sulfur removed by the sorbent tested. Regeneration was carried out in oxygen mixed with nitrogen. The sulfur compound (SO_2), released from the reactor during regeneration, was analyzed to evaluate the regenerability of the sorbent.

Table I. Synthesis details and the specifications of the desulfurization sorbent.

Precipitation method	Homogenerous
Precipitant	Urea
Calcination temp.	973K
Zn:Fe:Si molar ratio	1.0 : 2.0 : 2.0
Compounds	$ZnFe_2O_4$-SiO_2
Weight ratio of $ZnFe_2O_4$-SiO_2 and support material (TiO_2)	3.0 : 7.0
Content of Zn / Fe	0.83 / 1.7 mmol/g-sorbent

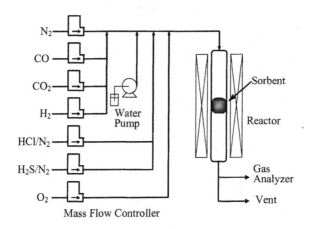

Figure 1. Fixed bed reactor used to evaluate sorbents.

Table II. Test conditions for fixed bed reactor.

	Sulfidation	Regeneration	Reduction
Temperature	723K	723K	723K
Pressure	0.98MPa	0.98MPa	0.98MPa
Gas composition	$CO : 20\%$	$O_2 : 1.5\%$	$CO : 20\%$
	$CO_2 : 5\%$	$N_2 :$ Balance	$CO_2 : 5\%$
	$H_2 : 8\%$		$H_2 : 8\%$
	$H_2O : 5\%$		$H_2O : 5\%$
	$H_2S : \begin{matrix} 1.1\% \\ 1000 \text{ ppm} \end{matrix}$		$N_2 :$ Balance
	$N_2 :$ Balance		

RESULTS AND DISCUSSION

Reactions with Zinc Ferrite in a Desulfurization Cycle

The reaction products associated with $ZnFe_2O_4$ in a desulfurization cycle were extensively studied as shown in Figure 2. During the sulfidation step, $ZnFe_2O_4$ can absorb sulfur compounds from the coal-derived gas, and convert them to ZnS and FeS. During regeneration, the sorbent used reacts with oxygen gas and converts to $ZnFe_2O_4$ and SO_2. Sulfates, such as $ZnSO_4$ and $Fe_2(SO_4)_3$, may also be produced. During the reduction step, these sulfates decompose into sulfides and remain in the sorbent. Thus the sulfur absorbing capacity of sorbent may drop after the desulfurization cycle. We examined data from fixed bed reactor test, and completed a chemical analysis of the residual sulfur in the spent sorbent.

Desulfurization Property Changes with Multiple Desulfurization Cycles

The sulfur absorbing capacity of the sorbent was determined from the breakthrough curves in cyclic desulfurization tests completed in the fixed bed reactor. Subsets of typical breakthrough curves up to 40 desulfurization cycles are shown in Figure 3.

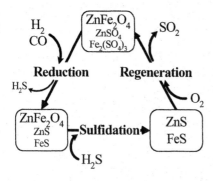

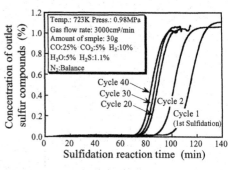

Figure 2. Desulfurization cycle for ZnFe$_2$O$_4$ sorbent.

Figure 3. Breakthrough curves for ZnFe$_2$O$_4$ sulfidation as a function of desulfurization cycle.

The results show that the sorbent removes sulfur compounds from the coal-derived gas. Figure 4 shows the change in sulfur absorbing capacity with the number of desulfurization cycles. The sulfur absorbing capacity decreases in the early cycles. This decrease is attributed to the increase in residual sulfur in the regenerated sorbent (Figure 4). The concentration of residual sulfur in spent sorbent is determined with infrared light absorption during combustion in oxygen. The sulfur absorbing capacity and the residual sulfur content become steady as the number of desulfurization cycles increases. Relative to the first cycle, the ratio of sulfur absorbing capacity in the 40th cycle is around 80%.

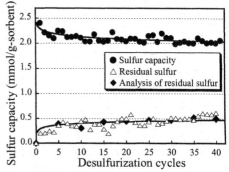

Figure 4. Change in the sulfur capacity of ZnFe$_2$O$_4$ as a function of desulfurization cycles.

The results of the sulfur reducing ability form desulfurization tests in the fixed bed reactor are shown in Figure 5. Fresh sorbent decreased the concentration of sulfur compounds in the simulated coal-derived gas to <1 ppm. Although the breakthrough time for the regenerated sorbent is shortened with more cycles, the ability to reduce the concentration of the sulfur compounds to <1 ppm is maintained through the 41st desulfurization cycle. This demonstrates that the zinc related sulfur removal performance is sustained after sorbent regeneration.

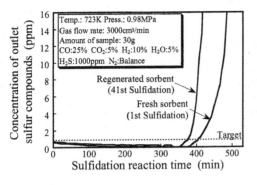

Figure 5. Effect of regeneration on the ability of $ZnFe_2O_4$ to remove sulfur from coal-derived gas.

Figure 6. Honeycomb $ZnFe_2O_4$ sorbent.

Reactivity of Honeycomb Sorbent

A large volume of coal-derived gas must be treated in the hot gas cleanup process in a large-scale advanced power generation system. Thus the desulfurization sorbent has to be molded to reduce pressure loss. We prepared extruded honeycombs of the $ZnFe_2O_4$ sorbent as shown in Figure 6. The cell pitch and wall thickness are 4.9 mm and 1.1 mm, respectively. The honeycomb sorbent also has the merits of high strength in the gas flow direction, and it promotes favorable contact between the gas and the sorbent. The breakthrough curves for the honeycomb and the particle $ZnFe_2O_4$ sorbent obtained from desulfurization test are shown in Figure 7. The results show that the honeycomb sorbent can reduce the concentration of sulfur compounds in coal-derived gas to <1 ppm; however the 1 ppm breakthrough time for the honeycomb sorbent is shorter than that for the particle sorbent.

Figure 8 shows the simulated breakthrough curve for the honeycomb $ZnFe_2O_4$ sorbent using the reaction model in Equation (1). This equation has been used to complete simulations of an IGCC honeycomb sorbent containing iron oxide (Fe_2O_3).[12]

$$\frac{dY}{dt} = \frac{Y \times (Cg_{S,in} - Cg_{S,e})}{f_F^{-1} + f_D^{-1} \times g_D(Y)^{-1} + f_R^{-1} \times g_R(Y)^{-1}} \quad (1)$$

$$Y : \text{Dimensionless concentration} = \frac{Cg_S - Cg_{S,in}}{Cg_{S,in} - Cg_{S,e}}$$

Cg_S : Outlet H_2S concentration
$Cg_{S,in}$: Inlet H_2S concentration
$Cg_{S,e}$: Equiliburium H_2S concentration
f_F : Conversion factor for gas film mass transfer
f_D : Conversion factor for diffusion
$g_D(Y)$: Conversion function for diffusion
f_R : Conversion factor for reaction rate
$g_R(Y)$: Conversion function for reaction rate

The simulated results are in good agreement with the experimental results. A decreasing reaction ratio in the honeycomb Fe_2O_3 sorbent due to the influence of gas film mass transfer and pore diffusion decrease its breakthrough time.[13] The decreasing reaction ratio in the honeycomb $ZnFe_2O_4$ sorbent may also contribute to the decrease in breakthrough time.

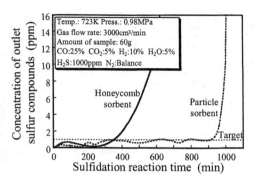

Figure 7. Breakthrough curves for $ZnFe_2O_4$ honeycomb and particle sorbents.

Figure 9 shows the influence of regeneration temperature on the sulfur balance, which indicates the behavior of absorbed sulfur in the regeneration step. The amount of remaining sulfur in the regenerated sorbent increases at lower temperatures because of sulfate formation. A high regeneration temperature is desirable to maintain sulfur removing capacity. However, regeneration above 823 K causes sintering, and decreases the surface area and performance of the sorbent. The regeneration temperature does not significantly affect the sorbent ability to remove sulfur below 1 ppm. Thus, the best regeneration temperature for desulfurization performance is 723 K, which is the same temperature as the desulfurization and reduction steps.

Figure 10 shows an example of an IG-MCFC power generation system with hot gas cleanup. The hot gas cleanup system consists of a dust removal facility with porous ceramic filters,[14] a dehalide unit with dehalide sorbent,[15] and a desulfurization unit with desulfurization sorbent. Impurities that have a harmful effect on MCFC performance in coal-derived gas are efficiently reduced with the hot cleanup system, and the resulting coal-derived gas is applicable for use in MCFC power generation.

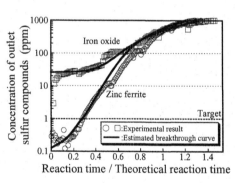

Figure 8. Comparison of simulated and experimentally-determined breakthrough curves for Fe2O3 and ZnFe2O4 honeycomb sorbents.

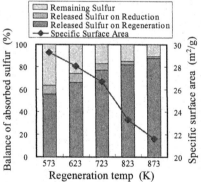

Figure 9. Influence of regeneration temperature on sulfur balance and surface area.

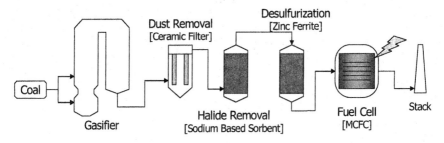

Figure 10. Example of an IG-MCFC power generation system with hot gas cleanup technology.

CONCLUSIONS

We developed a desulfurization sorbent containing zinc ferrite ($ZnFe_2O_4$) as the reactive material. This sorbent can reduce the concentration of sulfur compounds in coal-derived gas to less than 1 ppm. Spent sorbent is regenerated by reaction with oxygen gas. The sorbent maintains the ability to reduce the concentration of sulfur compounds in coal-derived gas below 1 ppm even after repeated desulfurization and regeneration cycles. An extruded honeycomb sorbent containing $ZnFe_2O_4$ was successfully fabricated and tested for use in a large-scale advanced power generation system.

REFERENCES

[1]"Fuel Cell Handbook, Sixth Edition", DOE/NETL-2002/1179 (2002).
[2]E.J. Vidt, G. Jablonski, J.R. Hamm, M.A. Alvin, R.A. Wenglarz, and P. Patel, "Evaluation of Gasification and Gas Cleanup Processes for Use in Molten Carbonate Fuel Cell Power Plants, Final Report," *U. S. DOE Report*, DOE/MC-16220/1306 (1982).
[3]M. Kobayashi, T. Nakayama, H. Shirai, H. Matsuda, and T. Tanaka, "A Review on Coal Gas Clean Up Technology for Molten Carbonate Fuel Cell," *Denryoku Cyuou Kenkyuuzyo Houkoku*, W90014 (1990).
[4]E. Sasaoka, Y. Iwamoto, S. Hirano, A. M. Uddin, and Y. Sakata, "Soot Formation over Zinc Ferrite High-Temperature Desulfurization Sorbent," *Energy & Fuels*, 9 344-53 (1995).
[5]V.S. Underkoffler, "Summary and Assessment of METC Zinc Ferrite Hot Coal Gas Desulfurization Test Program, Final Report Volume I," *U. S. DOE Report*, DOE/MC/21098-2247 (1986).
[6]S.E. Lyke, R.S. Butner, and G.L. Roberts, "Development of a Solid Adsorption Model and Improved Sorbent for Solid Supported Molten Salt Hot Gas Cleanup Process," *U. S. DOE Report*, DOE/RL/01830-2279 (1986).
[7]G.L. Andersen, F.O. Berry, A.H. Hill, E. Ong, R.M. Laurence, R. Shah, and H. L. Feldkirchner, "Development of Hot Gas Cleanup System, Final Report," *U. S. DOE Report*, DOE/MC/22144-2722 (1988).
[8]M. Kobayashi, M. Nunokawa, H. Shirai, and M. Watanabe, "Activation of Desulfurization Performance of Zinc Ferrite Sorbent for use in High Temperature Sulfur Removal of Coal Derived Gas for Application to Molten Carbonate Fuel Cell Power Plant," *CRIEPI Report*, EW93005 (1994).
[9]M. Kobayashi, H. Shirai and M. Nunokawa, "Investigation on Desulfuriza-

tion Performance and Pore Structure of Sorbents Containing Zinc Ferrite," *Energy & Fuels*, **11** [4] 887-96 (1997)

[10]M. Kobayashi, H. Shirai, and M. Nunokawa, "High-temperature Sulfidation Behavior of Reduced Zinc Ferrite in Simulated Coal Gas Revealed by In Situ X-Ray Diffraction Analysis and Mossbauer Spectroscopy," *Energy & Fuels*, **16** [3] 601-07 (2002).

[11]M. Kobayashi, H. Shirai, and M. Nunokawa, "Measurements of Sulfur Capacity Proportional to Zinc Sulfidation on Sorbent Containing Zinc Ferrite-Silica Composite Powder in Pressurized Coal Gas," *Industrial & Engineering Chemistry Research*, **41** [12] 2903-09 (2002).

[12]H. Shirai, M. Kobayashi, and T. Tanaka, "The performance of Desulfurization Sorbent at Pressurized Conditions for Coal Gas," *Denryoku Cyuou Kenkyuuzyo Houkoku*, W88033 (1989).

[13] H. Shirai, M. Kobayashi, and M. Nunokawa, "Modeling of Desulfurization Reaction for Fixed Bed System using Honeycomb Type Iron Oxide Sorbent and Desulfurization Characteristics in Coal Gas," *Kagaku Kougaku Ronbunshu*, **27** [6] 771-78 (2001).

[14] S. Ito, T. Tanaka, and S. Kawamura, "Dust Properties and Cleaning Systems as Factors in Pressure Loss of Hot Coal Gas Filtration," *Twelfth Annual International Pittsburgh Coal Conference*, 379-84 (1995).

[15] M. Nunokawa, M. Kobayashi, and H. Shirai, "Hydrogen Chloride Removal from Hot Coal-Derived Gas with Sodium-Based Sorbents," *High Temperature Gas Cleaning II*, 685-695 (1999)

Interface Control

Sic-COATED DIAMOND SYNTHESIS, AND PROPERTIES IN DIAMOND DISPERSED CEMENTED CARBIDES

Y. Miyamoto and Y. Morisada
Joining and Welding Research Institute
Osaka University
Ibaraki
Osaka 567-0047, Japan

H. Moriguchi, K. Tsuduki, and
A. Ikegaya
Itami Research Laboratories
Sumitomo Electric Industries, Ltd.
Itami 664-0016, Japan

ABSTRACT

A simple coating method has been developed to deposit nanometer-sized SiC on diamond using SiO vapor. The SiC coating improves the oxidation resistance of the diamond, and eliminates deleterious reactions between diamond and cobalt metal at high temperature. Cemented-carbide (WC/Co) composites with SiC coated diamond particles were fully densified without graphitization by pulsed-electric current sintering. Compared to conventional cemented carbides, these diamond dispersed cemented-carbides (DDC) show 50 % higher hardness (15.7-24.5 GPa), two times higher toughness (14-20 MPa m$^{1/2}$), five times lower friction coefficient (0.05), and an order of magnitude higher wear resistance. Large size DDCs (150 mm diameter and 5 mm thick) and functionally graded DDCs have been produced. Products have been commercialized as high performance wear-resistant tools.

INTRODUCTION

Sintered diamond is widely used for cutting, grinding, and polishing rocks, concrete, ceramics, and nonferrous metals. However, diamond oxidizes at over 530°C in air, converting to CO_2. It also reacts with transition metals such as iron, nickel, and cobalt at about 1150°C and converts to graphite.[1] To prevent the conversion of diamond to graphite, sintered diamond is produced under ultra-high pressure (e.g., 5.5 GPa) using cobalt binder.[2] This high pressure process, however, is very costly and limits the size and shape of products. To control interfacial reactions between diamond and molten metals, we have developed a simple method to coat nanocrystalline SiC on diamond using SiO vapor.[3] The SiC coating can protect the interface between the diamond and the molten cobalt so that the a dense SiC-coated diamond dispersed cemented carbide can be successfully fabricated at low pressure by pulsed-electric current sintering (PECS).[4] Because these diamond-dispersed cemented carbides have extremely high wear resistance, high toughness, and low friction coefficient, the products have been commercialized for various wear-resistant tools.

SYNTHESIS OF NANOCRYSTALLINE SiC COATED DIAMOND

The method for coating SiC on diamond is very simple and easy. Diamond powder (Allied Materials Co., Osaka, Japan) is sandwiched between carbon felt on top of solid granules of SiO (Nacalai Tesque Co., Kyoto, Japan) in an alumina crucible, and heated at 1350°C in vacuum (Figure1). The coating was identified as β -SiC by X-ray powder diffraction (XRD). Figure 2 shows SEM images of the SiC coated diamond treated at 1350°C for 1-90 min. The diamond surface is covered gradually with nanometer sized SiC by increasing the coating time. After coating for 1 min coating, a very thin SiC layer of about 10 nm is formed, and the majority of the diamond surface looks smooth. Tiny granules of SiC are deposited and aggregate with an increasing time.

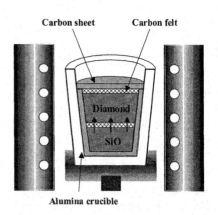

Figure 1. An assembly for the SiC coating of diamond particles.

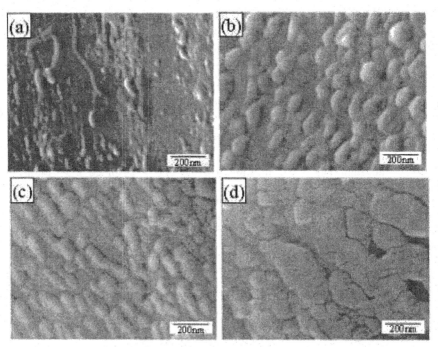

Figure 2. SEM images of the SiC coated diamond treated at 1350°C for: a) 1 min; b) 15 min; c) 30 min; and d) 90 min.

Growth Mechanism of SiC on Diamond

We analyzed the reaction process using thermogravimetric analysis (TGA) in flowing air. Figure 3 plots the mass gain ratio of the SiC coated diamond particles as functions of the coating temperature and time. The mass of the specimen increases linearly with coating time at each coating temperature. This result was used to calculate the activation energy of the reaction using the Arrhenius relation. Figure 4 shows the Arrhenius plot for the linear rate constant, k, of the coating reaction. From the slop of the line, we determined the activation energy to be 100 kJ/mol. This value agrees well with a reported value of 97 kJ/mol for the formation of SiC by the vapor phase reaction of $SiO(g)$ and $CO(g)$.

From the reaction analysis and SEM observation of the SiC coating, a coating mechanism can be proposed. The experimental result on the formation of a SiC layer as thin as 10 nm within one min of coating suggests a direct reaction occurs between SiO gas and the diamond surface to convert the diamond surface to SiC in the initial stage of the process. After forming a thin SiC layer on the diamond, it becomes difficult for the silicon to diffuse through the SiC layer to react with the diamond, or for carbon from the diamond to diffuse through SiC to react with the SiO. From morphological observations, SiC crystallites deposit and coagulate on diamond, suggesting that this process is entirely different from the earlier stage. In the later stage, the following reactions can be considered. At first, SiO gas reacts with CO gas and deposits SiC crystallites on diamond, and releases CO_2. This CO_2 reacts with the carbon felt or sheet to produce more CO. The CO reacts with the SiO. The cycle repeats itself, resulting in the deposition of SiC by the vapor phase reaction.

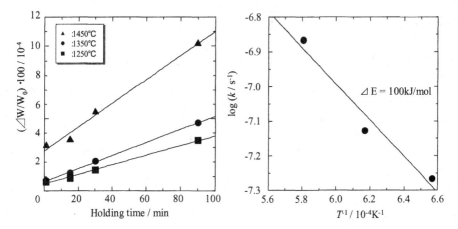

Figure 3. Mass gain with reaction between $SiO(g)$ and $C(dia)$. W_o is the initial weight of diamond before SiC coating. ΔW is the weight gain after coating.

Figure 4. Arrhenius plot of the liner rate constant, k, for the reaction of $SiO(g)$ with $C(dia)$.

SiC-COATED DIAMOND/CEMENTED CARBIDE COMPOSITES

Fabrication

The SiC-coated diamond powders produced were mixed with fine WC and cobalt powders (10 wt% cobalt, Allide Materials Co., Osaka, Japan) and sintered in a vacuum at 1250°C and 30 MPa for 5 min by PECS. A high pulsed current of 1000-3000 A was passed through the mixed powders charged into a graphite mold under uniaxial loading to complete the sintering. The diamond powders were coated at 1400°C for 30 min. The coated particles were 8-16 μm, and the diamond content was 20 vol%.

Microstructure and Mechanical Properties

Figure 5 shows a SEM micrograph of a DDC. The product is almost fully densified. The diamond particles are uniformly dispersed in the microstructure. They were not attacked by molten cobalt, and no graphitization at the interface between diamond and the matrix is observed. Some mechanical properties of DDCs are listed in Table 1, along with those of ordinary cemented carbide for comparison. The particle size of the WC is 2-3 μm, and it has 2-15 wt% Co, similar to the DDC. Additionally, the WC was sintered under high pressure.[5]

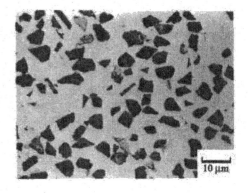

Figure 5. SEM image of a DDC.

The hardness of the DDC depends on the Co content, and ranges from 15-24 GPa. This is little higher than the hardness of the cemented carbide. The toughness measured by an indentation method under 50 N/Load is $14-20$ GPa m$^{1/2}$ which is almost two times higher than the toughness of the cemented carbide. The transverse rupture strength, measured using a three point bending test, is 1.5 to 2.5 GPa which is little lower than that of cemented carbide, but higher than that of polycrystalline sintered diamond. Young's modulus was measured using an ultrasonic method to be 60 to 69 GPa, which is higher than that

Table I. Mechanical properties of DDC, PDC, and WC-Co.

	DDC	PDC	WC-Co
Vickers hardness (GPa)	15.7~24.5	78.4~117.6	11.8~17.6
Fracture toughness (MPam$^{1/2}$)	14~20	−	8.5~12
Transverse rupture strength (GPa)	1.5~2.5	1.0~2.0	2.0~3.0
Young's modulus (GPa)	60~69	88~98	54~59
Coefficient of dynamic friction	0.05	0.05	0.2~0.3
Coefficient of thermal expansion ($\times 10^{-6}$/K)	3.7~4.1	3.1	4.8~5.3

of cemented carbide, but lower than that of the polycrystalline diamond. The coefficient of dynamic friction of the DDC is 0.05, which is very low. The friction coefficient was measured with a sliding speed of 3 m/min under 10 N load in air using a pin on disk method with an alumina ball. A low friction coefficient below 0.2 is very important for abrasion. The coefficient of thermal expansion of the DDC is about 4, which is little lower than that of WC/Co. The mechanical properties of the DDC show a large potential of tough wear resistant tools and parts.

Figure 6 shows the crack propagation in cemented carbide with and without dispersed diamond. The crack propagates straight in the cemented carbide. In contrast, in the DDC, the crack strongly deviates towards diamond particles and stops. As such, the toughness increases remarkably.

The machinability was evaluated by measuring the grinding force with a surface grinder using a #230 diamond grinding-stone. The DDC has five times better machinability than a diamond compact, and ten times higher wear-resistance.[6]

Diamond Dispersed WC/Co FGMs

It is possible to fabricate a functionally graded material of SiC coated diamond dispersed WC/Co.[6] Figure 7 shows the graded structure from a diamond dispersed cemented carbide to WC-20% Co, 30% Co, and 40% Co. The bottom part is easily welded to steel. The top part is highly toughened due to the compressive residual stress from the thermal expansion mismatch between the top and bottom compositions.

We evaluated the impact resistance of the DDC-FGM by applying an impact energy of 30 J to test pieces using an impact resistance tester. The WC/Co breaks by 15-33 impacts. The DDC shows no breakage after even 50 impacts.

Applications

Figure 8 shows an example of a commercial product made from a DDC. Zirconia rods for optical guide connectors are ground on a centerless blade between a grinding stone and a regulating wheel. The centerless blade must be long and must have high wear resistance. Large disks of DDC 150 mm in diameter and 5 mm thick have been produced by PECS. The long edge of the centerless blade is obtained by cutting this large disk.

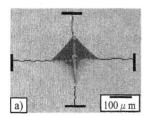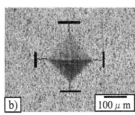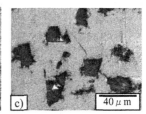

Figure 6. Optical micrograph showing crack propagation in cemented carbide: a) without diamond; and b) with diamond; and c) an SEM image showing crack deviation in a DDC.

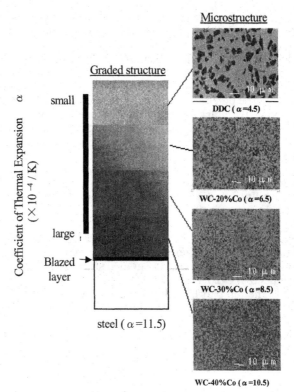

Figure 7. The graded structure from diamond dispersed cemented carbide to WC-20% Co, 30% Co, and 40% Co.

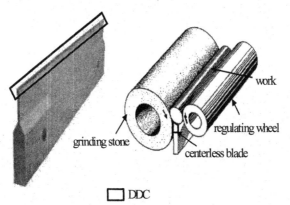

Figure 8. A centerless blade.

SUMMARY
1. Interfacial reactions between diamond and molten metal are eliminated using an easy and cost effective nano-crystalline SiC coating on the diamond particles.
2. The mechanisms to coat SiC on diamond are proposed in terms of the reactions of SiO (g) with carbon and CO.
3. Nearly fully dense SiC coated diamond dispersed cemented carbide composites were fabricated by pulsed electric current sintering (PECS).
4. SiC-coated DDCs show extremely high toughness and wear resistance.
5. A SiC coated diamond dispersed cemented carbide was commercialized as a high performance wear resistant tool.

REFERENCES
[1]N. Kinoshita, *Diamond Tool*, (1987).
[2]T. Hukaya, J. Shiraishi, and T. Nakai, *New Diamond*, **47**, 16-22 (1997).
[3]Y. Miyamoto, J. Lin, Y. Yamashita, T. Kashiwagi, O. Yamaguchi, H. Moriguchi, and A. Ikegaya, "Reactive Coating of SiC on Diamond Particles," *Ceramic Engineering and Science Proceedings*, **21** (4) 185-192 (2000).
[4]Y. Miyamoto, T. Kashiwagi, K. Hirota, O. Yamaguchi, H. Moriguchi, K. Tsuduki, and A. Ikegaya, "Fabrication of New Cemented Carbide Containing Diamond Coated with Nanometer Sized SiC Particles," *J. Am. Ceram. Soc.*, **86** (1) 73-76 (2003).
[5]K. Tsuduki and H. Moriguchi, "A Trend of Hardmetal for Wear-resistant Tools," *Journal of the Japan Society for Technology of Plasticity*, **43** (493) 5-9 (2002).
[6]H. Moriguchi, K. Tsuzuki, H. Itozaki, A. Ikegaya, K. Hagiwara, M. Takasaki, Y. Yanase, and T. Fukuhara, "Fabrication and Applications High-toughness, Highly Wear-resistant Diamond- and cBN-dispersed Cemented Carbide," *Sci Technical Review*, Sumitomo Electric Industries Ltd. **51** 121-125 (2001).

STRUCTURE OF THE INTERFACE IN SiC-BASED CERAMIC COMPOSITES FORMED BY TRIBO-OXIDATION UNDER DRY SLIDING CONDITIONS

Rolf Wäsche, Dieter Klaffke, Mathias Woydt, and Ilona Dörfel
Federal Institute for Materials Research and Testing (BAM)
D-12200 Berlin
Germany

ABSTRACT

The surface film formed at room temperature from oxidized wear debris in the wear scar of Al_2O_3/SiC-TiC-TiB_2 sliding couples has been analyzed by transmission electron microscopy. Although the sliding speed between oscillating and unidirectional sliding differs by more than a factor of 500, the basic structure and composition of the film are the same. The film is mainly an amorphous SiO_2 – B_2O_3 matrix with embedded crystalline particles. These are, on the one hand, wear debris of fractured bulk material and carbon, and on the other hand, newly formed silica, rutile, and substoichiometric Magnéli phases. The existence of these Magnéli phases, and the presence of carbon as a result of the SiC and TiC oxidation process show that the conditions in the interface during the wear process are partly reducing. This is attributed in part, to a sealing effect, and a thus hindered oxygen transport through the amorphous matrix.

INTRODUCTION

Depending on the ambient conditions, tribo-oxidation can either promote or reduce wear. For SiC, wear rate increases with decreasing relative humidity[1] or increasing temperature.[2] At a relative humidity greater than 5%, both the friction and the wear coefficient decrease.[3,4] These phenomena are associated with the tribochemical oxidation of the SiC surface under tribological load in the presence of water vapor[5-7] to form SiO_2, $Si-O_xC_y$,[8] and $(SiO_2)_n$ xH_2O. In a series of papers, it has been shown that non-oxide titanium compounds such as TiC or TiB_2, together with SiC, or in the system Si_3N_4-TiN,[9] can enhance the wear resistance significantly. The improvement is not only at room temperature, but at higher temperatures and sliding velocities as well.[8,10-14] The reason for this improvement is explained by the formation of titanium oxide phases and related Magnéli-type phases[15] due to tribo-oxidation. This forms a soft and malleable film within the tribo-interface on a hard substrate, which defines the micro-contact area. In this paper, the analysis of the structure and composition of this reaction film is presented.

EXPERIMENTAL

For the tribological experiments, an oscillating tribometer was used that works in the low displacement regime (amplitudes 5 μm to 1000 μm) with a ball-on-disk configuration. The principle of this test rig is described in more detail elsewhere.[16] In the experiments, the ball material was Al_2O_3 with a purity of 99.9%. The disk material had a composition of 50 mol% SiC 25 mol% TiC 25 mol% TiB_2. The normal force used was 10N. High speed experiments at 5 m/s sliding velocity where conducted on a continuous sliding tribometer using a wheel on disk configuration with a normal force of 20 N. After the tribology tests, the samples were characterized using transmission electron microscopy (TEM). The tribological experiments were carried out at room temperature. The TEM investigations were done with a Philips CM 20 microscope, and with a 4000 FX microscope from JEOL. The oxide layer formed by tribo oxidation in the wear scar was prepared by cutting the upper part of the bulk material carrying the wear scar. This sample was further thinned by ion milling from the backside to produce an oxide layer thin enough for the TEM investigation. This preparation procedure is schematically shown in Figure 1.

RESULTS AND DISCUSSION

Figure 2 shows the surface of a wear scar on a $SiC-TiC-TiB_2$ disk. Besides the phases of the non-oxide bulk material, tiny pads of oxide wear debris can be seen. A thin film of this oxide material covers the grains in the wear scar, and can be considered as the former sliding interface. The formation of the film has already been described in detail.[11] Figure 2 shows that the whole sliding surface is covered with a thin oxide film. The bright areas in the oxygen map show where oxide wear debris has collected. These act as oxide reservoirs during the ongoing sliding movement to continuously feed and maintain the oxygen content in the film Figure 3 shows a transmission electron micrograph (TEM) micrograph of the film obtained from a sliding experiment under oscillating sliding conditions (stroke length 0.8 mm) at a relative humidity of 5%. In this test the sliding speed was relatively low, 0.01 m/s.

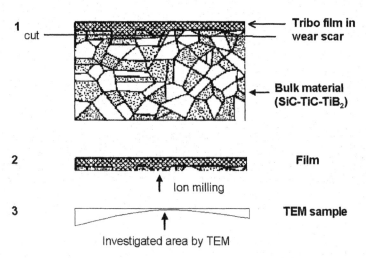

Figure 1. The steps used to prepare a sample of tribofilm from the wear scar for TEM investigation.

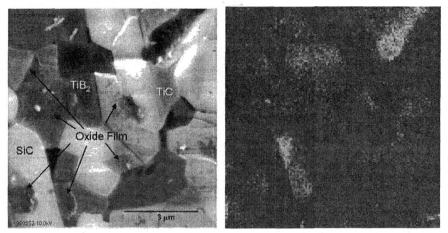

Figure 2. A SEM micrograph of the Sic-Tic-TiB_2 wear scar with the oxide film (left) formed due to tribo-oxidation of the bulk material during oscillating sliding. The corresponding oxygen map obtained by EDX (right) shows the oxide film and reservoirs.

Figures 4 - 6 show TEM micrographs of a similar film obtained from a sliding experiment under unidirectional, continuous sliding at a high sliding speed (5 m/s). The sliding distance was 5000 m. Similar to the observations at low sliding speed (Figure 3) the main part of the film is an amorphous matrix. Crystalline particles are embedded in the amorphous oxide matrix. TiO_2 (rutile, Magnéliphases), and monoclinic SiO_2 have been identified in Figures 4, 5, and 6 respectively.

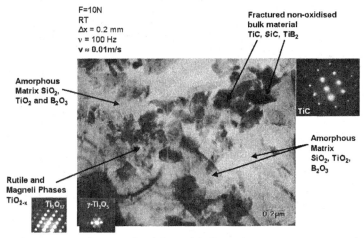

Figure 3. A TEM micrograph of the oxide layer from a wear scar produced using oscillating sliding. The maximum sliding velocity was about 0.01 m/s. The oxide matrix contains fractured bulk material, and newly formed substoichiometric Ti_nO_{2n-1} (Magnéli-phases). The matrix is mainly amorphous.

Characterization & Control of Interfaces for High Quality Advanced Materials   403

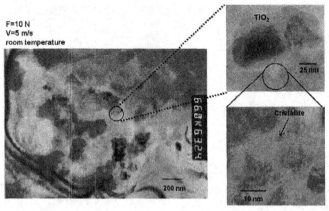

Figure 4. A TEM micrograph of the oxide layer from a wear scar using a unidirectional sliding velocity of 5 m/s. A magnified region of a TiO_2 crystallite in a partly crystalline matrix is also shown.

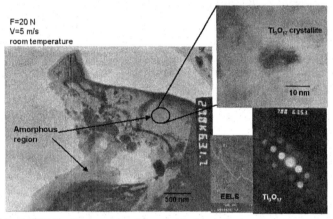

Figure 5. A TEM micrograph of the oxide layer from a wear scar produced using a unidirectional sliding velocity of 5 m/s after a sliding distance of 5000 m. The magnified area shows a Ti_9O_{17} crystallite in an amorphous matrix. The Ti_9O_{17} Magnéli phase is identified by the diffraction pattern.

The size of the fractured bulk material is relatively large, up to 1 μm. In contrast, the newly formed titanium oxide phases are on the nano meter size scale, from a few nm up to around 100 nm. The newly formed crystalline oxide phases comprise monoclinic SiO_2, TiO_2, γ-Ti_3O_5, Ti_5O_9, and Ti_9O_{17}. These were also previously analysed on self-mated (Ti, Mo)(C, N) dry sliding couples.[15]

However, the unoxidized wear debris of bulk material points to the fact that micro cracking, surface fatigue, and grain pull out are competing wear mechanisms with tribo-oxidation. In the ongoing wear process, most probably not all of the tiny non-oxide wear debris is oxidized. Hindered oxygen transport through the amorphous layer may cause an oxygen deficient condition on the sliding interface.

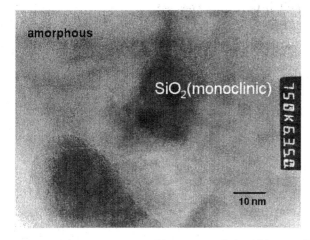 

Figure 6. A nanocrystalline monoclinic SiO_2 particle embedded in the amorphous matrix of the oxide layer from a wear scar, and the corresponding electron diffraction pattern.

The reasons for the coexistence of rutile and Magnéli phases are twofold, as follows:
1. The formation of an amorphous matrix effectively seals the wear scar and prevents oxygen vapor transport.
2. Oxygen deficiency as a consequence if (1) leads to competing oxidation reactions so that oxygen concentration gradients exist that allow for local equilibrium phases to form.

Although rutile is the most stable form of titania, the difference between the energy of formation between rutile and the Magnéli type titania is relatively small.[11] The formation of an amorphous matrix of SiO_2-B_2O_3 leads to a sealing effect due to it covering the wear scar and thus, hindering the oxygen transport. As a consequence, slight differences in the oxygen concentration or oxygen partial pressure can influence the stabilization of either form of titania. Subsequently this would lead to competing oxidation reactions in the non-oxide bulk material, and wear debris and the most stable oxides would be formed to their stoichiometric homologues at first. Titania however is a rather weak oxide in this sense.

The oxidation process of SiC is complex, and in a series of steps, different oxycarbides, SiC_xO_y, and even carbon forms.[17,18] Similar aspects are valid during the oxidation of TiC,[19,20] TiB_2, and B_2O_3 as well. The formation of carbon during the oxidation however, points strongly to the fact that part of the sliding interface remains in a reduced condition. SiO_2 and B_2O_3 are rather stable oxides compared to TiO_2, so that Magnéli phases are formed under these conditions. The formation of Magnéli-phases is most probably further supported by the high stress conditions[21] under the shearing of the upper region of the micro-asperities. This aspect, however, cannot be discussed in detail in the scope of this paper.

Figure 7 shows a schematic model of structure and composition of the interface formed based on the experimental results obtained in this investigation.

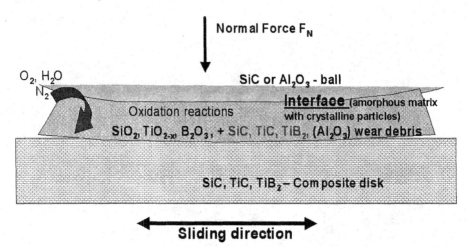

Figure 7. A model of the sliding interface structure and composition for an Al_2O_3/SiC-TiC-TiB_2 sliding couple in air.

SUMMARY AND CONCLUSIONS

Surface films formed at room temperature from oxide wear debris in the wear scar of Al_2O_3/SiC-TiC-TiB_2 sliding couples have been analyzed by transmission electron microscopy. The oxide film is mainly an amorphous SiO_2-B_2O_3 matrix in which crystalline particles are embedded. The particles are fractured bulk material and newly formed mono clinic SiO_2 and titanium oxide phases. Both rutile and Magnéli phases (Ti_9O_{17}, Ti_3O_5 and Ti_5O_9) are formed. The existence of the Magnéli phases points to the partly reducing conditions due to an oxygen deficiency during the wear process.

ACKNOWLEDGEMENT

The authors are grateful to Birgit Strauß for SEM investigations and furthermore to Christine Neumann and Manfred Hartelt for carrying out the tribological measurements. The financial support from Deutsche Forschungsgemeinschaft (DFG) under the contract Wa942/2 is greatly acknowledged.

REFERENCES

[1] D. Klaffke and K.-H. Habig, "Fretting Wear Tests of Silicon Carbide," *Wear of materials* **Vol. 1** 361-70 (1987).

[2] K.-H. Habig and M. Woydt, "Sliding Friction and Wear of Al_2O_3, ZrO_2, SiC and Si_3N_4," Proc. 5th Int. Congress on Tribology, **Vol. 3** 106-13 (1989).

[3] Y. Yamamoto and K. Okamoto, *Proceedings Jap. Int. Trib. Conf.*, Nagoya, 1455 (1990).

[4] P. Anderson, "Water-lubricated Pin-on-Disk Tests with Ceramics," *Wear* **154** 37-47(1992).

[5] S. Sasaki, "The Effects of the surrounding Atmosphere on the Friction and Wear of Alumina, Zirconia, Silicon Carbide and Silicon Nitride," *Wear* **134** 185-00, (1989).

[6]K. Mizuhara and S.M. Hsu, "Tribochemical Reactions of Oxygen and Water on Silicon Surfaces," *Proc. 19th Leeds-Lyon Conf.* 323-28 (1992).

[7]K. Mizuhara and S. M. Hsu, "Tribochemistry of Silicon Based Ceramics Under Aqueous Solutions," *Proc. 6th Int. Congress on Trib.*, Budapest, Hungary, ed. by M. Kozma, **Vol.3** 52-57 (1993).

[8]A Skopp and M. Woydt: "Ceramic and Ceramic Composite Materials with Improved Friction and Wear Properties," *Tribology Transactions*, **Vol. 38(2)** 233-42 (1995).

[9]A. Skopp, M. Woydt, and K.-H. Habig, "Tribological Behavior of Silicon Nitride Materials under Unlubricated Sliding Between 22 °C and 1000 °C," *Wear* **181-183** 571-80 (1995).

[10]R. Wäsche and D. Klaffke, "Schwingungsverschleiß an Keramischen Kompositwerkstoffen - eine Vergleichende Untersuchung," Proc. 10th Int. Colloquium Tribology, ed. by W. Bartz, Ostfildern TAE, **Vol. 3** 2371-80, (1996).

[11]R. Wäsche and D. Klaffke, "In-situ Formation of Tribologically Effective Oxide Interfaces in SiC-based Ceramics during Dry Oscillating Sliding," *Tribology Letters* **5** [2,3] 173-90 (1998).

[12]R. Wäsche and D. Klaffke, "Wear of multiphase SiC based Ceramic Composites containing free carbon," *Wear* **249** [3,4] 220-28 (2001).

[13]R. Wäsche and D. Klaffke, "High Temperature Tribological Behaviour of Particulate Composites in the System SiC-TiC-TiB$_2$ During Dry Oscillating Sliding," *Korean J. Ceram.* **5** [2] 155-61 (1999).

[14]R. Yarim, R. Wäsche, and M. Woydt, "Wear Behaviour of SiC-TiC-TiB2 Composites Sliding Against SiC up to 800°C and 2 m/s," Abstracts, Papers, Posters World Tribology Congress 2001, Vienna, Sept. 3- 7, published on CD by Austrian Tribology Society (2001).

[15]M. Woydt, A. Skopp, I. Dörfel, and K. Wittke, "Wear Engineering Oxides/Anti-wear Oxides," *Tribology Transactions* **42** 21-31 (1999).

[16]D. Klaffke, "Verschleißuntersuchungen an Ingenieurkeramischen Werkstoffen," *Tribologie und Schmierungstechnik* **34** [3] 139-47 (1987).

[17]C. Radtke, R.V. Brandao, R.P. Pezzi, J. Morais, I.J.R. Baumvol, and F.C. Stedile, "Characterisation of SiC Thermal Oxidation," *Nuclear Instruments and Methods in Physics Research*, **B 190** 579-82 (2002).

[18]D. Schmeißer, D.R. Batchelor, R.P. Mikalo, P. Hoffmann, P. and A. Lloydt-Spetz, "Oxide Growth on SiC (0001) Surfaces," *Applied Surface Science* **184** 340-45 (2001).

[19]M. Reichle, and J.J. Nickl,"Untersuchungen über die Hochtemperaturoxidation von Titankarbid," *J. Less-Common Metals* **27** 213-36 (1972).

[20]S. Shimada, "Interfacial Reaction on Oxidation of Carbides with Formation of Carbon," *Solid State Ionics* **141-142** 99-04 (2001).

[21]M.G. Blanchin, P. Faisant, C. Picard, M. Ezzo, and G. Fontaine, "Transmission Electron Microscopy Observations of Slightly Reduced Rutile," *Phys. Stat. Sol. (a)* **60** 357-64 (1980).

CONTACT DAMAGE BEHAVIOR OF TiN-COATED Si_3N_4 CERAMICS

Shin Suzuki, Junichi Tatami, Katsutoshi Komeya, Takeshi Meguro, and Akira Azushima
Yokohama National University
Hodogaya-ku,
Yokohama 240-8501, Japan

Young-Gu Kim and Do Kyung Kim
Department of Materials Science and Engineering
Korea Advanced Institute of Science and Technology
Taejon, Korea

ABSTRACT
    In this study, TiN coated silicon nitride was fabricated, and contact damage behavior was investigated with two types of indentation, including a single and a cyclic test using a spherical indenter. TiN was deposited on commercial silicon nitride ceramics by a cathodic arc ion plating method. The thickness of the TiN was varied from 1.0 to 3.5 μm according to coating time. After a single indentation, no crack was detected in the 1.0μm TiN coated specimen; however, a crack was observed in the uncoated silicon nitride at the same load. TiN coated silicon nitride is expected to have better performance than monolithic silicon nitride. The critical load for indentation cracking of TiN-coated silicon nitride decreases with increasing TiN thickness. Adhesive strength was measured by the scratch test and compared with the critical load of indentation test. A similar trend between adhesive strength and critical load is obtained for TiN thickness. Contact damage resistance is improved with the TiN coating. A 1 μm TiN-coated silicon nitride shows the highest critical load in single and cyclic indentation test.

INTRODUCTION
    Cutting and grinding tools used for machining must have high hardness and elastic modulus, low friction coefficient, and good corrosion resistance. Recently, requirements for cutting and grinding tools that enable high speed, high efficiency machining have increased. In response, tools made of ceramics, which have superior characteristics, are attracting attention and finding practical use.[1-3] Silicon nitride has been used as bearings and cutting tools because it has high hardness, high toughness, and excellent impact resistance. However, it has low abrasion resistance due to the high reactivity of silicon, its main component, with the material being cut. Therefore, coating silicon nitride with TiN, which is used to improve friction and abrasion resistance of ultra-hard metals, can effectively improve abrasion resistance and prolong service life.[4-8] Though silicon nitride is currently coated with TiN by CVD, the CVD coating requires a high temperature around 1000°C. Cracks that reduce the strength are thus generated in the coating during cooling.[9] PVD has the advantage of lower temperature processing that results in fewer cracks in the coating. However, there are few reports on TiN

coatings applied to silicon nitride using PVD, and the contact damage of silicon nitride coated with TiN has not been determined. The scratch test is usually used to evaluate thin film adhesion.[10-13] Contact damage by sphere indentation also has been extensively studied on monolithic silicon nitride and ceramic bilayered structures.[14-16] Sphere indentation is also expected to be a good evaluation tool for silicon nitride, especially for bearings. This study seeks to fabricate TiN-coated silicon nitride using arc ion plating PVD, and to evaluate the contact damage of PVD-TiN coated silicon nitride by single and cyclic indentation using a spherical indenter.

EXPERIMENTAL PROCEDURES

In this study, commercial silicon nitride sintered bodies were used as substrates. Y_2O_3, Al_2O_3, AlN, and TiO_2 were used as the sintering aids to produce the commercial Si_3N_4 substrates. TiN was coated onto these sintered bodies using the arc ion plating method, a type of PVD. During TiN coating, the film thickness was controlled by changing the coating time from 7.5 to 30 min. After coating, X-ray diffraction (XRD) was used to confirm the TiN phase on the coating. The TiN film thickness was measured by SEM. The adhesion between the TiN film and the Si_3N_4 substrate was evaluated using a continuous scratch test machine.

Single indentation (Pc) tests were used to evaluate contact damage and critical load for cracking. A WC spherical indenter with a radius of 1.98 mm was used. Cyclic indentation was conducted to evaluate fatigue strength, which is defined as the initial load for ring cracking on the surface. Indentation loads in a range from 1500 to 3000 N were used. Fatigue strength was estimated from 10^3 to 10^4 in cyclic indentation tests with a press fit load of 2000 N. The specimens subjected to the indenter tests were examined for ring cracks on the surface using an optical microscope.

RESULTS AND DISCUSSION

The film coating was confirmed to be TiN by XRD. The TiN film thickness is linearly proportional to deposition time. Figure 1 shows the adhesive strength as a function of TiN film thickness. The adhesion decreases as the thickness of the

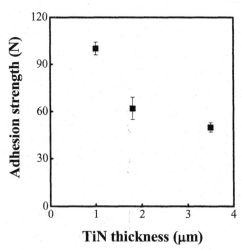

Figure 1. Adhesion strength of TiN coated silicon nitride as a function of TiN film thickness.

film increases, and the maximum adhesion is obtained for a film thickness of 1.0 µm. The coefficient of thermal expansion (CTE) of silicon nitride is

3.5×10^{-6}/K, and the CTE of TiN is 9.4×10^{-6}/K. Because the TiN has a higher thermal expansion coefficient than silicon nitride, the TiN film is in tension after coating. The CTE mismatch is always present, regardless of the film thickness, so there is more tensile stress as film thickness increases. Additionally, with a thicker film, the film becomes more rigid such that the CTE mismatch becomes more important. Figure 2 shows the surface damage on monolithic silicon nitride and on TiN-coated silicon nitride after a single indentation at a load of 2500 N. No cracks are observed on the specimen with a 1.0 µm TiN thick film. A ring crack is observed on a 1.8 µm TiN coated specimen. Double ring cracks are present on the 3.5 µm TiN-coated silicon nitride. The contact damage by sphere indentation is different depending on TiN coating thickness. The 1 µm TiN-coated specimen shows the best damage resistant behavior.

Figure 3 shows the relationships between TiN film thickness and the critical load for crack generation (P_c). This plot confirms that the critical load for cracking decreases as film thickness increases. The 1 µm TiN coated specimen shows the highest value, and TiN-coated silicon nitride has a higher P_c than Si_3N_4 alone. The compressive stress in the silicon nitride due to the thermal expansion difference between the TiN film and the silicon nitride may increase the critical load for cracking. The tensile stress in the TiN film increases as the film thickness increases, resulting in a decrease in critical load for cracking.

Figure 4 shows optical micrographs of the surface of specimens after cyclic indentation. 10^3 or 10^4 cycles were applied at a load of 2500 N. It becomes easier to generate cracks as the thickness of a TiN film increases, and the critical load for cracking decreases after a single indentation.

The contact damage behavior of TiN-coated silicon nitride after a single indentation and cyclic loading can be explained by the difference in thermal

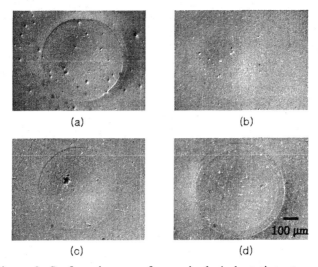

(a)　　　　　　　　　(b)

(c)　　　　　　　　　(d)

Figure 2. Surface damage after a single indentation at a load of 2500 N on: a) monolithic Si_3N_4; and b) 1.0 µm; c) 1.8 µm; and d) 3.5 µm TiN-coated silicon nitride.

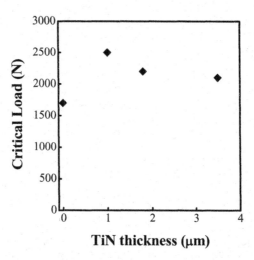

Figure 3. Critical load for cracking during single sphere indentation as a function of TiN coating thickness on Si_3N_4.

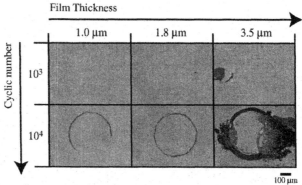

Figure 4. Surface damage after 10^3 or 10^4 cycles of cyclic indentation at a load of 2500 N on 1.0 μm, 1.8 μm, and 3.5 μm TiN coated silicon nitride.

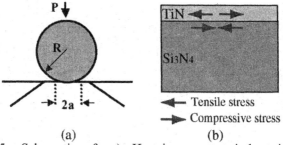

(a)   (b)

Figure 5. Schematic of: a) Hertzian contact indentation; and b) the residual stress near the interface of TiN coated silicon nitride.

expansion between the TiN film and the silicon nitride. Figure 5 shows a schematic of the Hertzian contact indentation and the residual stress near the TiN/Si$_3$N$_4$ interface in the TiN-coated silicon nitride. Tensile stress is applied in the TiN film and compressive stress in the silicon nitride substrate. Residual stress is produced inevitably during the cooling procedure in PVD coating. It is usually measured by XRD or by a laser scan method.[17,18] In this study, residual stress, σ_r, can be calculated from the force balance between the coating and substrate bilayers from their thermal expansion mismatch.[17]

$$\sigma_r = (\alpha_c - \alpha_s) \cdot E_c \cdot \Delta T / \{(1 - \upsilon_c) + (1 - \upsilon_s)(E_c / E_s)(2d_c / d_s)\} \quad (1)$$

where α is the thermal expansion coefficient, E is Young's modulus, ν is Poisson's ratio, and d_c is coating thickness. $(\alpha_c - \alpha_s) = 5.9 \times 10^{-6}$, $E_c = 600$ GPa, $\Delta T = 400$, $\nu_c = 0.25$, $\nu_s = 0.27$, $d_c = 1 \times 10^{-3}$mm, $d_s = 3$mm, $E_s = 300$ GPa, The estimated σ_r is 1.88 GPa, approximately.

From the Hertzian contact model, the maximum tensile stress at the surface with normal load P is[18]

$$\sigma_m = \frac{(1 - 2\upsilon) \cdot P}{2\pi a^2} \cdot \left(\frac{a}{r}\right)^2 \quad (2)$$

$$a^3 = \frac{4kPR}{3E} \quad (3)$$

where P is the applied load, R is the radius of the spherical indenter, a is the contact radius, and $k = [(1 - \upsilon^2) + (1 - \upsilon^{*2}) \cdot E / E^*]$, where E^*, ν^* is Young's modulus and Poisson's ratio of the indentor, respectively.

The Critical load for cracking in the monolithic silicon nitride is 1700 N. For the 1 μm TiN- coated specimen, it is 2500 N. When a load of 1700 N is applied to the monolithic silicon nitride surface, the maximum stress is 1.64 GPa. A 2500 N load induces 1.86 GPa stress from the Hertzian contact model. The contact radius, a, is approximately 275 μm when a load of 1700 N is applied to monolithic silicon nitride using a sphere with a radius of 1.98 mm. The TiN coating thickness is much smaller than the contact radius. The difference in the maximum stress level for an indentation load of 2500 N and 1700 N is about 0.22 GPa. By

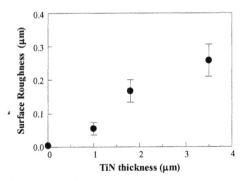

Figure 6. Surface roughness of TiN coated silicon nitride as a function of TiN thickness.

comparison, the calculated residual stress from the thermal expansion mismatch between TiN coated Si_3N_4 (1.86 GPa) is higher than the value from the Hertzian indentation (0.22 GPa). Stress relaxation during cooling usually results in cracking. However cracking of thin films with residual tensile stresses has been the focus of intensive research.[21-22] Residual tensile stresses in a thin film can cause film cracking when the film thickness is larger than a critical value. The reason why the adhesion strength decreases with TiN thickness seems to be caused by TiN surface roughness. As shown in Figure 6, the surface roughness increases with increasing film thickness. The decrease in the critical load for a thick TiN coating is considered to be due to the increase in the local stress on the rougher surface of the thicker TiN-coated specimen.

CONCLUSION

TiN was coated on silicon nitride to improve contact damage resistance. The critical cracking load in static and cyclic loading was investigated. Adhesion strength determined by a scratch test shows similar trends with single and cyclic indentation. A simple sphere indentation test can be a useful tool to evaluate a brittle thin film. It was clearly demonstrated that TiN-coated silicon nitride shows higher critical load than monolithic silicon nitride. Thus the contact damage resistance of silicon nitride ceramics is improved with a TiN coating. The highest critical load for cracking for a single indentation and for cyclic loading is observed in a 1.0 μm TiN-coated specimen.

REFERENCES
[1]M. Leoni, P. Scardi, S. Rossi, L. Fedrizzi, and Y. Massiani, "(Ti,Cr)N and Ti/TiN PVD Coatings on 304 stainless steel substrates: Texture and residual stress," *Thin Solid Films*, **345** 263-69 (1999).
[2]S.J. Bull, D.G. Bhat, and M.H. Stasia, "Properties and Performance of commercial TiCN coatings", *Surface and Coatings Technology*, **163-164** 499-06 (2003).
[3]D.M. Mattox, "Ion Plating Past, Present and Future", *Surface Coatings and Technology*, **133-134** 517-21 (2000).
[4]P.J. Burnett, and D.S. Rickerby, "The Scratch adhesion test: an Elastic-plastic indentation analysis," *Thin Solid Films*, **157** 233-54 (1998).
[5]P.A. Steinmann, Y. Tardy and H.E. Hintermann, "Adhesion Testing by the Scratch Test Method: the Influence of Intrinsic and Extrinsic Parameters on the Critical Load," *Thin Solid Films*, **154** 333-49 (1987).
[6]W.S. Baek, S.C. Kwon, J.Y. Lee, S.R. Lee, J.J. Rha, and K.S. Nam, "The Effect of Ti Ion Bombardment on the Interfacial Structure Between TiN and Iron Nitride," *Thin Solid Films*, **323** 146-52 (1998).
[7]M.T. Laugier, "An Energy Approach to the Adhesion of Coatings using the Scratch Test", *Thin Solid Films*, **117** 243-49 (1984).
[8]T. Hayashi, A. Matsumoto, M. Muramatsu, Y. Takahashi, and K. Yamaguchi, "Synthesis of TiN thin Films Prepared by Dynamic Ion Mixing Technique and their Mechanical Properties," *Thin Solid Films*, **349** 199-04 (1999).
[9]M. Stoiber, E. Badisch, C. Lugmair, and C. Mitterer, "Low Friction TiN Coatings Deposited by PACVD", *Surface and Coatings Technology*, **163-164** 451-56 (2003).
[10]A. Thobor, C. Rousselot, C. Clement, J. Takadoum, N. Martin, R. Sanjines, and F. Levy, "Enhancement of Mechanical Properties of TiN/AlN Multilayers by Modifying the Number and the Quality of Interfaces", *Surface and Coatings Technology*, **124** 210-21 (2000).
[11]N.X. Randall, G. Favaro and C.H. Frankel, "The Effect of Intrinsic Parameters on the Critical Load as Measured with the Scratch Test Method,"

Surface and Coatings Technology, **137** 146-51 (2001).

[12]A. Rodrigo and H. Ichimura, "Analytical Correlation of Hardenss and Scratch Adhesion for Hard Films," *Surface and Coatings Technology*, **148** 8-17 (2001).

[13]H. Ollendorf and D. Schneider, "A Comparative Study of Adhesion Test Methods for Hard Coatings," *Surface and Coatings Technology*, **113** 86-02 (1999).

[14]S.K. Lee, K.S. Lee, B.R. Lawan, and D.K. Kim, "Effect of Starting Powder on Damage Resistance of Silicon Nitrides," *J. Am. Ceram. Soc* **81** [8] 2061-70 (1998).

[15]K.S. Lee, S.K. Lee, B.R. Lawan, and D.K. Kim, "Contact Damage and Strength Degradation in Brittle/Quasi-plastic Silicon Nitride Bilayers", *J. Am. Ceram. Soc.*, **81** [9] 2394-04 (1998).

[16]D.K. Kim, Y.G. Jung, and I.M. Peterson, "Cyclic Fatigue of Intrinsically Brittle Ceramics in Contact with Sphere", *Acta Mater.*, **47** [18] 4711-25 (1999).

[17]F. Vaz, L. Rebouta, Ph. Goudeau, J.P. Riviere, E. Schaffer, G. Kleer, and M. Bodmann, "Residual Stress States in Sputtered Ti1-xSixNy Films," *Thin Solid Films*, **402** 195-02 (2002).

[18]N.B. Thomsen, A. Horsewell, K.S. Mogensen, S.S. Eskildsen, C. Mathiasen, and J. Bøttiger, "Residual Stress Determination in PECVD TiN Coatings by X-ray Diffraction: A Parametric Study", *Thin Solid Films*, **333** 50-59 (1998).

[19]H. Wang and X. Hu, "Surface Properties of Ceramic Laminates Fabricated by Die Pressing", *J. Am. Ceram. Soc.*, **79** [2] 553-56 (1996).

[20]B.R. Lawn, "Indentation of Ceramics with Spheres: A Century after Hertz", *J. Am. Ceram. Soc.*, **81** [8], 1977-94 (1998).

[21]M.H. Zhao, R. Fu, D. Lu and T.Y. Zhang, "Critical Thickness for Cracking of Pb(Zr0.53Ti0.47)O3 thin films deposited on Pt/Ti/Si(100) substrates", *Acta Mater.*, **50** 4241-54 (2002).

[22]Z.C. Xia and J.W. Hutchinson, "Crack Patterns in Thin Films", *J. Mechanics and Physics of Solids*, **48** 1107-31 (2000).

EFFECTS OF INTERFACE CONTROL ON THE FORMATION AND PROPERTIES OF CARBON NANOTUBES COMPOSITES

Lian Gao[*], Linqin Jiang, and Jing Sun
State Key Lab of High Performance Ceramics
Shanghai Institute of Ceramics
Chinese Academy of Sciences
1295 Dingxi Road
Shanghai 200050, P.R. China
[*]E-mail: liangaoc@online.sh.cn

ABSTRACT

Carbon nanotubes (CNTs)/alumina composites were formed by simple colloidal processing, and CNTs/magnetite composites were prepared using in situ solvothermal synthesis. Additionally, gold nanoparticles were selectively attached onto CNTs through various modifications of the nanotubes. The surface properties of CNTs were changed with a dispersant or by heat treatment with ammonia. Through the electrostatic interaction between components, CNTs were successfully coated with particles of oxide ceramic or metal. The addition of only 0.1 wt% CNTs in the alumina composite increases the fracture toughness from 3.7 to 4.9 MPa·m$^{1/2}$; this is partly due to a crack bridging effect between the CNTs and the alumina matrix, and the pullout of the CNTs from the matrix. CNTs/magnetite nanocomposites show improved electrical conductivity, owing to the percolation effect of the CNTs, and there is a good dispersion of the CNTs in the matrix.

INTRODUCTION

Since the discovery of CNTs in 1991 by Iijima,[1] a worldwide research effort has been triggered to develop their unique structural, electrical, mechanical, electro-mechanical, and chemical properties.[2-4] At present, CNTs are available in kilogram quantities at a price that is acceptable compared to that at the initial stage of their discovery. It costs about 2-20 U.S. dollars per gram for multiwalled carbon nanotubes (MWNTs). This makes the large-scale production of CNT composites possible and practical. Theoretical and experimental results have shown extremely high elastic modulus as high as 1 TPa, high tensile strength of 200 GPa, and superior electric-current-carrying capacity 1000 times higher than copper wires.[5,6] Therefore, it can be expected that CNTs can be used as strong, light, and tough fibers for nanocomposite structures, and as additives to improve the electrical properties of nanocomposite materials.[7]

Presently, two methods are commonly used to incorporate nanotubes into ceramics. One method involves mixing ceramic nanopowders directly with CNTs, and hot pressing.[8] A 10% improvement in the strength and fracture toughness has been reported with only 10 wt% CNTs as compared to SiC monolithic ceramics. In contrast, Peigney et al.[9,10] reported a novel catalytic route for the in situ formation of a composite powder based on alumina and SWNTs or MWNTs. The microstructure and mechanical properties of dense CNTs-Fe-Al$_2$O$_3$ materials prepared by hot-pressing these nanocomposite powders have been investigated.[11,12] It has been shown that CNTs confer electrical conductivity to ceramic-matrix composites that retain the mechanical properties of the ceramics. So in situ synthesis may be an effective way to produce homogeneous composites.

The homogeneous dispersion of CNTs throughout the matrix is critical to promote the effective utilization of nanotubes in composite applications. Moreover, good interfacial bonding across the CNT-matrix is significant for improving the mechanical and electrical properties of ceramic composites. Many researchers have reported recent progress to optimize the dispersion of CNTs in the matrix, and to strengthen the interfacial bonding between the nanotubes and the matrix. Hwang et al.[13] synthesized silicon glass rods as additives to reinforce ceramics using surfactant-CNTs co-micelles as templates. With the addition of ~6 wt% CNTs, ~100% enhancement of hardness was observed. Jin et al.[14] used polymer (polyvinylidene fluoride) to promote the dispersion of MWNTs, and to serve as a bridge to increase the interfacial adhesion between MWNTs and polymer (polymethyl methylacrylate). This resulted in an obvious increase in the storage moduli of the composites at low temperatures. Above all, a colloidal heterocoagulation process, using a dispersant to disperse the CNTs and interlink the components, or in-situ synthesis based on CNTs and a ceramic matrix could be potential ways to produce homogeneous composites.

In this paper, we summarize previous work from our group on the synthesis and properties of CNTs/composites.[15-19] Information related to the interface control and the effect on the properties of CNT-based composites will be discussed.

The surface chemistry of CNTs heat-treated in an ammonia atmosphere was examined.[15] Four different dispersants; poly acrylic acid (PAA), polyethyleneimine (PEI), citric acid (CA) and sodium dodecyl sulfate (SDS) were used to disperse the CNTs. Then the inherent surface potentials of the CNTs could be altered on demand.[15,16,18,19] Ceramic powders of alumina were successfully bonded with the nanotubes by a colloidal heterocoagulation process, which not only improved the homogeneous distribution of the CNTs in the ceramic matrix, but also made the bonding between the two phases tighter after sintering. The mechanical properties of CNTs-Al$_2$O$_3$ were investigated.[15] The CNTs modified by various treatments are also useful for selective attachment of gold nanoparticles.[18] Attaching metal nanoparticles to nanotube sidewalls is of interest to produce nanotube/nanoparticle hybrid materials with useful properties.[20] Moreover, a novel, simple, solvothermal method was successfully employed to synthesize CNTs-Fe$_3$O$_4$ nanocomposites from the precursor of Fe(III)-urea coordination compound and CNTs in situ. The electrical properties of the nanocomposites were studied.[17]

EXPERIMENTAL

MWNTs prepared by the catalytic decomposition of CH$_4$[21] were kindly provided by Chengdu Institute of Organic Chemistry, Chinese Academy of Sciences. Figure 1 shows a TEM micrograph of pristine MWNTs, which have an inner diameter around 10 nm, and which gave lengths ranging from several to tens

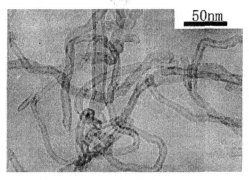

Figure 1. TEM micrograph of pristine MWNTs.

of micrometers.

The CNTs were placed inside a hollow tube within a laboratory furnace and heated to 600°C for 3 h in NH_3 gas flowing at 1 L/min. Then the samples were cooled to a room temperature. Treatment with NH_3 would effectively change the surface properties of the CNTs.[15,18] Four dispersants were used to disperse the CNTs into aqueous suspensions: a) PAA (Polymer Sciences, PA) with a molecular weight of 50000; b) PEI (BDH Laboratory) with a molecular weight of 50000; c) CA (Shanghai Chemical Co.) with a molecular weight of 210.1; and d) SDS (Shanghai Chemicals Co.) with a molecular weight of 288.4.

Commercial α-Al_2O_3 powder with a particle diameter of 30~100 nm (Taimei Chemical Co., Ltd., Japan) was used. In a typical experiment,[15] a certain amount of NH_3-treated CNTs was put into a 0.1 wt% solution containing PEI dispersant. Alumina (0.04 g) was dispersed in 100 mL of deionized water, and then 300 mg/L PAA was added to this very dilute alumina suspension. Sodium hydroxide was used to adjust the pH. PAA was added drop wise to the vigorously stirred as-prepared CNT suspension with PE, and ultrasonicated for 30 min. The coated CNTs collected from the mixed suspension were subsequently added to a concentrated suspension of about 50 wt% alumina in ethanol. Ultimately, the content of CNTs was only 0.1 wt% of the alumina content. Further drying and grinding produced the CNTs- Al_2O_3 composite powder.

Selective attachment of gold nanoparticles on the modified CNTs was achieved via the adsorption of dispersants on the nanotubes, followed by in situ reduction of the Au nanoparticles.[18] 0.01 g of PEI or CA-coated CNTs were added to 50 mL of 0.1 g citric acid aqueous solution. All of the CNT suspensions were ultrasonicated for 3-5 min. Then the prepared 50 mL solution of 0.2 g $HAuCl_4$ was added drop wise to the vigorously stirred, as-prepared CNT suspension at 70°C. After adding the $HAuCl_4$ solution, vigorous stirring was continued for 1 h. The mixed suspension was kept at 80°C for 8 h. The composite powder was obtained after further filtering and drying

A novel and simple solvothermal method was used to synthesize CNT-magnetite nanocomposites from a Fe(III)-urea complex precursor.[17] 0.2 g of CNTs were dispersed in 10 mL of ethylenediamine ($C_2H_8N_2$, >99.0%) by ultrasonication. Then the suspension and 2.0 g of the complex were mixed together with 20 mL of ethylenediamine in a 40 mL Teflon-lined autoclave. The autoclave was maintained at 200°C for 50 h and then cooled to room temperature. A black precipitate was obtained by filtration. After further drying and grinding, CNTs-magnetite nanocomposite powders were obtained.

Zeta-potential measurements (Zeta plus, Brookhaven, USA) were made to evaluate the surface properties of pristine, treated, and coated CNTs. Transmission electron microscopy (TEM, JEM 2010, JEOL, Japan) was used to observe the microstructure of the CNTs-oxide composites. The single-phase Al_2O_3 powder and the 0.1 wt% CNTs-Al_2O_3 composite powder were sintered by spark plasma sintering (Dr Sinter 1050, Sumitomo Coal Mining Co., Ltd.) in a graphite die at 1300°C, with a pressure of 50 MPa for 5 min in an Ar atmosphere. The CNT/Fe_3O_4 composite powders were pressed into $\Phi20 \times 1$ mm pellets by applying a 20 N force, and then the pellets were sintered at 500°C for 2 h in Ar. The fracture surfaces of the composites were observed by a scanning electron microscope (SEM, JSM-6700F, JEOL, Japan). Densities of the sintered composites were measured using the Archimedes technique. The hardness of the composites was measured using a Vickers hardness tester (AVK-A) with an applied load of 10 kg on the surface for 10 s. The fracture toughness was calculated by the Evans & Charles equation ($K_{IC} = 0.00824 \, P/C^{1.5}$), where P is the applied load in newtons and C is the crack length in meters.[22] The electrical conductivities of the composites were measured using a Hall effect measurement system (HL5500PC, U.K.).

RESULTS AND DISCUSSION

The zeta potential of the pristine CNTs, NH_3-treated CNTs, PEI, CA, PAA, and SDS-coated CNTs was measured as function of pH in 1 mM NaCl solutions. The results are shown in Figure 2. The isoelectric point (pH_{iep}) of pristine CNTs is at pH 3.8, but moves to 6.0 after heat treatment at 600°C in NH_3 for 3 h.

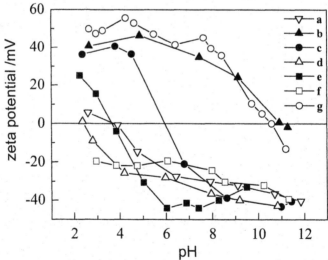

Figure 2. Zeta potential with pH of: a) pristine CNTs; b) PEI-coated CNTs; c) NH_3-treated CNTs; d) CA-coated CNTs; e) PAA-coated CNTs; f) SDS-coated CNTs; and g) NH_3-treated CNTs with PEI.

According to much fundamental research on the surface chemistry of activated carbon,[23,24] treatment with ammonia removes the acid oxygen-containing functional groups. Furthermore, it may also introduce basic nitrogen-containing groups (e.g. amine) onto the carbon surface.[25] Very similar results are obtained for CNTs heat treated in ammonia. The addition of PEI to NH_3-treated CNTs makes a more positive zeta potential value, which leads to a better dispersion after the ultrasonication (Figure 2g).

Four different dispersants, PEI, PAA, CA, and SDS were used to alter the inherent surface potentials of the pristine CNTs as shown in Figure 2. The addition of cationic PEI makes the zeta potential of the CNTs more positive, with the isoelectric point moving from 3.8 to 11.0. The three anionic dispersants make the isoelectric point move to a lower pH value. The changes in zeta potential are easily explained by the specific adsorption of the dispersants.[16]

A colloidal heterocoagulation method was adopted to coat the surface of CNTs with alumina powder.[15] It is well known that when two sols of opposite charge are mixed, mutual coagulation may occur. If the proper pH range is selected, ceramic particles will be adsorbed onto the CNTs by electrostatic forces. A previous study[26] demonstrated that the specific adsorption of PAA leads to a more negative charge on the alumina surface. Once the PAA-treated alumina suspension is mixed with the dilute solution of NH_3-treated CNTs with PEI, particles of alumina will bind to the CNTs because of the strong electrostatic attractive force between each other. The composite powders containing 0.1 wt% CNTs and the single-phase alumina were fully sintered by SPS in 5 min. at 1300°C. The hardness of the alumina is 16.9 GPa, and that of the CNT reinforced composite is 17.6 GPa.[14] The addition of 0.1 wt% CNTs to the alumina increases the fracture toughness[15] by about 31% from 3.7 to 4.9 MPa·m$^{1/2}$. There is an effective and significant improvement in the mechanical properties of the CNT composites prepared using the simple colloidal processing method as compared to previous results.[27]

Figure 3 shows SEM micrographs of the fracture surface of a CNT-alumina composite.[15] Carbon nanotubes are located inside alumina grains, or pulled out

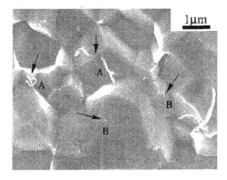

Figure 3. SEM micrograph of a fracture surface of the 0.1 wt% CNT/Al_2O_3 composite (Several CNTs ruptured in the fracture surface, as pointed out by the arrows in the A areas, and several CNTs pulled out from the matrix, as shown by the arrows in the B areas). Reprinted with permission from (15). Copyright (2003) American Chemical Society.

from the alumina matrix. Typical examples of interactions between cracks and the nanotubes are shown in Figure 4.[15] Some CNTs align perpendicular to the crack direction and bridge the crack (Figure 4A), or some CNTs are trapped in the crack (Figure 4B). Thus, it can be concluded that pullout of CNTs at the interface is a possible mechanism contributing to the improvement in the fracture toughness. Figure 5 shows the selective attachment of gold nanoparticles to the CA- or PEI-coated CNTs. The adsorption of the CA dispersant produces acidic groups on the CNTs, and the PEI generates basic nitrogen-containing groups around the nanotubes. The electrostatic interaction on the outer layer of the CNTs, and the subsequent in situ reduction of $HAuCl_4$ produces the Au nanoparticles. The surface modification of the CNTs is critical to anchor the metal particles, and it results in a strong interaction between the particles and the nanotubes.

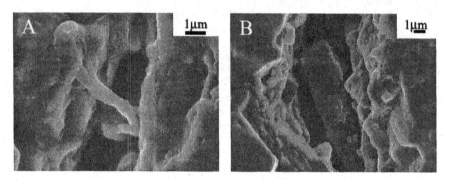

Figure 4. SEM micrographs showing typical interactions between cracks and CNTs in a CNT/alumina composite. Reprinted with permission from (15). Copyright (2003) American Chemical Society.

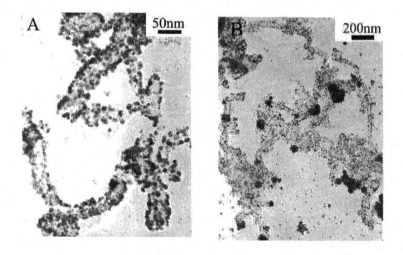

Figure 5. TEM micrograph of gold nanoparticles attached to: A) CA-coated CNTs; and B) PEI-coated CNTs (right).

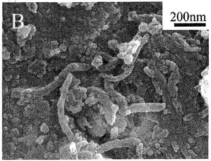

Figure 6. SEM micrograph of a fracture surface of: A) a magnetite composite without CNTs; and B) a magnetite-CNTs composite.

In the magnetite-CNT composite, the electrical conductivity (σ) is 2.5 S cm^{-1}. In the composite without CNTs, it is only 1.9 S cm^{-1}.[17] Figure 6 shows the fracture surface of the composite with and without CNTs. It is obvious that the CNTs introduce conducting pathways in the magnetite, viz., the percolation of the CNTs.[27,28] The relative densities are approximately identical in the two composites.[17] Thus, we can conclude that the influence of the conducting CNT pathways is the main factor that improves the electrical conductivity of the composite.

CONCLUSIONS

A colloidal processing route and an in situ solvothermal synthesis process have been shown to be effective ways to improve the mechanical and electrical properties of CNTs/oxide composites. By adjusting the surface properties of the oxide powder and of the CNTs, it is feasible to produce strong cohesion between the two materials via electrostatic interaction. Future work, including a detailed investigation on the adhesion between CNTs and the matrix, the optimization of the CNT concentration in composites for different applications, and the preparation of other CNTs/oxide composites, is in progress.

REFERENCES
[1]S. Iijima, "Helical Microtubules of Graphitic Carbon," *Nature*, **354** [6348] 56-8 (1991).
[2]W.A. De Heer, A. Chatelain, and D. Ugarte, "A Carbon Nanotube Field-Emission Electron Source," *Science*, **270** [5239] 1179-80 (1995).
[3]R.F. Service, "Materials Science: Mixing Nanotubes Structures to Make a Tiny Switch," *Science*, **271** [5253] 1232 (1996).
[4]P. Ball, "The Perfect Nanotube," *Nature*, **382** [6586] 207-8(1996).
[5]R.S. Ruoff, J. Tersoff, D.C. Lorents, S. Subramoney, and B. Chan, "Radial Deformation of Carbon Nanotubes by Van Der Waals Forces," *Nature*, **364** [6431-8] 514-6 (1993).
[6]P.G. Collins and P. Avouris, "Nanotube for Electronics," *Sci. Am*, **283** [6] 62-3 (2000).
[7]C.N.R. Rao, B.C. Satishkumar, A. Govindaraj, and M. Nath, "Nanotubes,"

ChemPhysChem, **2** [2] 78-105 (2001).

[8]R.Z. Ma, J. Wu, B.Q. Wei, J. Liang and D.H. Wu, "Processing and Properties of Carbon Nanotube-Babi-Sic Ceramic," *J Mater Sci*, **33** [21] 5243-6 (1998).

[9]A. Peigney, Ch. Laurent, O. Dumortier, and A. Rousset, "Carbon Nanotube-Fe-Alumina Nanocomposites. Part I: Influence of the Fe Content on the Synthesis of Powders," *J Eur Ceram Soc*, **18** [14] 1995-2004 (1998).

[10]A. Peigney, Ch. Laurent, F. Dobigeon and A. Rousset, "Carbon Nanotubes Grown In Situ by a Novel Catalytic Method ," *J Mater Res*, **12** [3] 613-5 (1997).

[11]A. Peigney, Ch. Laurent, E. Flahaut, and A. Rousset, "Carbon Nanotubes in Novel Ceramic Matrix Nanocomposites," *Ceram Inter*, **26** [6] 667-83 (2000).

[12]Ch. Laurent, A. Peigney, O. Dumortier, and A. Rousset, "Carbon Nanotubes-Fe-Alumina Nanocomposites. Part II: Microstructure and Mechanical Properties of the Hot-Pressed Composites," *J Eur Ceram Soc*, **18** [14] 2005-13 (1998).

[13]G.L. Hwang and K.C. Hwang, "Carbon Nanotube Reinforced Ceramics," *J Mater Chem*, **11** [6] 1722-5 (2001).

[14]Z. Jin, K.P. Pramoda, S.H. Goh, and G. Xu, "Poly(vinylidene fluoride)-Assisted Melt-Blending of Multi-Walled Carbon Nanotube/Poly(methyl methacrylate) Composites," *Mater Res Bull*, **37** [2] 271-8 (2002).

[15]J. Sun, L. Gao, and W. Li, "Colloidal Processing of Carbon Nanotube/Alumina Composites," *Chem Mater*, **14** [12] 5169-72 (2002).

[16]J. Sun and L. Gao, "Development of a Dispersion Process for Carbon Nanotubes in Ceramic Matrix by Heterocoagulation," *Carbon*, **41** [5] 1063-8 (2003).

[17]L.Q. Jiang and L. Gao, "Carbon Nanotube-Magnetite Nanocomposites from Solvothermal Processes: Formation, Characterization, and Enhanced Electrical Conductivity," *Chem Mater*, **15** [14] 2848-53 (2003).

[18]L.Q. Jiang and L. Gao, "Modified Carbon Nanotubes: an Effective Way to Selective Attachment of Gold Nanoparticles," *Carbon*, **41** [15] 2923-2929 (2003).

[19]L.Q. Jiang and L. Gao, "Production of Aqueous Colloidal Dispersions of Carbon Nanotubes," *J Colloid Inter Sci*, **260** [1] 89-94 (2003).

[20]J. Kong, M. Chapline, and H. Dai, "Functional Carbon Nanotubes for Molecular Hydrogen Sensors," *Adv Mater*, **13** [18] 1384-6 (2001).

[21]Q. Liang, B.C. Liu, S.H. Tang, Z.J. Li, Q. Li, L.Z. Gao, B.L. Zhang, and Z.L. Yu, "Carbon Nanotubes Growth Catalyzed By La_2NiO_4," *Acta Chim Sinica*, **58** [11] 1336-9 (2000).

[22]G.R. Anstis, P. Chantikul, B.R. Lawn, and D.B. Marshall, "A Critical Evaluation of Indentation Techniques for Measuring Fracture Toughness: I, Direct Crack Measurements," *J. Am. Ceram. Soc.* **64** [9] 533 (1981).

[23]R.J.J. Jansen and H.V. Bekkum, "Amination and Ammoxidation of Activated Carbons," *Carbon*, **32**[8] 1507-16 (1994).

[24]J.A. Menendez, J. Philips, B. Xia, and L.R. Radovic, "On the Modification and Characterization of Chemical Surface Properties of Activated Carbon: in the Search of Carbon with Stable Basic Properties," *Langmuir*, **12** [18] 4404-10 (1996).

[25]S. Biniak, G. Szymanski, J. Siedlewski, and A. Swiatkowski, "The Characterization of Activated Carbons with Oxygen and Nitrogen Surface Groups," *Carbon*, **35**[12] 1799-810 (1997).

[26]J. Cesarano III, and I.A. Aksay, "Stability of Aqueous α-Al_2O_3 Suspensions with Poly(methacrylic acid) Polyelectrolyte," *J Am Ceram Soc*, **71** [4] 250-5 (1988).

[27]E. Flahaut, A. Peigney, Ch, Laurent, Ch, Marlière, F. Chastel, and A. Rousset, "Carbon Nanotube-Metal-Oxide Nanocomposites: Microstructure, Electrical

Conductivity and Mechanical Properties," *Acta Mater*, **48** [14] 3803-12 (2000).

[28]E. Kymakis, I. Alexandou, and G.A.J. Amaratunga, "Single-Walled Carbon Nanotube-Polymer Composites: Electrical, Optical and Structural Investigation," *Synth Met*, **127** [1-3] 59-62 (2002).

COMPOSITE POWDERS SYNTHESIS BY SPRAY PYROLYSIS

S. Ohara, M. Umetsu, S. Takami,
and T. Adschiri
Institute of Multidisciplinary Research
for Advanced Materials
Tohoku University
2-1-1 Katahira, Aoba-ku
Sendai 980-8577, Japan

M. Itagaki and J.-H. Lee
Japan Fine Ceramic Center
2-4-1 Mutsuno, Atsuta-ku
Nagoya 456-8587, Japan

H. Abe, M. Naito, and K. Nogi
Joining and Welding Research Institute
Osaka University
11-1 Mihogaoka, Ibaraki
Osaka 567-0047, Japan

T. Fukui
Hosokawa Powder Technology
1-9 Shodai, Tajika, Hirakata
Osaka 573-1132, Japan

ABSTRACT

A variety of spray pyrolysis techniques have been developed to directly synthesize advanced powders from aqueous solutions. The purpose of this study is to control the morphology of composite particles by determining the effect of the starting materials in the solution. Examples that are discussed include composite powders of SiO_2-Al_2O_3, SiO_2-ZrO_2 and Al_2O_3-ZrO_2. SEM and TEM observations reveal that the morphology of composite powders is altered by the differences in starting materials. In the case of syntheses from a solution of two sols, the composite powders consist of well-dispersed, fine primary particles of both precursors. In contrast, encapsulated composite powders are formed from solutions of a sol and an ionic solute.

INTRODUCTION

The physical and chemical properties of ceramic materials are highly dependent on the morphology and the chemical composition of the starting powder. Therefore, it is expected that various properties of ceramic materials can be improved by controlling the morphology and composition of the starting powder. Promising techniques to produce ceramic powders with controlled morphology are sol-gel, co-precipitation, and pyrolysis. Among these techniques, spray pyrolysis is the most suitable one to control the morphology and the chemical composition of the powders. The spray pyrolysis technique integrates the evaporation, precipitation, decomposition, and sintering stages of powder synthesis into a single continuous process.[1]

Spray pyrolysis also has a great advantage in preparing composite powders with controlled morphology. It is considered a useful method to fabricate starting powders for composite materials with advanced properties and property combinations. Ceramic composite powders synthesized by spray pyrolysis are often used to fabricate composite ceramics with improved mechanical and electronic properties.[2-5] Synthesis of metal composite powders and metal-ceramic composite powders by spray pyrolysis has been also applied to design materials with unique combinations of properties.[6-9]

The aim of our work is to develop powder processing understanding of spray pyrolysis to control the morphology of composite powders to create composite materials with advanced properties and unique property combinations. In this paper, the synthesis of SiO_2-Al_2O_3, SiO_2-ZrO_2 and Al_2O_3-ZrO_2 composite powders, and the effect of the starting materials in the aqueous solutions on the morphology evolution of the powders will be discussed. SiO_2, Al_2O_3 and ZrO_2 are representative ceramics, and composites composed of these materials are promising for various applications as highly functional materials.

EXPERIMENTAL

The spray pyrolysis apparatus used in this study is shown in Figure 1. This apparatus consists of an ultrasonic vibrator, a reactor (a reaction furnace with four zones and a quartz reaction tube), and an organic filter.

SiO_2 sol (Nissan Chemical Product Ltd., ST-20, particle size: 10-20 nm), Al_2O_3 sol (Nissan Chemical Product Ltd., AS-520, particle size: 10-20 nm), ZrO_2 sol (Nissan Chemical Product Ltd., NZS-20A, particle size: 70 nm), $Al(NO_3)_3\cdot9H_2O$ (Nakalai Tesque Inc., purity: 98%) and $ZrO(NO_3)_2\cdot2H_2O$ (Nakalai Tesque Inc., purity: 95%) were used as the starting materials. These materials were dissolved in distilled water to produce the spray solution. The solution was atomized with an ultrasonic vibrator operated at 1.7 MHz. Droplets were introduced into a horizontal reaction furnace with air as the carrier gas (flow rate: 0.5 L/min). The droplets underwent evaporation, precipitation, decomposition, and finally, sintering. To have a thermal gradient in the reactor, the temperature of each heating zone – zone 1, zone 2, zone 3 and zone 4 – was set at 200°C, 400°C, 600°C and 800°C, respectively.

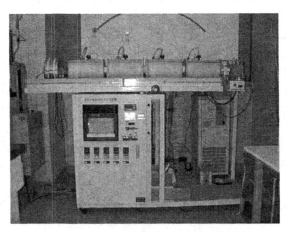

Figure 1. Spray pyrolysis apparatus.

The morphology of the synthesized powders was characterizing using a scanning electron microscopy (SEM; Hitachi, S-800). A transmission electron microscope (TEM; JEOL, JEM2000EX) equipped with an energy dispersive analysis X-ray spectrometer (EDS) was also employed to examine the morphology of the powders. TEM samples were prepared by embedding the particles in epoxy resin, and cutting into thin sections (thickness: about 70 to 80 nm) with a microtome (LKB, LKB4800). The different phases were analyzed by X-ray diffraction (XRD; Philips PW1877).

RESULTS AND DISCUSSION

First, we synthesized single component SiO_2, Al_2O_3 and ZrO_2 particles from the solutions of each sol. As shown in Figure 2, a spherical powder morphology is obtained by spray pyrolysis processing of the ceramics. The shape of all of the different ceramic powders is equiaxial and uniform. To examine the morphology of the powders in detail, TEM observations of each powder was carried out (Figure 3). It is obvious that the dense spheres consist of fine primary particles. The size of the primary particles for the SiO_2, Al_2O_3 and ZrO_2 powder is 10-20 nm, 50-100 nm, and about 100 nm, respectively. The shape is spherical, rectangular, and disk-like, respectively. From XRD, the crystalline structure of the powder is amorphous-SiO_2, γ-Al_2O_3 and monoclinic-ZrO_2, respectively.

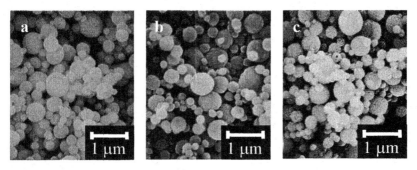

Figure 2. SEM picture of the: a) SiO_2; b) Al_2O_3; and c) ZrO_2 powder produced by spray pyrolysis.

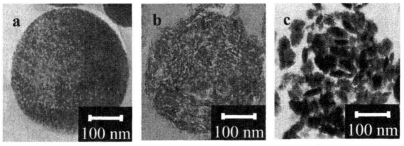

Figure 3. TEM image of the: a) SiO_2; b) Al_2O_3; and c) ZrO_2 powder produced by spray pyrolysis.

Figures 4 and 5 show SEM and TEM pictures of the SiO_2-Al_2O_3, SiO_2-ZrO_2, and Al_2O_3-ZrO_2 composite powders prepared from SiO_2 and Al_2O_3 sols, SiO_2 and ZrO_2 sols, and Al_2O_3 and ZrO_2 sols, respectively. The ratio of the two components in the solution was kept at 50:50 vol.% for all powders. It is clear from Figure 5 that the composite powders consist of well-dispersed, fine primary particles of each of the two components, i.e., dispersions of SiO_2 and Al_2O_3, SiO_2 and ZrO_2, and Al_2O_3 and ZrO_2 primary particles, respectively. The size and shape of each of the primary particles in the composite powders is similar to those produced with the single component powders, as seen in Figure 3. XRD confirms that the crystalline phases for SiO_2-Al_2O_3, SiO_2-ZrO_2 and Al_2O_3-ZrO_2 composite powders are mixture of amorphous-SiO_2 and γ-Al_2O_3, amorphous-SiO_2 and tetragonal-ZrO_2, and γ-Al_2O_3 and tetragonal-ZrO_2, respectively. Therefore, it was found that the well-dispersed composite powder mixtures are obtained when powders are synthesized from the solution of two sols. By the way, the crystal structure of ZrO_2 in the composite powders is tetragonal, while that in the single component ZrO_2 particle is monoclinic. The reason for the change in the crystal structure of the ZrO_2 is yet an open question. It may be possible that the tetragonal phase of ZrO_2 is favored by the mechanical stress in the composite powder (e.g., due to the thermal expansion mismatch between the different materials).

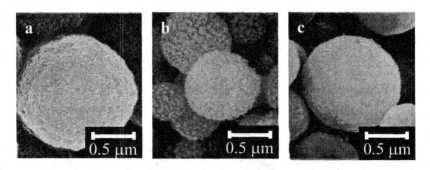

Figure 4. SEM picture of the: a) SiO_2-Al_2O_3; b) SiO_2-ZrO_2; and c) Al_2O_3-ZrO_2 composite powder synthesized from two sols.

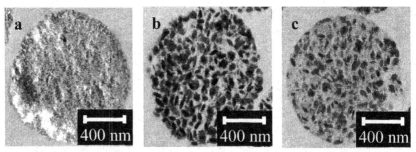

Figure 5. TEM images of the: a) SiO_2-Al_2O_3; b) SiO_2-ZrO_2; and c) Al_2O_3-ZrO_2 composite powder synthesized from two sols.

Figures 6 and 7 present the SEM and TEM photographs for the SiO_2-Al_2O_3 and SiO_2-ZrO_2 composite powders prepared from solutions of SiO_2 sol and $Al(NO_3)_3 \cdot 9H_2O$, and from SiO_2 sol and $ZrO(NO_3)_2 \cdot 2H_2O$. The ratio of SiO_2 and Al_2O_3, and of the SiO_2 and ZrO_2 was maintained at 50:50 vol.%. No reaction phase is observed in either composite powder. TEM-EDS analysis determined that the weakly contrasted particles in the composite powder are SiO_2, and that the strongly contrasted particles are Al_2O_3 and ZrO_2. As evident from Figure 7a, the SiO_2-Al_2O_3 composite powder has a morphology in which Al_2O_3 is encapsulated by SiO_2. In the SiO_2-ZrO_2 composite powder (Figure 7b), SiO_2 particles tend to be forced out toward the powder surface. Therefore, it was determined that encapsulated composite powders are obtained when synthesized from a solution of the sol and the ionic solute.

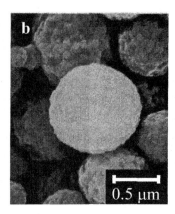

Figure 6. SEM picture of the: a) SiO_2-Al_2O_3; and b) SiO_2-ZrO_2 composite powder synthesized from sol and ionic solute.

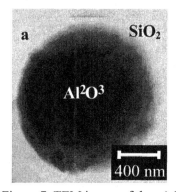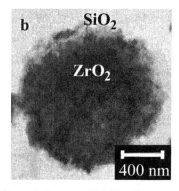

Figure 7. TEM image of the: a) SiO_2-Al_2O_3; and b) SiO_2-ZrO_2 composite powder synthesized from sol and ionic solute.

Finally, we discussed the formation mechanism of the composite powders by spray pyrolysis. Recently Fukui et al.[9] reported the formation mechanism for ZrO_2 encapsulated NiO composite particles synthesized from a solution of ZrO_2 sol and $Ni(CH_3COO)_2 \cdot 4H_2O$. According to them, during the evaporation stage up to 200°C, a solid particle of $Ni(CH_3COO)_2$ surrounded by ZrO_2 fine particles is formed from an atomized droplet containing the Ni ion dispersed in the ZrO_2 sol by volume precipitation. Then, ZrO_2 particles move to the surface of the composite powder during the outgassing and the oxidation of $Ni(CH_3COO)_2$ during the thermolysis stage up to 400°C. Based on the findings of Fukui et al., it is presumed that the SiO_2 encapsulated Al_2O_3 and ZrO_2 composite powders are formed by the outgassing and the oxidation of $Al(NO_3)_3$ and $ZrO(NO_3)_2$, respectively. In contrast, when powder is synthesized from a solution of two sols, outgassing due to decomposition of the starting material does not occur. Therefore, composite powders composed of dispersed individual particles are obtained. We are now examining the morphology of composite powders synthesized at temperatures of 200°C, 400°C, and 600°C, to determine the mechanism of composite powder formation. The detailed results will be published in the near future.

CONCLUSIONS

SiO_2-Al_2O_3, SiO_2-ZrO_2 and Al_2O_3-ZrO_2 composite powders were synthesized from aqueous solutions of different starting materials by ultrasonic spray pyrolysis. The effect of starting materials on the morphology of the composite powders was investigated. In the case of synthesis from a solution of two sols, the composite powders consist of well-dispersed, fine primary particles. On the other hand, when synthesized from a solution of a sol and an ionic solute, encapsulated composite powders are obtained.

ACKNOWLEDGMENTS

This work was partly supported by the Ministry of Education, Science, Sports and Culture, a Grant-in-Aid for the COE project, Giant Molecules and Complex Systems, 2003, and NEDO International Joint Research Grant supervised by the Ministry of Economy, Trade and Industry of Japan.

REFERENCES

[1]G.L. Messing, S.C. Zhang, and G.V. Jayanthi, "Ceramic Powder Synthesis by Spray Pyrolysis," *Journal of American Ceramic Society*, **76** [11] 2707-26 (1993).

[2]T. Suzuki, K. Itatani, M. Aizawa, F.S. Howwel, and A. Kishioka, "Sinterbility of Spinel ($MgAl_2O_4$)-Zirconia Composite Powder Prepared by Double Nozzle Ultrasonic Spray Pyrolysis," *Journal of European Ceramic Society*, **16** 1171-78 (1996).

[3]T. Fukui, S. Ohara, and K. Mukai, "Long-Term Stability of Ni-YSZ Anode with a New Microstructure Prepared from Composite Powder," *Electrochemical and Solid-State Letters*, 1 [3] 120-22 (1998).

[4]S. Ohara, R. Maric, X. Zhang, K. Mukai, T. Fukui, H. Yoshida, T. Inagaki, and K. Miura, "High Performance Electrodes for Reduced Temperature Solid Oxide Fuel Cells with Doped Lanthanum Gallate Electrolyte; I. Ni-SDC Cermet Anode," *Journal of Power Sources*, **86** 455-58 (2000).

[5]T. Fukui, S. Ohara, M. Naito, and K. Nogi, "Morphology Control of the Electrode for Solid Oxide Fuel Cells by using Nanoparticles," *Journal of Nanoparticles Research*, 3 171-74 (2001).

[6]T.C. Pluym, T.T. Kodas, L.M. Wang, and H. D. Glicksman, "Silver-Palladium Ally Particle Production by Spray Pyrolysis," *Journal of Materials Research*, **10** 1661-73 (1995).

[7]S. Che, O. Sakurai, K. Shinozaki, and N. Mizutani, "Effect of Starting Materials on the Preparation of Spherical Pd Powders by Ultrasonic Spray Pyrolysis," *Journal of Ceramic Society of Japan*, **105** [3] 269-71 (1997) [in Japanese].

[8]M. Matsumoto, K. Kaneko, Y. Yasutomi, S. Ohara, and T. Fukui, "Synthesis of TiO$_2$-Ag Composite Powder by Spray Pyrolysis," *Journal of Ceramic Society of Japan*, **110** [1] 60-62 (2002).

[9]T. Fukui, S. Ohara, M. Naito, and K. Nogi, "Synthesis of NiO-YSZ Composite Particles for an Electrode of Solid Oxide Fuel Cells by Spray Pyrolysis," *Powder Technology*, **132** 52-56 (2003).

AEROSOL SYNTHESIS AND PHASE DEVELOPMENT IN Ce-DOPED NANOPHASED YTTRIUM-ALUMINUM GARNET ($Y_3Al_5O_{12}$:Ce)

Olivera Milosevic and Lidija Mancic
Institute of Technical Sciences of
Serbian Academy of Sciences and Arts
11000 Belgrade, K. Mihajlova 35/IV
Serbia & Montenegro

Satoshi Ohara
Institute of Multidisciplinary Research
for Advanced Materials
Tohoku University
2-1-1 Katahira, Aoba-ku
Sendai 980-8577, Japan

Gilberto del Rosario
University Rey Juan Carlos I. Tulipan
s/n. Móstoles
Madrid 28933, Spain

Predrag Vulic
Faculty of Mining and Geology
Djusina 7, 11000
Belgrade, Serbia & Montenegro

ABSTRACT
Nanophase phosphor particles in the complex $Y_3Al_5O_{12}$:Ce system were synthesized from aerosols of 0.06 mol/dm^3 nitrate solution ultrasonically generated with frequency of $8 \cdot 10^5$ s^{-1} and thermally treated at 1273 to 1473 K. The secondary particles obtained are spherical, and range in size from 300-800 nm. TEM reveals a composite secondary particle structure comprised of clusters of < 60 nm primary particles. XRD of the as-synthesized powders indicates the presence of poorly crystallized phases. The phase development and structural changes, followed using a Koalariet-Xfit program, identifies the target cubic YAG - $Y_3Al_5O_{12}$ being prevalent after annealing at 1073 K, with The YAG content increases with temperature. Both CeO_2 and monoclinic $Y_4Al_2O_9$ are present in lower concentration. EDX analysis verifies a uniform distribution of the constitutive elements inside the secondary particles. Photoluminescence measurements show an emission spectrum peak maximum at 537 nm for the thermally treated particles, which is attributed to the phase content of the powder.

INTRODUCTION
The fundamental aspects and processing of solid-state functional materials that exhibit ionic and/or electronic conduction must consider microstructure refinement to achieve good homogeneity and high reliability.[1] Controlled and/or fine grain size, grain and grain boundary chemistry, and chemical homogeneity through the uniform distribution of the constituent phases are opportunities that nanostructures provided to improve functional properties. Decreasing the grain size to the nanometer scale could have a huge effect on structure disordering, increasing interfacial area and grain boundary diffusion, interconnected porosity,

quantum confinement effects, and the stabilization of metastable structures that include a wide range of nonstoichiometry. Several methods are currently being used for nanophase synthesis.[2] Among them, synthesis through a dispersion phase (aerosol) enables the generation of fine size single-phase or complex powders with improved properties.[3, 4] The short residence time in these processes enables the suppression of compositional segregation, which is especially important for the case of multicomponent systems.[5]

The opportunities for the synthesis of spherical, nonaglomerated particles with uniformly distributed components and phases are of special importance when considering materials for photonic applications.[6,7] Among them, the garnet phase in the complex yttrium-aluminium oxide system is a suitable material for solid-state lasers that is widely used for various display applications when doped with rare-earth ions.[8] There are several stable phases in the Y_2O_3–Al_2O_3 system, including $Y_3Al_5O_{12}$ (YAG) with the garnet structure, $YAlO_3$, (YAP) and $Y_4Al_2O_9$ (YAM).[9] YAG is the phase that is typically produced by solid-state synthesis above 1873 K. Some authors have implied that the high heating rates and short residence times associated with spray pyrolysis cause the formation of the kinetically stable phases, such as $Y_4Al_2O_9$, rather than the thermodynamically stable $Y_3Al_5O_{12}$;[10] however the target $Y_3Al_5O_{12}$ (YAG) phase doped with europium has been obtained after annealing as-prepared aerosol synthesized powder from the amorphous Y:Al = 3:5 phase using a filter-expansion aerosol generator.[8] Similarly, luminescence properties are revealed after annealing ultrasonically-derived YAG:Ce particles at 1573-1873 K.[11] We speculate that the aerosol synthesis process may favor lower decomposition temperatures that allow particle spheroidization, and that maintain a uniform distribution of yttrium and aluminum in the absence of phase crystallization. The subsequent thermal treatment then produces the thermodynamically stable phase, while preserving the main characteristics of a spray pyrolysis powder (i.e. homogeneous, spherical, unagglomerated particles). This paper examines low-temperature synthesis of spherical, nonaglomerated YAG:Ce particles with uniformly distributed constitutive elements using the aerosol route. Additionally, structure and phase development during aerosol synthesis are evaluated.

EXPERIMENTAL
The experimental work involved *in situ* synthesis of submicron powders through reactions in aerosols (spray pyrolysis). Composite particles based on YAG:Ce were prepared by combining $0.081 \cdot 10^3$ mol m^{-3} (water solution) of $Y(NO_3)_3 \cdot 6H_2O$ (Aldrich, p.a., 99.9%) with $0.136 \cdot 10^3$ mol m^{-3} $Al(NO_3)_3 \cdot 9H_2O$ (Merck, p.a., 99.9%) and $0.0054 \cdot 10^3$ mol m^{-3} $Ce(NO_3)_3 \cdot 6H_2O$ (Aldrich, p.a., 99.9%) in the desired 3:5 Y:Al molar ratio. The solubility of the precursors, and the precursors solution chemistry were carefully controlled and monitored. The concentration ($0.06 \cdot 10^3$ mol m^{-3}), pH(2.7), viscosity ($1.0436 \pm 0.0007 \cdot 10^3$ Nm^{-2} s, MLW Viskosimeter B3), density ($1.0642 \pm 0.0002 \cdot 10^3$ kg m^{-3}, AP PAAR density meter DMA55) and surface tension (112.295 ± 0.001mN m^{-1}, tensiometer K10T KRUSS) were controlled and monitored to control aerosol synthesis. Measurements were made at room temperature, (293 K) and at elevated temperature (323 K). Based on these values, the average droplet size (2316 nm), the mean particle size (745 nm), and the aerosol droplet number density ($<10^{12}$ m^{-3}) were estimated in accordance with a previously proposed procedure.[3]

The precursor solutions were atomized ultrasonically (atomizing frequency, $8 \cdot 10^5$ s^{-1}). The experimental set-up employed for the synthesis is explained elsewhere.[5] The aerosol was introduced into a heated zone using air as a carrier gas. The flow rate was set at $0.028 \cdot 10^3$ m^{3}s^{-1}, and the droplet/particle residence time was 3 s. Aerosol decomposition was performed in a high temperature tubular flow reactor at 1173 K. The as-prepared powders were thermally treated at 1273-1473 K for 7200 s.

X-ray automated powder diffraction (XRD, Philips PW 1710 and 1877), with CuKα radiation and typical scan parameters: $10°-90°2\theta$ scan, 0.02 step size and 15 s step time, was used to characterize the crystal structure of the phases produced. Peak positions were used to determine microstructure parameters. Structural refinements were carried out using the Rietveld program Koalariet-Xfit.[1] Compositional homogeneity and particle morphology were characterized using scanning electron microscopy (SEM, Hitachi, S-800 and JSM 5300-JEOL) and energy dispersive x-ray spectroscopy (EDX, HORIBA, QX-2000 and Philips XL). Particle structure was examined using transmission electron microscopy (TEM, TECNAI 20 Philips) operated at 200 kV, with small angle electron diffraction (SAED). Photoluminescence measurements were done using sharp excitation spectra generated with an excitation wavelength of 351 nm from an Ar laser (7 mW). The measurements were done in the wavelength range of 400-700 nm using a conventional lock in amplitude and photomultimeter detector.

RESULTS AND DISSCUSSION:

SEM reveals that the as - prepared ("secondary particles"[3]) produced by spray pyrolysis are spherical, range in size from 300-800 nm, and have smooth particle surfaces (Figure 1A). After the thermal treatment, the particle size remains mostly unchanged (Figure 2A, C, E). The secondary particle morphology shows a composite structure that is an aggregate of "primary particles" (Figure 2 B). It is believed that these clusters arise during the thermally induced   processes of crystallization and growth.

In some cases, particle bonding and neck formation are evident (Figure 2D). EDAX spot and square area analysis verifies that there is a uniform distribution of the constitutive elements inside the as-prepared particles: aluminum (Al, Kα, 1.48 KeV), yttrium (Y, Lα, 1.921 KeV), cerium (Ce, Lα 4.843 KeV, Mα 0.882 KeV) and oxygen (O, Kα 0.517 KeV). The results of semi quantitative analysis and particle profiling imply that compositional homogeneity persists after annealing as well.

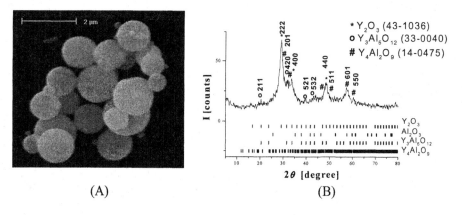

(A)   (B)

Figure 1. YAG:Ce powder produced by spray pyrolysis: a) SEM micrograph; and b) XRD pattern.

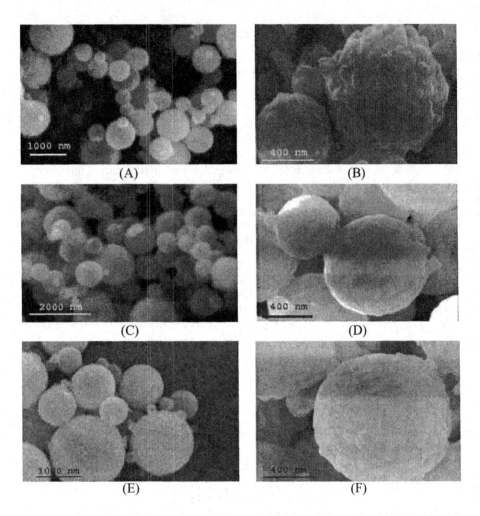

Figure 2 SEM micrographs of spray pyrolized powders after heat treating at 1273 (A,B), 1373 (C,D) and 1473 K (E,F)

XRD of the as-synthesized powders indicates the presence of poorly crystallized phases: Y_2O_3 (JCPDS file card 43-1036), orthorhombic Y_4Al_2O9 (JCPDS file card 14-0475) and cubic $Y_3Al_5O_{12}$ (YAG, JCPDS file card 33-040) (Figure 1B). Peaks at 1.7864, 1.5881 and 1.5106 nm might be connected with the (024), (211) and (018) (hkl) planes of Al_2O_3 (file card 31-0026), respectively.

The phase development at 1273-1473 K were followed in detail, and the peak fitting using the Koalariet-XFit is summarized in Table I together with the weight % of the respective phases and their structural parameters. Note that satisfactory peak fitting is obtained even after proposing that the phases obtained are twin-domains, where one domain relates solely to the crystalline component and the other to the intercrystalline component that has a high defect content. Taking into consideration the high heating rates during the aerosol synthesis, and the

Table I. $Y_3Al_5O_{12}$:Ce structural parameters (Koalariet-Xfit[12])

Temp K	Phase wt %	Unit cell parameters nm	Crystallite size, nm	Microstrain %
1273	$Y_3Al_5O_{12}$ C, 43.40	Cub. $Ia3d$ 1.20412±0.000075	39.5 ± 2.1	0.173 ± 0.082
	$Y_3Al_5O_{12}$ IC, 40.72	Cub. $Ia3d$ 1.20517±0.00011	-	0.951 ± 0.013
	CeO_2 C, 6.79	Cub. Fm-$3m$ 0.53976±0.00004	23.4 ± 0.8	0.603 ± 0.068
	$Y_4Al_2O_9$ C, 8.34	Mon. P_21/c	46.4±5.0	0.249 ± 0.221
1373	$Y_3Al_5O_{12}$ C, 68.06	Cub. $Ia3d$ 1.20438±0.00002	136.7 ± 5.2	0.556 ± 0.009
	$Y_3Al_5O_{12}$ IC, 24.71	Cub. $Ia3d$ 1.20489±0.00008	-	1.576 ± 0.288
	CeO_2 C, 6.42	Cub. Fm-$3m$ 0.54067±0.00002	35.0 ± 0.8	0.443 ± 0.032
	$Y_4Al_2O_9$ C, 0.80	Mon. P_21/c	-	-
1473	$Y_3Al_5O_{12}$ C, 71.27	Cub. $Ia3d$ 1.20318±0.000004	197.5 ± 6.5	0.476 ± 0.005
	$Y_3Al_5O_{12}$ IC, 22.94	Cub. $Ia3d$ 1.20414±0.000014	-	1.446 ± 0.019
	CeO_2 C, 5.08	Cub. Fm-$3m$ 0.54128±0.00004	41.9 ± 3.9	0.241 ± 0.089
	CeO_2 IC, 0.71	Cub. Fm-$3m$ 0.54139±0.00016	-	0.559 ± 0.107

C-crystalline; IC-intercrystalline

nanophase inner structure associated with a large interfacial area (i.e., from the fine particle size) (Figure 3), the results seem reasonable. A similar approach has been applied previously to a single-phase metal system comprised of different sized fraction particles.[13] The XRD results show that the target cubic YAG phase is present at all temperatures, and that YAG content increases with annealing temperature.

Low-magnification TEM analysis (Figure 3, left) reveals the composite inner particle structure, comprised of differently oriented primary particles (< 60 nm) and nanoporosity. A higher density of the primary particles is evident at the surface of the secondary particle. This is believed to be a consequence of surface precipitation, which is prevalent during the aerosol droplet evaporation/precipitation step. A high-magnification TEM image of a sample treated at 1273 K (Figure 3, middle) shows two domains with different orientations and different inter-planar spacing. Details of these domains are shown in the inset. An inter-planar spacing of 0.405 and 0.841 nm is measured in zone a and zone b, respectively. The former value is presumably attributed to the (220) plane of the YAG ($Y_3Al_5O_{12}$) phase ($d_{(220)} = 0.424700$ nm). The later can be assigned to the finely dispersed aluminium oxide (d = 0.80400 nm, JCPDS file card 31-0026) that has short distance order. The cubic symmetry in both domains, which is implied by the SAED patterns (Figure 3, right) is attributed to the cubic YAG phase. A typical high-magnification image of a sample treated at 1373 K indicates that the

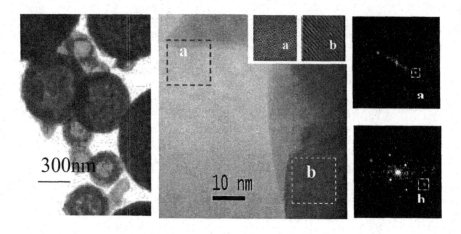

Figure 3. Particles annealed at 1373 and 1273 K, respectively. Low-magnification (left), and high-magnification (middle) TEM image, and the corresponding SAED image (right).

inter-planar spacing of 0.486 nm corresponds to the x-ray value for the (211) plane of the YAG phase ($d_{(211)}$ = 0.490500 nm).

The average crystallite sizes, and the microstrains (Table I) indicate an increase in crystallinity and structure ordering with temperature. The presence of an intermediate monoclinic phase $Y_4Al_2O_9$ is evident in low concentration at 1273 (~8 wt%) and 1373 K (<1 wt%). The increase in the lattice parameter of cubic CeO_2 (theoretical: 0.541134 (12) nm, JCPDS file card 34-0394), and the decrease in concentration with temperature (Table I) might be connected with the tendency for Ce^{4+} to be reduced to the larger Ce^{3+} ion to stabilize the fluorite structure.[14] This would be followed by the diffusion of Ce^{3+} into the garnet structure, since there is close ionic radius matching (0.101 nm Ce^{3+} and 0.09 nm for Y^{3+}). The larger lattice parameters measured for YAG (Table I) compared with the theoretical value (1.20089 (3) nm) is presumably the consequence of substituting cerium (Ce_Y) for Y in the YAG matrix. This substitution is further confirmed by the broad emission from the particles in the range from 450-700 nm with a luminescent peak (maximum) at 537 nm, which is attributed to the Ce^{3+} intershell transition (5d→4f) in the YAG lattice[15]. The results obtained are in agreement with the ultrasonically synthesized (f = $1.75 \cdot 10^6$ s^{-1}) Y:A l= 3:5 system containing 0.5 – 3 at% Ce^{11}. However, the relatively broad luminescence spectrum obtained, and the peak shifting in comparison to a single crystal[12] are presumably associated with the primary particle nano-structure.

CONCLUSIONS:

$Y_3Al_5O_{12}$:Ce spherical phosphor particles (300-800 nm), with luminescence properties (emission spectra 450-700 nm, peak at 537nm), were synthesized from ultrasonically (frequency $8 \cdot 10^5$ s^{-1}) generated aerosol solution. Detailed phase and structural analyse were completed using various methods. TEM reveals a composite particle structure comprised of primary nano particles-clusters (below 60 nm after annealing at 1373 K). High resolution TEM and XRD confirm the synthesis of cubic $Y_3Al_5O_{12}$.

ACKNOWLEDGEMENTS:
This research was financially supported through the NEDO International Joint Research Grant Program 01MB7: "Wetability of solid by liquid at high temperatures", the Ministry of Education, Science, Sports and Culture, through Grant-in-Aid for the COE project, Giant Molecules and Complex, 2003 as well as Republic of Serbia Science Foundation project 1832.

REFERENCES:
[1]H. Tuller, "Solid State Electrochemical Systems - Opportunities for Nanofabricated or Nanostructured Materials," *J. Electroceramics*, **1-3** 211-18, (1997).
[2]A.Gurav, T. Kodas, T. Pluym, and Y. Xiong, "Aerosol Processing of Materials," *Aerosol science and technology*, **19** 411-52, (1993).
[3]G.L.Messing, S.-C. Zhang, and G. V. Jayanthi, "Ceramic Powder Synthesis by Spray Pyrolysis," *J. Am. Ceram. Soc.*, **76** 2707-05, (1993).
[4]O. Milosevic, M. Mirkovic, and D. Uskokovic, "Synthesis of BaTiO$_3$ and ZnO Varistor Precursor Powders by Reaction Spray Pyrolysis", *J. Am. Ceram. Soc.*, **79** [6] 1720-22, (1996).
[5]Z.V. Marinkovic, L. Mancic, R. Maric,and O. Milosevic, "Preparation of Nanostructured Zn-Cr-O Spinel Powders by Ultrasonic Spray Pyrolysis," *J.Europ.Ceram.Soc.*, **21** 2051-55, (2001).
[6]Y.C. Kang, S.B. Park, I.W. Lengorro, and K. Okuyama, "Morphology Control of Multicomponent Oxide Phosphor Particles Containing High Ductility Component by High Temperature Spray Pyrolysis," *J. Electrochemical Society*, **146** [7] 2744-47, (1999).
[7]O. Milosevic, R. Maric, S. Ohara, and T. Fukui, "Aerosol Synthesis of Phosphor Based on Eu Activated Gadolinium Oxide Matrices," in Ceramic Processing Science VI, Ed. S. Hirano et al., *Ceram. Trans.* **112** 101-06, (2001).
[8]Y.C Kang, Y.S. Chung, and S.B. Park, "Preparation of YAG:Europium Red Phosphors by Spray Pyrolysis Using a Filter-expansion Aerosol Generator," *J. Am. Ceram. Soc.*, **82** [8] 2056-60, (1999).
[9]K.M. Kinsman, J. McKittrick, "Phase Development and Luminescence in Chromium-doped Yttrium Aluminium Garnet (YAG:Cr) Phosphors", *J. Am. Ceram. Soc.*, **77** [11] 2866-72, (1994).
[10]M. Nyman, J. Caruso, M.J. Hampden-Smith, and T.T. Kodas, "Comparison of Solid-state and Spray Pyrolysis Synthesis of Yttrium Aluminate Powders," *J. Am. Ceram. Soc.*, **80** [5] 1231-38, (1997).
[11]Y.C., Kang, I.W. Lengorro, S.B. Park, K. Okuyama, "YAG:Ce Phosphor Particles Prepared by Ultrasonic Spray Pyrolysis," *Materials Research Bulletin*, **35** 789-98, (2000).
[12]R.W. Cheary and A.A.Coelho, Programs XFIT and FOURYA, CCP14 Powder Diffraction Library-Daresbury Laboratory, Warnington, England, (1996).
[13]T.Ungar,"The Dislocation Based Model for Strain Broadening in X-ray Line Profile Analysis," in IUCR-Oxford University Press Inc New York (1999).
[14]R. Li, S. Yabe, M. Yamashita, S. Momose, S. Yoshida, S. Yin, T. Sato, "Synthesis and UV-shielding Properties of ZnO- and CaO-doped CeO$_2$ via Soft Solution Chemical Process", *Solid State Ionics*, **151** 235-41, (2002).
[15]E. Zych, C. Brecher, "Temperature Dependence of Ce-emission Kinetics in YAG:Ce Optical Ceramic, *J. Alloys Compounds*, **300-301** 495-99, (2000).

PHASE EVOLUTION IN Ag:$(Bi,Pb)_2Sr_2Ca_2Cu_3O_x$ COMPOSITE POWDER

L. Mancic
Institute of Technical Sciences
of the Serbian Academy of Sciences
& Arts, Knez Mihailova 35/IV
11000 Belgrade, Serbia & Montenegro

Z. Marinkovic
Center for Multidsciplinary Studies
Kneza Viseslava 1a,
11000 Belgrade, Serbia & Montenegro

B. Marinkovic, P.M. Jardim, and
F. Rizzo
Departamento de Ciência dos Materiais
e Metalurgia
Pontificia Universidade Catôlica do
Rio de Janeiro, 22453-900, Brasil

O. Milosevic
Institute of Technical Sciences of the
Serbian Academy of Sciences & Arts
Knez Mihailova 35/IV,
11000 Belgrade, Serbia & Montenegro

ABSTRACT

The phase evolution in submicron superconducting composite particles prepared by spray pyrolysis was followed from the droplet-particle formation stage through the powder calcination stage. Aerosol was generated ultrasonically (1.7 MHz) from a urea-modified nitrate precursor solution containing 20 wt% silver. The salt decomposition process was aided by self-combustion of the droplets due to the presence of the urea. A uniform distribution of all of the constitutive chemical elements of the (Bi,Pb)-2223 phase was confirmed by energy dispersive spectroscopy. Ag is present together with (Bi,Pb)-2212, 2201, and $(Sr,Ca)_{14}Cu_{24}O_{41}$ in submicron equiaxial particles that are aggregates of primary nanoparticles. Initial thermal treatment of these particles increases the 2212 phase content, and forms Ca_2PbO_4.

INTRODUCTION

Although the phenomenon of superconductivity was discovered a long time ago, a break-through in this field was the discovery of high T_c oxide superconducting ceramics.[1] The correct phase and precise stoichiometry precursor powders are required to make high T_c superconducting ceramics. It is generally accepted that fine and homogenous powders are required to produce a uniform microstructure and a high current density (J_c) superconducting devices. Specially, for Bi-based tapes, the precursor powders must be composed of a reactive mixture of phases that readily forms the desired final product on firing.[2] According to the Bi-Sr-Ca-Cu-O phase diagram, the phase with the highest T_c value, i.e., $Bi_2Sr_2Ca_2Cu_3O_{10}$ or the 2223 phase, exists in a extremely limited monophasic field. Phase pure synthesis is additionally complicated by the need to partially substitute Pb for Bi to stabilize the 2223 phase.[3] Various attempts have been made to develop a convenient technology for high T_c phase synthesis. Solution-based routes for precursor powder synthesis have shown promise to substantially reduce the heat treatment time required to form the (Bi,Pb)-2223 phase. The use of submicron powders, characterized by a large surface area, also can increase the surface energy and decrease the diffusion distance to promote the formation of this phase.[3] Ultrasonic spray pyrolysis, as a dispersion phase powder processing method, is capable of producing such particles.[4] Moreover, this method has some advantageous in preparing multicomponent or composite powders for tape processing.[5] In addition to the particle size and homogeneity effects on phase evolution in spray pyrolyzed particles, there is also a benefit of having urea present during the (Bi,Pb)-2223 synthesis process.[6]

There are some preliminary results on Ag:(Bi,Pb)-2223 composite particle synthesis.[7] Silver is indispensable for superconducting tape production, both due to its resistance to oxidation at the tape processing temperature, and because it is oxygen permeable. Also, Ag assists the 2212 to 2223 phase transformation, and promotes grain alignment.[8] All this gives rise to the production of high quality composite powders, which are of considerable interest as precursor powders for superconducting tapes, wires or bulk materials.

Silver particles mixed with the superconducting precursor powders by hand result mainly in inhomogeneous mixing.[9,10] Silver additions during evaporative decomposition of solutions results in a more uniform 2223 phase distribution across the thickness of the core.[11] However, no detailed study has been made regarding the effect of silver on the phase conversion during precursor powder synthesis. Therefore, the aim of the present investigation is to explore and establish the influence of Ag on high T_c phase evolution during spray pyrolysis, and during initial thermal treatments.

EXPERIMENTAL

Composite particles of Ag:(Bi,Pb)-2223 stoichiometry were synthesized from a mixed nitrate solution with the appropriate ratio of the cations necessary to produce the 2223 phase (Bi:Pb:Sr:Ca:Cu=1.8:0.2:2:2:3). Nominally, 100 g of mixed metal oxides were combined with 1000 ml of a 5 wt% HNO_3, 2 wt% urea and 20 wt% Ag (in the form of $AgNO_3$). The precursor solution characteristics (density, pH, viscosity, and surface tension) were measured and used to calculate[12] the average droplet size (2725 nm), the mean particle size (318 nm) and the aerosol droplet number density ($<10^{14}$ droplets/m^3). The solution was atomized using an ultrasonic atomizer with a working frequency of $1.7 \cdot 10^6$ s^{-1}. Aerosol decomposition was carried out in a tubular flow reactor in argon gas at temperatures up to 1113 K. The details of the synthesis process are given elsewhere.[6] The aerosol flow rate is determined by the argon flow rate ($1.7 \cdot 10^{-6}$ m^3 s^{-1}) and the reactor geometry (quartz tube characteristics: inner diameter $3.6 \cdot 10^{-2}$ m, length $6 \ 10^{-1}$ m), so the residence time of a droplet/particle at the maximum reaction temperature was 5 s. The powder produced was heat treated in air at 1023 K (3600 s) and 1053 K (10800 s).

Thermal analysis was completed on the dehydrated precursor nitrates mixture (3600 s at 423 K) after the urea addition, and on the heated powders using differential-scanning calorimetry (DSC). The analysis was performed in a platinum crucible using nitrogen as a purge gas (Shimadzu DSC-50).

Phase analysis of the as-prepared and heat-treated powders was completed using a Siemens D-5000 diffractometer. CuKα radiation and graphite mono-chromator were used with a 0.02 scanning step and a step time of 25 s. Quantitative phase analysis was done using the Rietveld method[13] Particle structure was characterized by transmission electron microscopy (TEM), using a

JEOL 2010 operated at 200 kv. Energy dispersive spectroscopy (EDS) was performed on different size particles to determine chemical homogeneity.

RESULTS AND DISSCUSION

During pyrolysis of the ultrasonically formed aerosol, additional heat is generated in each droplet/particle as a result of urea. The nature and the temperature of these processes were determined using DSC analysis of partially dehydrated nitrate mixtures containing 2 wt% urea (Figure 1). According to a previous investigation,[6] the intense evolution of heat and gases caused by the decomposition of the urea and the urea intermediate products leads to a volume reduction of the droplet/particle. This reduction is beyond the levels calculated for shrinkage by water evaporation and thermal decomposition of the metal salts alone. Thus, the unexpected formation of smaller dense particles results from the rapid explosion of individual droplets when they enter the intense temperature field in the tubular flow reactor. Due to the additional heat released, all of the nitrates are decomposed below 773 K to produce spheroid particles with a uniform distribution of the chemical components. Because of the high evaporation rate of the volatile metal species, one might not expect this result.

According to x-ray diffraction (XRD) analysis the decomposition of spray pyrolized precursor gives both binary and multicomponent oxides. The calcination process leads to the formation of the several typical phases (Figure 2), consistent with the results of Rath et al.[14] Table I summarizes the structural parameters and the phases determined by Rietveld refinement. The results indicate that a short reaction time and a high heating rate lead to the formation of the (Bi,Pb)-2212 phase. Providing sufficient time during the calcination process promotes the formation of the thermodynamically favored Ca_2PbO_4 phase (Fig 3).

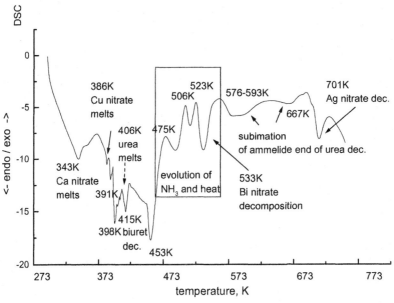

Figure 1. DSC of the partially dehydrated nitrate precursor containing 2 wt% urea.

Table I Structural parameters and the phases of composite Ag:(Bi,Pb)-2223 particles produced by spray pyrolysis.

Sample	Phase	wt%	Unit cell type & parameters (10 nm)				Crystal. size, nm
				b	a	c	
As-prepared	2212	25.5	Amaa	5.417	5.420	30.896	51
	14:24	27.5	Cccm	11.24	12.87	27.61	12
d	2201	16.8	Amaa	5.358	5.397	24.423	25
	CuO	10.7	C2/c	4.63	3.455	5.063	52
	Ag	19.5	Fm3m	4.089	4.089	4.089	177
Calcined	2212	62.1	Amaa	5.405	5.405	30.855	78
	14:24	10.2	Cccm	11.17	12.55	27.82	207
	Ca$_2$PbO$_4$	4.8	Pbam	5.719	9.902	3.461	70
	Ag	23	Fm3m	4.090	4.090	4.090	148

Pb release from the 2212 crystal structure during the initial thermal treatment is indicated by the decrease in the orthorhombic distortion for the (200)/(020) reflections at $2\theta = 28.52$ (Figure 4), as well as by the additional satellite diffraction lines at $2\theta = 28.52; 30.02;$ and $37.02°$ (marked as 2* in Figure 2B). These reflections are related to incommensurate structural modulation of the 2212 phase[15] characteristic of Bi-2212 without Pb.

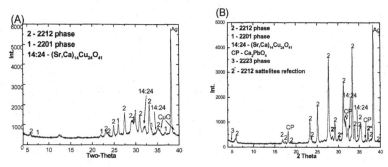

Figure 2. XRD of the composite Ag:(Bi,Pb)-2223 powder produced by spray pyrolysis: A) as-prepared; and B) calcined.

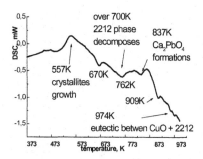

Figure 3. DSC of as-prepared powder.

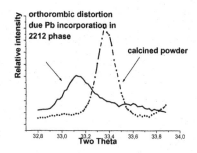

Figure 4. Orthorombic distortion of (200/020) planes.

The orthorhombic distortion factor, z, calculated from unit cell parameters as

$$\frac{b-a}{b+a}$$

is 2.6×10^{-4} in the as-prepared powder, and 4×10^{-5} in the calcined powder. The absence of alkaline-earth bismuthates, typical for precursor powders synthesized through spray pyrolysis,[14,16] should be treated as an advantage, since these indicate a heterogeneous phase composition in thermally-treated powders. The weak reflection visible at $2\theta = 4.8°$ (marked a 3 on Figure 2B) could be the first indication of the 2223 phase appearing, but additional measurements of magnetic susceptibility must be performed to verify this. The improved homogeneity and favorable assembly of phases could be a consequence of the Ag present. Figure 5a shows an equiaxial composite particle with a size around 600 nm. According to the EDS analysis (Figure 5b), a high Ag content is detected together with the other precursor cations. The particle morphology implies primary particle clustering, and the formation of secondary particles during the synthesis process. Presumably, as a result of the droplet/particle combustion, many different particle shapes are formed.

In Figure 6, a polycrystalline rectangular particle with a size 190 x 145 nm (marked A) is shown together with a rod-shaped particle marked B (210 x 145 nm). Electron diffraction analyses of the rectangular particle indicates that the superconducting 2212 and the 2201 phase coexist together with the $(Ca,Sr)_{14}Cu_{24}O_{41}$ phase. New crystallite formation, together with a phase change in the assembly, are noticeable after the crystallites interact with the electron beam, i.e., a result of local heating due to the poor heat dissipation. A new network of 2212 crystallites appears in particle B due to the evolution of a glassy hollow area into superfine crystallites (distinguished as the black dots in Figure 6). Pronounced 2212 formation, as well as crystallite nucleation and growth, are clearly indicated by DSC (Figure 3) and XRD structural analysis. This is consistent with the predicted higher reactivity of the composite nanoparticles due to the larger interfacial area related to the finer particle size. In this context, cation

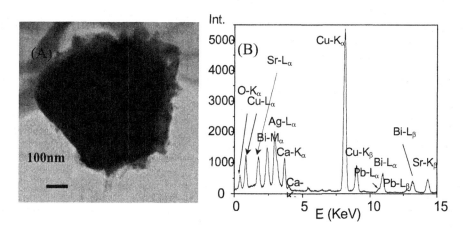

Figure 5. TEM image of a typical: A) as-prepared particle with the corresponding EDS analysis (B).

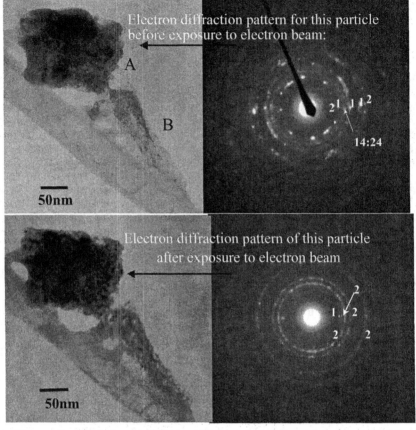

Figure 6. TEM images (left) with corresponding electron diffraction patters (right): 2 - 2212; 1 - 2201; 14:24 - $(Ca,Sr)_{14}Cu_{24}O_{41}$

diffusion will be enhanced, not only by liquid phase formation under controlled heating conditions in subsequent thermal cycles, but also by the increased defect concentration in grain boundary area.

CONCLUSION

Phase assemblies in as-prepared and calcined powders were investigated to determine the effect of Ag on spray pyrolysed (Bi,Pb)-2223 particles. An intimate mixture of all of the precursor components exists in submicron equiaxial particles. The 2212 and 2201 phases coexist in different shape polycrystalline nanoparticles. In-situ formation of (Bi,Pb)-2212, together with the absence of alkaline-earth bismuthates in the precursor powder after initial heating, should give rise to rapid formation of the (Bi,Pb)-2223 phase during subsequent thermal treatment.

ACKNOWLEDGEMENT
This research was partially financially supported through the NEDO International Joint Research Grant Program 01MB7: "Wetabilitty of solid by liquid at high temperatures".

REFERENCES
[1]K.A. Muller, J.G. Bednorz, "The Discovery of a Class of High Temperature Superconductors," Science, **273** 1134-1135 (1987).
[2]F.S. Guenther, M. Bernd, K. Mathias, R. Dietmar, and R. Rodney, "Bi(Pb)SrCaCuO Superconductors," U.S. Pat. No. 5 814 585, Sep.29, 1998.
[3]U. Endo, S. Koyama, and T. Kawai, Preparation of the High Tc phase of BiPbSrCaCu Superconductors," Jpn. J. Appl. Phys., 27, L1476-1479 (1988).
[4]Y.-W. Hsueh, S.C. Chang, R.S. Liu, L. Woodall and M. Gerards, "A Comparison of the Properties of Bi-2223 Precursor Powders Synthesized by Various Methods," *Materials Research Bulletin, 36* 1653-1658 (2001).
[5]M. Matsumoto, K. Kaneko, Y. Yasutomi, S. Ohara, T. Fukui, and Y. Ozawa, "Synthesis of TiO$_2$-Ag Composite Powder by Spray Pyrolysis," *Journal of the Ceramic Society of Japan*, **110** [1] 60-62 (2002).
[6]L. Mancic, O. Milosevic, B. Marinkovic, M.F. de Silva Lopes, and F. Rizzo, "The Influence of Urea on the Formation Process of BiPbSrCaCuO Superconducting Ceramics Synthesized by Spray Pyrolysis Method," *Materials Science and Engineering B*, 76 127-132 (2000).
[7]L. Mancic, B. Marinkovic, P. Vulic, O. Milosevic, "Aerosol Processing of Fine Ag:(Bi,Pb)2223 Composite Particles," *Physica C*, in press.
[8]S.K. Xia, M.B. Lisboa, A. Polasek, M.A. Sens, E.T. Serra, F. Rizzo, and H. Borges, "Preparation of Bi-2223/Ag Tapes by Controlling Precursor Powders," *Physica C, 354* 467-471 (2001).
[9]I. Karaca, S. Celebi, A. Varilci, and A.I. Malik, "Effect of Ag$_2$O Addition on the Intergranular Properties of the Superconducting Bi-(Pb)-Sr-Ca-Cu-O System," *Supercond. Sci. Technol., 16* 100-104 (2003).
[10]A. Sobha, R.P. Aloysius, P. Guruswamy, K.G.K. Warrier, and U. Syamaprasad, "Phase Evolution in Ag, Ag$_2$O and AgNO$_3$ Added (Bi,Pb)-2223 Phase Superconductor," Physica C, **307** 277-283 (1998).
[11]Z.Yi, L. Law, C. Beduz, R.G. Scurlock, and R. Riddle, "The Local Phase Conversion and Critical Current Density of Silver-Doped (Bi/Pb)$_2$Sr$_2$Ca$_2$Cu$_3$O$_x$ Tapes," *Cryogenics, 37* [10] 605-608 (1997).
[12]G.L.Messing, S.-C. Zhang, and G.V. Jayanthi, "Ceramic Powder Synthesis by Spray Pyrolysis," *J. Am. Cer. Soc.,* 76 2707-2805 (1993).
[13]R.J. Hill, 61-101 in *The Rietveld Method*, Ed. R.A. Young, International Union of Crystallography, Oxford University Press, Oxford, 1993.
[14]S. Rath, L. Woodall, C. Deroche, B. Seipael, F. Schwaigerer, and W.W. Schmahl, "Quantitative Phase Analysis of PBSCCO 2223 Precursor Powders – An XRD/Rietveld Refinement Study," *Supercond. Sci. Technol., 15* 543-554 (2002).
[15]F. Hai-Fu, W. Zheng-Hua, L. Jian-Qi, Γ. Zheng-Qing, M. You-De, L. Yang, S. Bing-Dong, C. Ting-Zhu, L. Fang-Hua, and Z. Zhong-Xian, 285-294 in *Electron Crystallography*, Ed. D.L. Dorset, S. Hovmoller, and X. Zou, Kluwer Academic Publishes, The Netherlands, 1998.
[16]L.Woodall, R.S. Liu, Y.-W. Hsueh, and W. Schmahl, "A Pb-Bi-Sr-Ca-Cu-Oxide Powder Mix with Enhanced Reactivity and Process for its Manufacture," W.O. Pat. No. 0 112 557, February 22, 2001.

Nanotechnology

NANOTECHNOLOGY- FROM PROMISING TO PRACTICAL
THE ROLE OF STANDARDS

Dr. Stephen Freiman
Deputy Director, Materials Science and Engineering Laboratory
National Institute of Standards and Technology
Gaithersburg, MD, USA

ABSTRACT
Products based upon nanotechnology represent a potential market for revolutionary new materials, particularly those where one or more dimensions are less than 50 nm, e.g., films, carbon nanotubes, nanoparticles. Commercialization of materials such as these will demand that consensus-based standards, e.g., common characterization and measurement techniques, be established. Such standards help facilitate commerce through more reproducible and consistent data, better specifications, and harmonized performance characteristics. In addition, standards can lend credibility to a new material, and help to educate the market as to the possibilities for applications. Examples illustrating a number of these points are presented. The needs for harmonized material measurement methods applicable to various segments of nanotechnology will be emphasized. The beneficial role of pre-normative collaborations will also be discussed.

INTRODUCTION
The ability to make common measurements on the same materials at various places on the globe is critical to world commerce. Worldwide users and suppliers of materials need the assurance that the property of the material obtained in one locality is obtained in the same way as in another. For new materials in emerging markets, such standards are particularly important. The path from the research laboratory to the commercial sector is often strewn with pitfalls of inconsistent data that cause confusion, inefficiency, and added costs. Recognition of the data inconsistency problem usually occurs when an innovative material has matured to the point that multiple sources or users are involved. Standards, in the context of this discussion, range from documentary, e.g., formal ASTM, ISO, and standards, to informally agreed upon test methods.
In this paper, I will attempt to provide specific examples where material measurement standards really made a difference, and to suggest where measurement standards at the nanoscale for materials could be important to the growth of markets for new and innovative applications.

EXAMPLES OF THE VALUE OF STANDARDS
Advanced ceramics ("fine ceramics") are a broad class of materials from which examples of the importance of standards related to market development can be

extracted. More information on the applicability of standards to advanced ceramics is provided by Freiman and Quinn.[1]

Harmonized Test Procedures Lead To Cost Savings

Because of their brittleness, whenever ceramics are employed, either in direct structural applications, e.g., as engine components, or elsewhere, e.g., optical fibers, it is important for the designer to be able to predict their safe, reliable operation. Therefore, being able to accurately and consistently measure the strength of these materials is particularly important. Confidence in manufacturers' reported strength data is also important.

The easiest way to measure the strength of a ceramic is to bend it in a flexure test. This type of test is the bread and butter method of the ceramics industry, and is much simpler than traditional tensile tests, for which specimens are costly to machine (>$100 per specimen), require significant quantities of material, and need meticulous alignment during the test. However, before the development of harmonized measurement procedures, everything about the flexure test could change from one laboratory to the next, giving rise to the reporting of different properties.

A primary benefit of the development of a unique specimen size and shape is the reduction of specimen machining costs. Prior to the development of standard test methods, preparation of a typical specimen cost $20 to $33. With the development of the ASTM flexural test standard C 1161-90, the cost of those tests dropped to $10. Why? Standard fixtures could be employed, and machine shops knew that they were going to make exactly the same specimen for everyone, resulting in savings of $0.8M to $1.5M a year to ceramic manufacturers.

Other benefits accrued. Prior to standardization, it was recognized that the myriad methods then in use were not optimized, and could be faulty. Data discrepancies led to confusion and even distrust. Furthermore, rudimentary quality control or materials development data often did not meet the more stringent requirements for design or materials specifications. This often led to costly, duplicative testing. The adoption of a simple, technically rigorous standard method solved the problem. Now almost everyone measures flexural strength in the same way. Data collected for quality control purposes is immediately acceptable for the most stringent design applications, and the costs of redundant testing have been eliminated. Intangible costs of doubt and distrust have also been reduced.

Standards Speed Acceptance by Regulatory Agencies

Biomaterials are a rapidly growing market segment, and artificial hips are one of the most prevalent uses of such materials. At present, most of the balls of such hip replacements are made of metal. But if one wants to replace hips in younger individuals, and leave them in for longer periods of time, we must look for materials, e.g. alumina, zirconia, etc., that are more inert, harder, and have better biocompatibility. However, in the U.S., the use of any new materials for such applications must have the approval of the Food and Drug Administration (FDA). The FDA would like to see consensus standards and specifications in place to enable them to more rapidly certify new materials. Although the FDA has the authority to write regulatory standards, they now rely on consensus standards developed in both national and international venues. Standards for biomaterials were developed through ASTM, originally in the committee on advanced ceramics, C-28, which wrote the material test standards for the material properties, namely flexural strength, elastic modulus, hardness, and fracture toughness. Committee F-04,

Surgical and Medical devices, used these standards as building blocks in new implant material specification standards.[2,3]

Standards Facilitate Purchasing
One especially relevant example of the importance of new materials to modern technology, and of where standards can be influential, is in wireless communication. Cellular telephones would not exist were it not for the unique dielectric properties of key ceramic components. The development of new materials for the wireless industry provides a good illustration of how a lack of standards can directly affect commerce in new materials. The following statements are paraphrased from comments made by a leading manufacturer of wireless materials:

- A significant problem with the lack of standards is that one company can promote its material over another, when in fact the only difference between the two materials is the fact that their properties are measured in two different ways. One sees apparent conflict; the buyer is not quite sure what is the true property of the material.
- There is potential confusion in interpreting data. If one isn't sure how a particular property is measured, there is clearly a problem in understanding what that property is.
- Thirdly, two vendors may supply a different product even though the material was ordered to the same specification.

All of the above conditions lead to the overall problem that customers may have to qualify each of their vendors'particular products. Property test standards lead to a harmonized set of a material's data, thereby relieving most of the above problems.

When Are Standards Needed for New Materials?
What should be the timing in the development of standards relative to the commercial development of new materials? When a new material is developed, and if there is only one manufacturer, specifications can result from a private agreement between the manufacturer and the end user. As the material matures, more manufacturers of ostensibly the same material appear, and there are more end-users. At this point, some kind of standard becomes important, because it will define the way the critical properties of the material should be measured.

I believe that we are now at this stage for a number of applications of nanomaterials. It is already clear that measurement procedures that work for "bulk" materials will not be valid for the nanoscale. I would suggest that it is time to seriously consider the kind of measurements that will be needed, and to work cooperatively in order to develop consensus on the most efficient measurement techniques. In the next section, I provide three examples of such measurement needs.

THE PROMISE OF NANOTECHNOLOGY
Devices based upon nanotechnology hold promise in a number of fields including data storage, healthcare, electronics, pharmaceuticals, and many more. A number of products based on nanosize particles, films, etc., are already on the market (Table I). Three examples of the need for measurement development for nanomaterials were chosen: nanoparticles, materials issues related to micro/nano-electromechanical (MEMS/NEMS) devices, and carbon nanotubes.

Table I: Current Products Incorporating Nanomaterials[1]
• Data storage - IBM -disk drives with nanostructured GMR materials • Materials enhancement - Nanophase Technologies -nanocrystalline particles for transparent sunblocks and catalysts • Semiconductor electronics - Intel, Motorola, etc., manufacturing using chemomechanical polishing • Drug delivery - Gilead Sciences -100 nm diameter lipid spheres encase anti-cancer drugs • Catalysts – Exxonmobil -zeolites with pore sizes < 1 nm to crack hydrocarbon molecules to form gasoline

Measurement Issues for Nanoparticles

Some of the existing and potential markets for nanoparticles include catalysts, magnetic data storage, chemomechanical polishing slurries, cosmetics, and pharmaceuticals. For some of these applications, knowledge and control of particle size distributions is critical to the reproducible manufacture of the product.

The primary factor that makes measurements of nanoparticles difficult is their increasing surface-to-volume ratio as particle diameters fall below about 50 nm. The increasing surface area leads to problems such as agglomeration. A consequence of the tendency to bond to one another is that nanoparticles cannot be taken out of suspension without altering the measured size distribution, which would include individual particles plus agglomerates. There is clearly a need to be able to differentiate individual particles from agglomerates, and to determine particle shapes.

In the particle size range 100 nm and larger, laser light scattering is used extensively to measure size distributions in concentrated suspensions. However, below a particle size of ≈ 50 nm surface scattering and multiple scattering from particles begin to take on increasing importance, for which there are currently no valid models. A further problem is the fact that the surface chemistry of these particles is different than that in their interior so that determining values of the refractive indices needed for the scattering models based on knowledge of average particle chemistry is not possible. The use of dynamic light scattering can alleviate some of these difficulties, but this technique is only applicable to very dilute suspensions, i.e., < 0.1 volume %.

Mechanical Properties at the Nanoscale

For nanoscale applications in which the mechanical properties of a material, e.g., hardness, modulus, strength, are important, a significant question is whether such properties can be extrapolated from bulk values to the point at which only a few atoms of material are involved. There is data to suggest that the character of failure-producing flaws in glasses is a function of flaw size,[4] and that failure

---

[1] "The identification of any commercial product or trade name does not imply endorsement or recommendation by the National Institute of Standards and Technology."

processes in silicon films used in MEMS can be quite different than those observed in the fracture of bulk silicon.[5,6]

An international workshop, organized on behalf of the Versailles Project on Advanced Materials and Standards (VAMAS) and the Standardization and Research Working Group of CEN (CEN-STAR) was held in 2002 at the National Physical Laboratory (NPL). The objective of the workshop was to identify measurement needs for a range of themes in the area of nanotechnology. It was agreed that the particular applications of MEMS/NEMS that would be most in need of knowledge of mechanical properties were those in which moving parts were involved, i.e., switches and motors. Durability/reliability of such devices is currently a limitation to their widespread development. There are currently no acceptable test methods to measure durability, and in fact, there are very few methods available to determine mechanical properties of any kind in this size range. While current test geometries, e.g., dogbone tensile tests, can be conducted, sample design and preparation is difficult, and obviously quite expensive. There are currently no models that allow the reliable extrapolation of data from large-scale tests to predict the behavior of material in the nanometer size range.

In NEMS and MEMS, the lack of understanding and control of friction and stiction at the nanoscale is a "log jam" to many exciting new developments. The most immediate high priority needs include tests for a range of properties, e.g., adhesion, deformation, stiffness, and fracture. For imaging at the nano-scale, the workshop highlighted the trend for real-time, in-vivo measurements. Furthermore, the importance of scanning probe tip characterization, critical to reliable measurements, was heavily stressed.

Today, the most widely used direct tests for mechanical properties of small-scale devices involve nano-indentation (or nano-scratch) procedures. However, there are no widely accepted guidelines for the use of such procedures. Standards for nano-indentation measurements of hardness and elastic modulus are currently under development.

The workshop concluded that some of the areas in need of attention (not in order of priority) are:

- Scanning Probe Microscopy (SPM) best practice and tip characterization.
- Development of constitutive relationships for modeling friction at the nanoscale.
- Development of scaling laws (to predict properties at the nanoscale based on data obtained at larger scales).
- Development of nanomechanical properties tests.

Single Wall Carbon Nanotubes

Single wall carbon nanotubes (SWCNTs) are an exciting new material with the potential for use in such widely diverse applications such as extremely stiff and strong structural composites, and electronic components. A workshop organized jointly by the National Aeronautics and Space Administration, Lyndon B. Johnson Space Center (NASA/JSC) and the National Institute of Standards and Technology (NIST) was held at NIST in May 2003. The topics of purity and dispersion were chosen as focal areas for the workshop because of their critical importance for the production of high quality SWCNTs. At the present time, a variety of measurement techniques employing significant differences in test methodology are utilized to assess purity and dispersion, resulting in unacceptable uncertainties in data

Characterization & Control of Interfaces for High Quality Advanced Materials   457

interpretation from one laboratory to another. The workshop findings are summarized below:

The purity of single walled carbon nanotubes is defined as the quantity of SWCNTs relative to the metal catalysts and other carbon-like materials present (amorphous, graphitic, and C_{60} carbons). The strengths, limitations, and research needs for most of the commonly used techniques, e.g., thermogravimetric analysis (TGA), transmission electron microscopy (TEM), scanning electron microscopy (SEM), and Raman spectroscopy, were discussed. TGA is the primary technique used to define purity, but there was general accord that unless TEM demonstrates a significant quantity of SWCNTs, no one will agree on the quality of the material. There was consensus that protocols for purity measurement techniques would be valuable even if they are incomplete.

Dispersion is defined as the distribution of nanotube bundles, the splitting of the bundles into individual tubes, and the agglomeration of SWCNTs in solvents or polymers. In macrodispersion, the focus is on dispersability (degree and ease of placing the nanotubes in suspension) and eliminating agglomerates. In nanodispersion, the focus is on debundling the SWCNT ropes. There was a suggestion that a standard dispersing liquid and a standard dispersed solid be distributed to researchers willing to perform characterization so that methods can be compared. Variables could include composition, particle size, and dispersability. An agreement on a solvent for dispersion is also needed. Optical microscopy was conceded to be the primary technique to determine dispersion, but other techniques of value include SEM, ultraviolet/visible spectroscopy, and atomic force microscopy (AFM). Workshop participants agreed there is an urgent need for sonication methods research to determine the effects of time, frequency, and power of the ultra-sonicator on the dispersion of SWCNTs.

Some of the presentations given at this workshop will be published in the Journal of Nanoscience and Nanotechnology. Potential future workshop topics were also discussed. These include: 1) size and chirality of nanotubes; 2) defects; 3) surface chemistry and functionalization; 4) functional, e.g., electronic, properties; 5) applications and performance measures; and 6) health and safety issues.

MATERIALS PRESTANDARDIZATION RESEARCH

Finally, I would like to suggest one approach to obtaining the harmonized measurement procedures needed by the growing field of nanomaterials. This is what I term as prestandardization. Prestandardization research may take many forms including investigating new measurement methods, clearing up gaps or inconsistencies in existing methods, preparing of reference materials, and conducting interlaboratory round robins. In this era of the global marketplace, it is particularly important that prestandardization research be coordinated on an international level. To that end, one effective forum for prestandardization research in materials has been the aforementioned VAMAS organization. The mission of VAMAS is to support world trade in products dependent on advanced materials technologies by providing the technical basis for harmonized measurements, testing, specifications, and standards. VAMAS promotes collaboration among materials laboratories throughout the world, bringing together experts in many materials fields who participate in the harmonization activities through Technical Working Areas (TWAs). TWAs address a wide variety of topics, e.g., surface chemistry, mechanical measurement of films, electroceramics, and ceramic powders, to name but a few.

VAMAS has formal linkages to both ISO and IEC, and perhaps of equal importance, the individuals who participate in VAMAS typically also participate in their national standards bodies and in international standards development. These individuals see each other frequently, work together, and ultimately develop a mutual trust, which facilitates the development of standards on an international basis.

I would point out that formal standards are not necessarily the only valuable output from the work of VAMAS. Other types of documents such as Technology Trends Assessments, published by ISO, describe measurement procedures in some detail, and provide valuable guidelines into carrying out these procedures. The work carried out under VAMAS collaborative activities can also find their way into Good Practice Guides published by National Laboratories such as NIST and the National Physical Laboratory.

SUMMARY

I have shown through examples, how measurement standards have had a direct economic effect on advanced materials. Benefits resulting from such standards include cost savings due to harmonized procedures, more reproducible data, better specifications, and enhancing the credibility of a new material. These standards can include formal documents published through standards development organizations, informally agreed-upon protocols, or recommended practice guides.

I have also suggested some of the measurement needs in three areas of importance in the upcoming field of nanotechnology, namely nanoparticles, MEMS/NEMS applications, and single wall carbon nanotubes.

I also suggested the benefits of international cooperation in measurement development through organizations such as VAMAS.

REFERENCES
[1]S.W. Freiman and G.D. Quinn, "How Property Test Standards Help Bring New Materials to the Market," *ASTM Standardization News,* **29** [10] 26-31 (2001).
[2]ASTM Standard F1538-03 Standard Specification for Glass and Glass Ceramic Biomaterials for Implantation.
[3]ASTM Standard F1873-98 Standard Specification for High-Purity Dense Yttria Tetragonal Zirconium Oxide Polycrystal (Y-TZP) for Surgical Implant Applications.
[4]B. Lin and M.J. Matthewson, "Inert Strength of Subthreshold and Post-threshold Vickers Indentations on Fused Silica Optical Fibers," *Philos. Mag.* **A74** 1235-44 (1996).
[5]C.L. Muhlstein, E.A. Stach, and R.O. Ritchie, "Mechanism Of Fatigue In Micron-Scale Films Of Polycrystalline Silicon For Microelectromechanical Systems," *App. Phy. Lett.,* **80** [9] 1532-34 (2000).
[6]A.M. Fitzgerald, R.S. Iyer, R. H. Dauskardt, and T.W. Kenny, "Subcritical Crack Growth In Single-Crystal Silicon Using Micromachined Specimens," *J. Mat. Res.,* **17** [3] 683-92 (2002).

NANO-SIZED HYDROXYAPATITE CRYSTALS GROWN IN PHASE SEPARATED MICROENVIRONMENTS

Kimiyasu Sato, Yuji Hotta, Yoshiaki Kinemuchi, and Koji Watari
National Institute of Advanced Industrial Science and Technology (AIST)
Anagahora 2266-98, Shimoshidami, Moriyama-ku
Nagoya, 463-8560, JAPAN

ABSTRACT

In the biological synthesis of inorganic solids, an organism creates the proper organic matrix, and the crystals precipitate onto the matrix due to chemical interaction at an inorganic/organic hetero-interface. We attempted to apply this mechanism in material synthesis to produce hydroxyapatite (HAp) nano-crystallites. Phase separated microenvironments of organic molecules, i.e., microemulsions, were employed as microreactors for HAp formation. The surfactant bound water mediated HAp crystal nucleation, and HAp nano-crystallites were obtained. The crystallographic properties are similar to those of HAp in natural bones.

INTRODUCTION

Due to similarities with the mineral constituents of bones and teeth, hydroxyapatite (HAp), $Ca_{10}(PO_4)_6(OH)_2$, is one of the most biocompatible materials. HAp contains only nontoxic species, and they should dissolve in a human body without causing any health problems.[1] Nano-sized HAp crystals are thought to be suitable for use in various medical materials; for instance, drug delivery systems, protein delivery systems, artificial organs, and scaffolds for tissue engineering. In addition, HAp crystals may be useful to immobilize environmentally harmful materials. Recently, the surface structure of the HAp crystal has been determined,[2] and the interfacial interactions between the HAp surface and organic substances have been discussed.[3] This accumulated knowledge should help promote the use of HAp nano-crystals. There are several reports concerning the fabrication of HAp nano-crystals through soft-chemical processing.[4,5] We would like to synthesize nano-particles without firing or melting. Processes requiring high temperatures consume a lot of energy, and they can result in environmental problems. Biomineralization process, the formation of inorganic crystals in living bodies, proceeds at ambient temperature. Mimicking biomineralization will be the friendliest for the environment. In biomineralization processes, organic supramolecular systems are used as pre-organized environments to control the formation of finely divided inorganic materials.

In the present paper, we attempt to synthesize HAp nano-crystals through a bio-inspired method. The HAp particles were grown in phase-separated microenvironments of organic molecules.

MATERIALS AND METHODS

Four milliliters of 2.5 M calcium hydroxide aqueous suspension was added to 50 ml of cyclohexane containing 11 g of sodium bis(2-ethylhexyl)sulfosuccinate (AOT) while vigorously stirring. AOT possesses a sulfo group in its structure, and acts as an anionic surfactant (Figure 1). This mixture was aged for 1 h, and then 2 milliliters of 1.5 M potassium dihydrogenphosphate aqueous solution was added to the reaction mixture. In surfactantoil-water systems, water is readily solubilized to form small water-pools surrounded by the surfactant molecules. These phase-separated microenvironments, i.e., microemulsions, were utilized as microreactors for nucleation and crystal growth. The reactions were conducted for 72 h under intense agitation. All the processes were conducted at a fixed temperature of 25°C. The resulting materials were separated by centrifugation, washed with water and ethanol, and then dried at ambient temperature. The products were identified using X-ray diffraction (XRD), Fourier transform infrared (FT-IR) spectroscopy, and transmission electron microscopy (TEM).

RESULTS AND DISCUSSION

Figure 2 shows the XRD pattern for the HAp powder in comparison to a calculated spectrum. Peaks that can be attributed to the apatite structure are observed in the XRD pattern. Other than apatite, no crystalline phase (such as tricalcium phosphate) is detected. The broadness of the diffraction peaks is attributed to both the fine size and low crystallinity of the powder obtained. In an FT-IR spectrum for the specimen, absorptions due to OH stretching and vibration were found. Stretching bands due to carbonate substituting into the phosphate positions in the apatite lattice are also observed. The carbonate ion comes from a reaction between atmospheric carbon dioxide and the solution. The precipitation product is revealed to be carbonate-containing HAp powder. Through TEM observations, the morphology of the HAp particles was investigated (Figure 3). The nano-crystals exhibit an elongated plate-like form

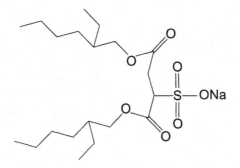

Figure 1. Molecular structure of AOT.

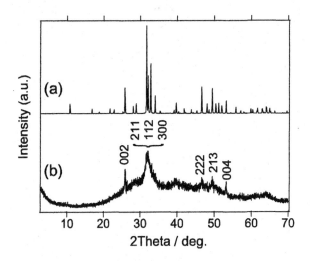

Figure 2. XRD spectra for HAp: a) calculated; and
b) measured nano-particles.

of about 50-100 nm in length, and about 10 nm thick (Figure 3a). They
assemble to form aggregates 500 nm to 1 μm in diameter. High-resolution
TEM (HRTEM) shows lattice fringes in the nano-crystals. The lattice-fringe
image of the nano-crystals observed is shown in Figure 3b together with its
Fourier transform. The periodicity of the lattice fringe is 0.82 nm, which can
be assigned to the interplanar spacing of {100} in the HAp structure. The
plate-like HAp single crystals are parallel to the {100} planes. It is reported
that carbonate ions incorporated in HAp affect the morphology to form flat
crystals with hexagonal symmetry.[6] This structural characteristic is similar to
that of HAp crystals in natural bones.

Throughout the reaction process, the microemulsion cavity of the AOT
molecule is stable. When KH_2PO_4 solution is added, ion exchange occurs
slowly via collisions between and dissociation of the microemulsions containing
Ca^{2+} and PO_4^{3-}. It is known that HAp nucleation can be induced by negatively
charged functional groups on organic matrices.[7,8] Heterogeneous nucleation
probably occurs within the water droplet at the molecular wall. Slow diffusion
of the reactants through the stable wall of the microemulsions will restrict
crystal growth. It is possible that coordination between the combination of
high frequency nucleation and slow diffusion results in the small primary
particles observed.

In various kinds of hard tissue in natural bodies, organic sheaths are
observed in the vicinity of inorganic crystals.[9] Inorganic ion condensation and
crystallization occur within the organic sheaths, which contribute to the
constancy of the shapes and sizes of the inorganic crystals. In our experimental
system, the AOT microemulsions act as the organic sheath that results in the
formation of the bone-like HAp crystals.

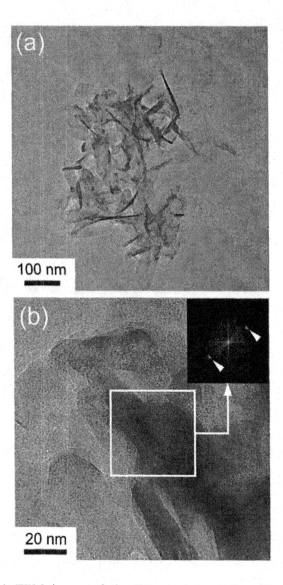

Figure 3. a) TEM image of the HAp nano-crystals. Note that the apparent elongated crystals are in a plate-like form viewed edge-on. b) HRTEM image of the HAp crystals. Fourier transforms and the corresponding areas are also shown. Arrows in the Fourier transforms indicate the spots corresponding to 0.82 nm, which is ascribed to the interplanar spacing of {100} in HAp. Note those spots are found in the direction vertical to the elongated form.

CONCLUSION

HAp nano-sized crystals were synthesized within AOT microemulsions. Heterogeneous nucleation can be induced using organic functional groups within the water droplet surrounded by the molecular wall. The particles obtained are aggregates composed of plate-like HAp crystals containing CO_3^{2-} ions. The plate-like HAp crystals are 50–100 nm in length, and about 10 nm thick, and the plate is parallel to specific {100} planes in the HAp structure. The combination of high frequency nucleation and the slow diffusion of the reactants results in the formation of HAp nano-crystals similar to those found in natural bones.

REFERENCES

[1]H. Aoki, H. Aoki, K. Kutsuno, W. Li, and M. Niwa, "An in Vivo Study on the Reaction of Hydroxyapatite-Sol Injection into Blood," *Journal of Materials Science: Materials in Medicine*, **11** 67-72 (2000).

[2]K. Sato, T. Kogure, H. Iwai, and J. Tanaka, "Atomic-Scale {10$\bar{1}$0} Interfacial Structure in Hydroxyapatite Determined by High-Resolution Transmission Electron Microscopy," *Journal of the American Ceramic Society*, **85** [12] 3054-58 (2002).

[3]K. Sato, Y. Suetsugu, J. Tanaka, S. Ina, and H. Monma, "The Surface Structure of Hydroxyapatite Single Crystal and the Accumulation of Arachidic Acid," *Journal of Colloid and Interface Science*, **224** 23-27 (2000).

[4]M. Yoshimura, H. Suda, K. Okamoto and K. Ioku, "Hydrothermal Synthesis of Biocompatible Whiskers," *Journal of Materials Science*, **29** 3399-02 (1994).

[5]T. Furuzono, D. Walsh, K. Sato, K. Sonoda, and J. Tanaka, "Effect of Reaction Temperature on the Morphology and Size of Hydroxyapatite Nanoparticles in an Emulsion System," *Journal of Materials Science Letters*, **20** 111-14 (2001).

[6]R.Z. LeGeros, "Biological and Synthetic Apatites"; 3-28 in *Hydroxyapatite and Related Materials*, Ed. P.W. Brown and B. Constantz. CRC Press, Boca Raton, 1994.

[7]K. Sato, Y. Kumagai, and J. Tanaka, "Apatite Formation on Organic Monolayers in Simulated Body Environment," *Journal of Biomedical Materials Research*, **50** 16-20 (2000).

[8]K. Sato, T. Kogure, Y. Kumagai, and J. Tanaka, "Crystal Orientation of Hydroxyapatite Induced by Ordered Carboxyl Groups," *Journal of Colloid and Interface Science*, **240** 133-38 (2001).

[9]K. Hoshi, S. Ejiri, and H. Ozawa, "Organic Components of Crystal Sheaths in Bones," *Journal of Electron Microscopy*, **50** [1] 33-40 (2001).

REDUCTION OF THIN FILM SURFACE ROUGHNESS BY SELF-ASSEMBLING OF ORGANIC MOLECULES

M. Itoh, K. Aota, R. Sugano, and H. Takano
Department of Chemical Engineering and Materials Science
Faculty of Engineering
Doshisha University, Kyo-Tanabe
Kyoto 610-0321, JAPAN

ABSTRACT

A self-assembled monolayer (SAM) of alkane-thiols was applied to improve the surface flatness of a gold thin film produced by the QDPD process to increase the interfacial mechanical strength of a hybrid thin film made of gold and an organic-layer synthesized by Langmuir-Blodgett technique. A gold thin film was synthesized on a silicon substrate by the QDPD process, and it was dipped in a 1-octadecanthiol solution. Within 3 h, the surface roughness was reduced to about 1% of the original condition, proving that an ultra-flat surface in the nano-scale range can easily be formed with a SAM.

INTRODUCTION

The surface roughness measured by atomic force microscopy (AFM) can be regarded as the variation of surface height. The roughness can be characterized in a statistical manner. The most obvious parameter is the average roughness

$$Ra = \frac{1}{N} \sum_{i=1}^{N} |\Delta ri,|$$ (1)

which is also called the centerline average (CLA) roughness. This is the average of the surface height deviations, Δri, of N data points. This parameter is often used in the manufacturing industry.

Organic thin films have been fabricated using the Langmuir-Blodgett method and/or other single but complex procedures. However, recently, industries require a much simpler and easier process to fabricate organic thin films on inorganic substrates. Some types of organic molecules form a monolayer on solid substrate by the self-assemble of molecules (SAM). Some organic molecules will make an ultra-flat surface by embedding themselves in the cavities of the surface by the method shown in Figure 1, which involves intermolecular interaction or adsorption between orientation-ordered molecules.

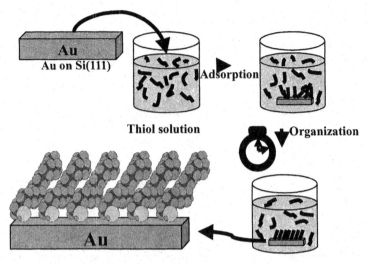

Figure 1. Preparation of SAM coating.

The adsorption of an S-H functional group on gold occurs when a gold substrate is dipped into a solution containing alkanethiol; the phenomenon occurs spontaneously by the following reaction.

$$-SH + Au \rightarrow -S - Au + \frac{1}{2}H_2 \qquad (2)$$

A highly oriented and self-assembled monolayer is formed on the gold substrate by the repulsive interaction between the S-H functional group and the gold substrate. Such a monolayer made of self-assembled organic molecules is generally stable and tightly bound, due to the strong chemical affinity of the functional group with the surface material. Such SAMs have a wide variety of applications, such as for sensors, corrosion inhibition, wetting control, and in bio-molecular, molecular electronic devices.

Water droplets deposited on a glass plate grow by merging with adjacent droplets. A small island made up of the water droplets will spread on a plate to form a water film by self-organization. Metallic particles less than 3 nm in diameter often display liquid-like behavior at room temperature due to the lowering of the melting point by the so-called quantum size effect. By a process analogous to the formation of the water film described above, the atoms in nano-size metallic particles in a liquid phase diffuse and spread on a substrate[2] to form islands of metal. A metallic thin film will organize automatically by the fusion of the adjacent islands, and the metal film will crystallize automatically at room temperature without any annealing or sintering. This method of fabricating a thin film via direct deposition of nano-phase particles (quantum dots) is called the Quantum Dot Physical Deposition (QDPD) process.

In this study, self-assembled monolayers of alkane-thiols were applied to improve the surface flatness of a gold thin film produced by the QDPD process.[1]

EXPERIMENTAL

A silicon substrate (111) was cleaned with Semico-Clean-23 in an ultrasonic cleaner (US-4), rinsed with deionized water, and air dried with filtered N_2 gas at 125°C. Then a gold thin film was produced on the clean silicon substrate by the Quantum Dot Physical Deposition (QDPD) process.

Figure 2 shows the experimental setup of the QDPD system. The system consists of a vacuum chamber, an evacuation system, and a film growth section. An oil rotary pump (PMB003 ULVAC) and diffusion pump (ULKO ULVAC) are used to establish an initial high vacuum condition inside the chamber.

To self-assembly molecules on the gold substrate, 1-octadecanthiol ($CH_3(CH_2)_{17}SH$ = 286.57: Aldrich Chemical Company, Germany) was dissolved in benzene (C_6H_6=78.11, Wako chemical company, Japan). The concentration was regulated to be 1.0mg/ml. The QDPD produced gold thin film was dipped into the 1-octadecanthiol solution for 1 to 240 min. The self-assembly of 1-octadecanthiol molecules on the gold surface was observed with SHIMADZU SPM-9500 atomic force microscope (AFM) as a function of the time immersed in the solution. The AFM was operated at room temperature in the contact mode using an e-scanner. The contact force was minimized during scanning with the set point by located just above the disengagement point. Height images were captured at the highest allowable integral and the proportional gain was usually set between 5 and 10. The scan rate was 1 to 5 Hz. Deflection images of molecular lattices were obtained with the gain set at 0.1 using a scan rate between 10-20 Hz. After flattening to remove underlying surface curvature, raw images of the AFM were converted into standard image files (such as TIF or JPEG) with Adobe Photoshop. The size and dispersibility of the gold particles were also characterized using transmission electron microscopy (TEM) (Hitachi, H-500H and H-8100).

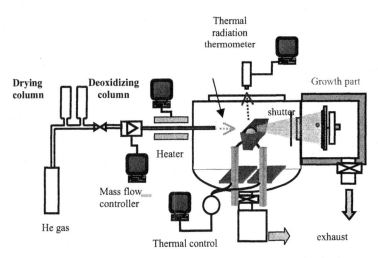

Figure 2. Experimental setup for the Quantum Dot Physical Deposition (QDPD) Process.

Figure 3. TEM photograph of nano-size gold particles formed by the
QDPD process (Deposition time = 60 s, D_g = 2.28 nm, σ_g = 1.36).

RESULTS AND DISCUSSION

The liquefaction of gold particles[3] at room temperature due to the quantum size effect is an essential process in our hybrid thin film fabrication process (QDPD process), and the particle size is the dominant factor to qualify the process itself. Figure 3 shows typical nano-size gold particles generated in the QDPD system; the geometric mean diameter of the particles is 2.28 nm, and the geometric standard deviation is 1.36. Figure 4 shows the growth behavior of the gold film on the substrate at 90, 120, and 240 s, respectively. The pictures make it clear that the liquefied Au nano-particles deposited on the substrate grow by fusing with the adjacent ones. The small islands spread on the silicon substrate and self-organize.

The formation of an ultra-flat surface by the self-assembly of molecules on the QDPD prepared gold surface is depicted in Figure 5. The surface roughness decreases with increasing immersion time, and it improves by a factor of 10 within 3 hours compared to the original substrate surface. In every case, the SAM coating reduces the average roughness (Ra) relative to the original gold substrate (Figure 6).

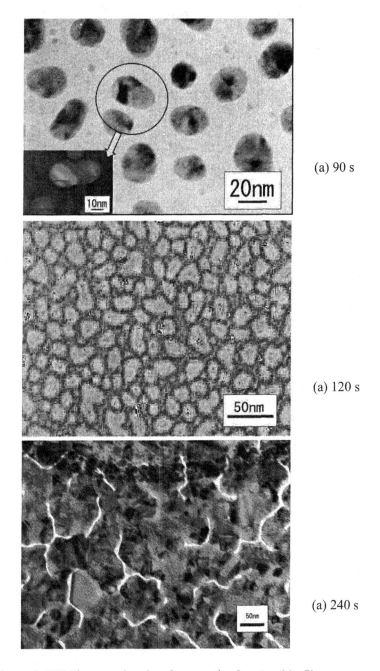

(a) 90 s

(a) 120 s

(a) 240 s

Figure 4. TEM images showing the growth of an Au thin film
with increasing QDPD time (sampled in front of buffer plate).

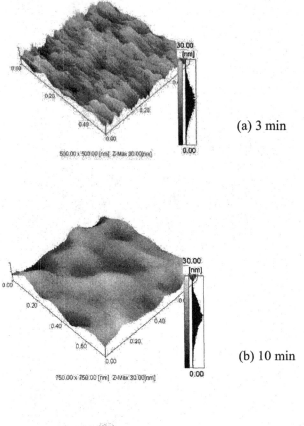

(a) 3 min

(b) 10 min

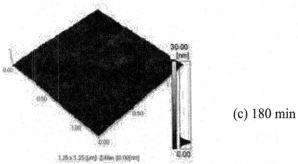

(c) 180 min

Figure 5. AFM images of surface roughness (by the SA method) of a SAM coated QDPP film as a function of coating time. (Ra: roughness parameter of surface).

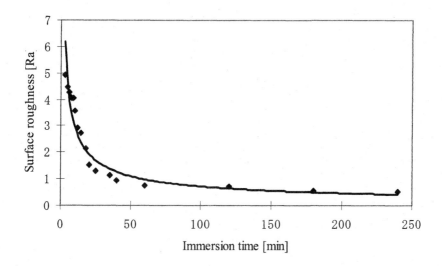

Figure 6. Relationship between immersion time (SAM coating time) and surface roughness (Ra) of a self-assembled organic film on a QDPD gold film.

The self-assemble method is a very simple and widely applicable coating process. Additionally, the time required to produce a coating by self-assembly could be shorter than for other similar processes, e.g., the Langmuir-Blodgett (LB) process. As shown in Figures 5 and 6, the surface becomes smoother with increasing coating time; to some extent, it will be possible to produce arbitrarily rough surfaces using the self-assembled method. This characteristic will be very helpful for hybrid thin film production[4] because it will provide a tight and smooth interface for organic films in a hybrid system without any trade-off in the synthesis process.

CONCLUSION
Surface roughness was improved by factor 10 within 3 hours compared to the original the gold substrate surface. It has been proved that an ultra-flat surface in the nano-scale range can easily be formed with self-assembled monolayers (SAMs). The surface roughness of the SAM coating can be controlled by controlling the coating (immersion) time.

ACKNOWLEDGMENTS
This study is funded by a part of the Research Project of the Ministry of Education, Sports, Culture, Science and Technology of Japan in 2002-2004, and by a part of the Research Promotion Fund of Aid at Doshisha University in 2001-2002.

REFERENCES

[1] M. Itoh, K. Sakiyama, S. Handoh and H. Takano, "The Film Synthesis by Self-organization of Physically Deposited Nano-sized Metallic Particles," in *Advanced Technologies for Particle Processing*, Eds. M.C. Roco, *et al.*, 246, American Institute of Chemical Engineers (1998)

[2] Y. Matsubara, M. Takahasi, H. Oku, S. Ogasawara, H. Takano, and M. Itoh, "Numerical Analysis on the Thin-Film Growth at Room Temperature by Cool-Liquefaction of Nano-Phase Aerosol Particles," *J. Aerosol Sci.*, **32** S229-30 (2001).

[3] S. Tohno, M. Itoh, S. Aono, and H. Takano, "Production of Contact-Free Nanoparticles by Aerosol Process: Dependence of Particle Size on Gas Pressure," *J. Colloid and Interf. Sci.*, **180** 547-77 (1996)

[4] K. Sakiyama, M. Izumi, K. Matsuoka, H. Takano, O. Yamaguchi, and M. Itoh, "Novel Method for the Synthesis of Metal-Organic Hybrid Thin Film at Room Temperature by Aerosol Process of Nano-Particles," *Proc. 6th-International Congress on Chemical Engineering 2001*, 879 (2001).

THE SYNTHESIS OF Nio-CGO POWDER, AND THE PROCESSING AND PROPERTIES OF Nio-CGO ANODES

Eisaku Suda and Bernard Pacaud
R&D Department
ANAN KASEI Co., Ltd.
210-51 Ohgata, Anan
Tokushima 774-0022, Japan

Mikio Itagaki and Satoshi Ohara
Japan Fine Ceramics Center (JFCC)
2-4-1, Mutsuno, Atsuta-ku
Nagoya 456-8587, Japan

Yvan Montardi
Aubervilliers Research Center,
Rhodia Electronics and Catalysis
52, rue de la Haie-coq-F-93308
Aubervilliers Cedex, France

Yasuo Takeda
Department of Chemistry
Faculty of Engineering
Mie University
1515 Kamihama
Tsu 514-0008, Japan

ABSTRACT
The synthesis and processing of 50 vol% Ni - $Ce_{0.9}Gd_{0.1}O_{1.95}$ (CGO) composite powder and an anode for solid oxide fuel cells (SOFCs) has been studied. Using sinterable CGO powder, three powder synthesis methods were examined, including spray drying, sol-gel, and traditional ball milling. NiO-CGO powder prepared by the spray drying process gave the best result. The powder consists of spherical particles with a narrow particle size distribution between 0.2 to 0.6 μm in diameter. Ni-CGO anodes made at 1300°C have low over potential, η_a, (100mV) and low ohmic loss, IR_a, (65mV) and a current density of 0.2 A/cm^2. A well-connected Ni network seems to form in the sintered anode, and the IR_a is low. The NiO particles, however, may impede the sintering of the CGO particles.

INTRODUCTION
Solid oxide fuel cells (SOFCs) are expected to be a new power generation system that can provide high efficiency when used in co-generation. In general, SOFCs operate at temperatures between 900 and 1000°C to take advantage of the oxygen conductivity in a yttria-stabilized zirconia (YSZ) electrolyte. In high temperature SOFCs, materials and cell manufacturing costs are issues. If the operation temperature could be decreased, various low-cost materials could be considered for the cell components in SOFCs.[1] Moreover, lowering the sintering temperature of the cell component materials will reduce manufacturing costs. Due to polarization losses at both the anode and the cathode, as well as ohmic loss in the electrolyte, it is necessary to develop high performance electrodes for lower

temperature SOFCs.[2]   The activity and stability of an electrode depend not only on chemical composition, but also on microstructure.   The latter is closely related to the dispersion and interconnectivity of the nickel particles and the electrolyte particles in the electrode.[3-5]

For the anode, especially, the use of doped-ceria has been suggested,[2-4,6-8] because Doped-ceria is both an oxygen ion and an electron conductor.   The conductivity of doped-ceria depends on the dopant cation.   In general Sm doped ceria shows the highest oxygen-ion conductivity.[3, 6-7,9-11]   However the isotope Sm 147 emits radioactive alpha (α) particles and is toxic.   The conductivity of Gd-doped ceria is close to that of Sm doped ceria; $Ce_{0.9}Sm_{0.1}O_{1.95}$ and $Ce_{0.9}Gd_{0.1}O_{1.95}$ is 0.07 S/cm and 0.06 S/cm, respectively.[3,11]   Additionally, the earth's crust is almost as rich in Gd as it is in Sm.[12]   Consequently Gd is a candidate dopant for ceria.

Ohara et al.[2] reported that the best anode properties are obtained for Ni-Samarium doped ceria with a Ni content of around 50 vol%.   In this study, 50 vol% $Ni-Ce_{0.9}Gd_{0.1}O_{1.95}$ (CGO) cermets are synthesized using spray drying, sol-gel, and traditional ball mill mixing, and the relationship between microstructure and anodic performance is discussed.

EXPERIMENTAL

For all of the Ni-CGO samples in this study, the volume ratio of Ni to CGO is 50:50.   Spray drying of a slurry was first investigated.   Sinterable CGO powder (HT-CGO: D50 = 0.2 μm) was prepared using a previously described heat-treatment process.[13]   Ni nitrate (99.9%, Wako pure chemicals) was dissolved in pure water to prepare a 1 mol/L of Ni nitrate solution.   Then a slurry consisting of Ni nitrate solution and CGO powder was prepared and spray dried at 240°C.   After the spray dry process, the Ni nitrate / CGO powder mixture was calcined at 700°C for 5 h.   Finally the calcined powder was ground using a mortar and pestle.   The NiO-CGO powder prepared by spray drying is denoted as SD.

For the sol-gel method, a slurry of Ni nitrate and CGO was prepared using the same procedure described above.   The slurry was decomposed in a stainless vessel at 400°C.[14]   After decomposition, the powder was calcined at 700°C to remove any remaining organic materials.   The calcined powder was ground using a mortar and pestle.   This NiO-CGO is denoted as SG.

Traditional ball mill mixing also was used to prepare powder.   NiO powder (D50 = 1 μm, Nikko Rika) and HT-CGO were thoroughly mixed with ethanol in a ball mill for 24 h.   Nylon balls with an iron core were used with a polypropylene lined milling jar.   After ball milling, the mixed slurry was dried at 80°C for 10 h.   The powder was calcined at 700°C, and ground using a mortar and pestle.   The NiO-CGO powder prepared by ball milling is denoted as BM.

The morphology and microstructure of the powders were characterized using a Hitachi scanning electron microscope (SEM).   Energy dispersive X-ray (EDX) elemental analysis was completed using a Kevex microanalyzer attachment. X-ray diffraction (XRD) analysis (Rigaku RINT2000) was used to identify the phases of the powders.

Anode characterization was studied as follows.   Powders were screen-printed onto dense 8 mol% yttria stabilized zirconia (8YSZ) electrolyte pellets 14 mm in diameter and 0.2 mm thick.   A pore-former (polyethylene glycol) was added, and the mixing weight ratio of powder to binder was 5:2 for all samples.   The anode area was 0.3 cm^2.   Samples were sintered at 1250 – 1350°C for 2 h.   After sintering, $La_{0.6}Sr_{0.4}CoO_3$ was screen-printed on as a counter electrode on the opposite side of the pellet, and sintered at 900°C for 4 h.

Figure 1 shows a schematic of the single cell test configuration used in this study. The as-prepared cell was operated at 800°C. Air and 3% wet hydrogen gas (H_2 + 3% H_2O) were supplied to the cathode and anode, respectively. A flow rate of 50 cm^3/min was used for both gases. For electrochemical characterization, the current-interruption technique was used.[15,16]

In this report, the anode polarization (Pa) is defined as the potential difference between the anode and the reference electrode, which includes the electrochemical overpotential (ηa) and the ohmic loss (IRa) between the anode and the reference electrode, i.e.,

$$P_a = \eta_a + IR_a$$

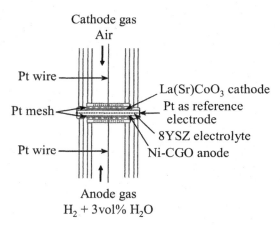

Figure 1. Schematic diagram of the single-cell test setup.

RESULTS AND DISCUSSION

No second phase is observed by XRD in the composites of NiO and CGO. Figure 2 shows SEM micrographs of the NiO-CGO powders synthesized using the different processes. Figure 2a shows that the SD powder consists of spherical particles with a narrow particle size distribution between 0.2 to 0.6 μm in diameter. Figure 2b shows that the SG powder consists of very small submicron particles, but most of the particles are significantly agglomerated. The BM powder is an agglomerated mixture of 1-2 μm diameter particles and submicron particles, because the average particle sizes of NiO and CGO are 1 μm and 0.2 μm, respectively.

(a) SD   (b) SG   (c) BM

Figure 2. SEM micrographs of NiO-CGO powders prepared by a) spray drying (SD), b) sol-gel (SG), and c) ball milling (BM).

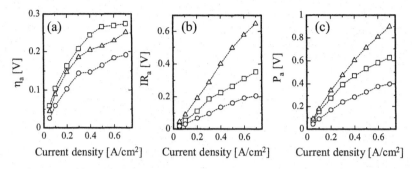

Figure 3. Electrical performance of NiO-CGO anodes made at 1300 °C from SD (○), SG (△) and BM (□) powder.

Figure 3 shows the anode performance for cells made at 1300°C using the SD, NS and BM powders. For the SD powder, the ηa and IRa, at 0.2 A/cm^2 are 100 mV and 65 mV, respectively. The SD powder gave the best results among the three powders examined. The SG and BM powder contained large particles that produce poor performance anodes.[5] The microstructure of the anode after the anode performance measurements is shown in Figure 4. The SG powder sintered microstructure shows large Ni-CGO agglomerates, large pores, and an inhomogeneous skeleton microstructure (Figure 4b). The BM powder sintered microstructure is also inhomogeneous, and contains small agglomerated particles (Figure 4c). It is believed that the IRa for the SG and BM powder is high because agglomerates can prevent the formation of the Ni network in the microstructure. The SD sintered powder microstructure consists of a homogeneous skeleton of Ni-CGO particles, and well-connected surfaces (Figure 4 a). The homogeneous dispersion of Ni and CGO in the microstructure of the SD sintered powder is confirmed by EDX analysis. Since the Ni-CGO particles are uniform, a well-connected Ni network seems to form, and a low IRa is measured.

Figure 4. SEM micrographs of cross section of anodes produced at 1300 °C from a) SD, b)SG, and c) BM NiO-CGO powder.

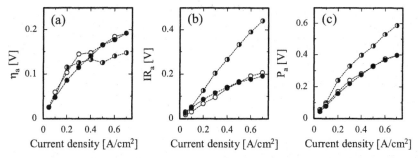

Figure 5. Electrical performance of anodes made from spray dried NiO-CGO sintered at 1250 °C (◑), 1300 °C (○) or 1350 °C (●).

To investigate the effect of sintering temperature on anode performance, cells were made using SD powder sintered at 1250°C, 1300°C and 1350°C for 2 h. Figure 5 shows the anode performance for these cells. The ηa for all sintering temperatures is almost same and low (Figure 5a). Anodes sintered at 1300°C and 1350°C have a lower IRa than the anode sintered at 1250°C (Figure 5 b). Although the anode sintered at 1350°C contains agglomerates (Figure 6 b), the performance is almost the same as for the anodes sintered at 1300°C. The inhomogeneous microstructure produced at 1350°C, however, isn't appropriate for the anode of a SOFC, because of the degradation in performance at longer time.[5,17]

The microstructure of the anode sintered at 1250°C consists of loosely bonded grains that are quite similar in morphology to the particles in the starting powder (Figure 2a and Figure 6a). The IRa of the anode sintered at 1250°C is high because sintering has not progressed sufficiently to get good interconnectivity of the particles. The relative densities of sinterable CGO and NiO-CGO at 1200°C are 94% and 85%, respectively. This reveals that NiO particles impede the sintering of the CGO particles. To lower the sintering temperature of anode for the investigation, it is necessary to optimize the Ni content in this system.

(a) 1250 °C   (b) 1350 °C

Figure 6. SEM micrographs of the microstructure of SD powder sintered at a) 1250 °C and b) 1350 °C.

CONCLUSION

Single cell tests on SOFCs with anodes made from different NiO-CGO powders determined that spray drying (SD) produces the best powder. The anode polarization for the SD powder is the lowest; at a current density 0.2 A/cm^2,

ηa and IRa are 100 mV and 65 mV, respectively. The SD starting powder has a narrow particle size distribution. Additionally, the Ni and CGO particles in the sintered microstructure showed good interconnectivity between the Ni-Ni, CGO-CGO and Ni-CGO particles in the anode. Cells sintered at 1300°C have a homogeneous microstructure, and have the best performance.

ACKNOWLEDGEMENT

The authors wish to thank Dr. Masashi Mori of Central Research Institute of Electric Power Industry for valuable suggestion and insight.

REFERENCES

[1]K. Atsuta, M. D. Civiello, B. Pacaud, T. Seguelong, and E. Suda, "Industrial Ceria-Based Powders for Solid Electrolyte Applications in SOFC: Physical-Chemical Characterization and Properties," *in Proceedings of Fifth European Solid State Fuel Cell Forum*, Edited by J. Huijsmans, ISBN 3-905592-10-X, 1-5, July 2002, Lucerne/Switzerland, 327-334 (2002)

[2]S. Ohara, R. Maric, X. Zhang, K. Mukai, T. Fukui, H. Yoshida, T. Inagaki, and K. Miura, "High Performance Electrodes for Reduced Temperature Solid Oxide Fuel Cells with Doped Lanthanum Gallate Electrolyte I. Ni-SDC Cermet Anode", *J. Power Sources*, **86** 455-458 (2000).

[3]Q.M. Nguyen and T. Takahashi, "Science and Technology of Ceramic Fuel Cells," Elsevier Science, ISBN044489568, (1995).

[4]T. Setoguchi, K. Eguchi, and H. Arai, "Effects of Anode Material and Fuel on Anodic Reaction of Solid Oxide Fuel Cells," *J. Electrochem. Soc.*, **139** 2875-2880 (1992).

[5]T. Fukui, S. Ohara, M. Naito, and K. Nogi, "Morphology and Performance of SOFC Anode Fabricated from NiO/YSZ Composite Particles," *J. Chem. Eng. Japan*, **34** [7] 964-966 (2001).

[6]B.C.H. Steele, "Appraisal of $Ce_{1-y}Gd_yO_{2-y/2}$ Electrolytes for IT-SOFC Operation at 500°C," *Solid State Ionics*, **129** 95-110 (2000).

[7]T. Kudo and H. Obayashi, "Mixed Electrical Conduction in the Fluorite-Type $Ce_{1-x}Gd_xO_{2-x/2}$," *J. Electrochem. Soc.*, **123** 415-419 (1976).

[8]M. Mogensen, N.M. Sammes, and G.A. Tompsett, "Physical, Chemical and Electrochemical Properties of Pure and Doped Ceria," *Solid State Ionics*, **129** 63-94 (2000).

[9]H.L. Tuller and A.S. Nowick, "Doped Ceria as a Solid Oxide Electrolyte," *J. Electrochem. Soc.*, **122** 255-259 (1975).

[10]R.T. Dirstine, R.N. Blumental, and T.F. Kuech, "Ionic Conductivity of Calcia, Yttria, and Rare Earth-Doped Cerium Dioxide," *J. Electrochem. Soc.*, **126** 264-269 (1979).

[11]H. Yahiro, Y. Eguchi, K. Eguchi, H. Arai, "Oxygen Ion Conductivity of the Ceria-Samarium Oxide System with Fluorite Structure", *J. Appl. Electrochem.*, **18** 527-531 (1988).

[12]S.R. Taylor, "Abundance of Chemical Elements in the Continental Crust: A New Table," *Geochim. Comochim. Acta*, **28**[8] 1273-1285 (1964).

[13]E. Suda, B. Pacaud, Y. Montardi, M. Mori, M. Ozawa, and Y. Takeda, "Low-Temperature Sinterable $Ce_{0.9}Gd_{0.1}O_{1.95}$ Powder Synthesized through Newly-Devised Heat-Treatment in the Coprecipitation Process," *Electrochemistry*, **71**[10] 866-872 (2003).

[14]M. Pechini, *US Patent*, No.3330697 (1967).

[15]D.Y. Wang and A.S. Nowick, "Cathodic and Anodic Polarization Phenomena at Platinum Electrodes with Doped CeO_2," *J. Electrochem. Soc.*, **126** [7] 1156-1165 (1979).

[16]T. Kawashima and Y. Matsuzaki, "Effect of Particle-Diameter Ratio of YSZ

to Ni on Polarization of Ni/YSZ Cermet Anode," *J. Ceram. Soc. Japan*, **104** [4] 317-321 (1996).

[17]H. Itoh, T. Yamamoto, M. Mori, T. Horita, N. Sakai, H. Yokokawa, and M. Dokiya, "Configurational and Electrical Behavior of Ni-YSZ Cermet with Novel Microstructure for Solid Oxide Fuel Cell Anodes," *J. Electrochem. Soc.*, **144** [2] 641-646 (1997).

Author and Keyword Index